U0275589

寰宇文献 Universal Library | SINOLOGY 系列

SELECTED WORKS OF BERTHOLD LAUFER

劳费尔著作集

第十一卷

[美] 劳费尔 著

黄曙辉 编

图书在版编目(CIP)数据

劳费尔著作集 / (美) 劳费尔著；黄曙辉编. —上海：中西书局，2022
(寰宇文献)
ISBN 978-7-5475-2015-4

Ⅰ. ①劳… Ⅱ. ①劳… ②黄… Ⅲ. ①劳费尔 – 人类学 – 文集 Ⅳ. ①Q98-53

中国版本图书馆CIP数据核字（2022）第207067号

第 11 卷

151

书评五则

AMERICAN ANTHROPOLOGIST

NEW SERIES

ORGAN OF THE AMERICAN ANTHROPOLOGICAL ASSOCIATION,
THE ANTHROPOLOGICAL SOCIETY OF WASHINGTON,
AND THE AMERICAN ETHNOLOGICAL
SOCIETY OF NEW YORK

VOLUME 21

LANCASTER, PA., U. S. A.
PUBLISHED FOR
THE AMERICAN ANTHROPOLOGICAL ASSOCIATION

1919

Man began by giving his best to the gods. At first the king or priest was sacrificed, and as he was often thought to be deity incarnate he was eaten by the worshippers in the belief that by doing so they became permeated with the divine spirit. Later an animal, such as a bull, was substituted. Bulls came to be regarded as too expensive and a goat or pig was sacrificed. Man became more niggardly still and fashioned a piece of dough to represent the victim and finally they did not even take the trouble to fashion the dough in any image.

The irony is not always maintained at this level and sometimes dips into sheer burlesque, as in the statement that

Some writers, but they are not up to date, assert that it is his reputation as the Father of Lies which entitles Herodotus to be hailed as the Father of Anthropology.

or the punning, under Cannibalism, on struggling missionaries, *piéces de resistance*, and mission furniture. Yet such passages, in juxtaposition with references to the Golden Bough, Tylor, Reinach's Orpheus, Folkways, Spencer and Gillen, and Robertson Smith, will bewilder and shock those whom a subtler sarcasm would have passed through without a scar. The skit is just broad enough to amuse any well-read person; but much of it is so finely sharpened, and its venom at once so gentle and so genial, as to make it doubly refreshing within the profession. Some of us may even temporarily succeed in inhibiting, under its ridicule, habits in which we have heretofore indulged without shame.

This being the first sustained notice which wit has deigned to give anthropology, the science must be arriving. There remain many pompousnesses, but thanks are due this professor of the classics for the clatter of deflations caused by the pricks which he has strewn between his compact dedication and his barbed last sentence.

A. L. Kroeber

ASIA

Collection Tovostine des antiquités préhistoriques de Minoussinsk conservées chez le Dr. Karl Hedman à Vasa. *Chapitres d'archéologie sibérienne par* A.-M. Tallgren, Conservateur-Adjoint au Musée Historique de Finlande. Société Finlandaise d'Archéologie, Helsingfors, 1917. 94 p. 4°. 12 plates. 90 text-figs.

Mr. Tallgren gives in this elegant volume an accurate and intelligent account of a collection of 1,053 antiquities gathered in the region of Minusinsk on the upper Yenisei in central Siberia by the Russian collector Tovostin. Of this number, 298 are illustrated on the plates and seventeen in the text; the reproductions are excellent. The archaeology of this region has for some time been the object of general interest, as it is apt to establish a connecting link between the Scythian area of southern Russia and the culture of ancient China. Klementz, Aspelin, Radloff, Martin, and others, have worked in this field, and Tallgren has utilized the labors and results of his predecessors with skill and laudable care. He is not content, however, with a mere description of his material, but endeavors to study it historically and chronologically and in its relation to the surrounding cultural provinces. He thus presents us with a lucid exposé of this vast field, which is of value to every archaeologist; and with the spirit of the true scholar he constantly points out lacunae in our knowledge and problems still awaiting solution.

The antiquities of the stone age along the banks of the upper Yenisei are but imperfectly known, and while many stations of the stone age have been located, rational excavations have not yet been carried out. Finds of stone implements are rather scarce; and many of these, particularly phallic emblems and ceremonial nephrite objects, are doubtless contemporaneous with the early bronze age. The author has justly recognized the similarity of these perforated nephrites with analogous specimens in ancient China, as described by me, but what the real historical connection is in this as well as in other groups still escapes our knowledge.

Bronze and copper socketed celts play a prominent rôle in the early bronze period of central Siberia; at least eight hundred of these are known at present. The author holds tentatively that, unless further information comes to light, this type may be regarded as due to European, more particularly, Hungarian influence. I feel somewhat sceptical about this point. In China this type is not so rare, as the author inclines to assume, but is rather common; even the spiral decorations on the celts are identical in China and Siberia. Assuredly we require more facts before we can hope to formulate positive conclusions. A very characteristic feature of the outfits of Minusinsk culture is represented by the short double-edged daggers provided with a guard, hilt, and blade being turned out in a single cast; exactly the same type prevailed at a certain time in China. Knives of a great variety of forms are still more abundant, and are regarded by Tallgren as thoroughly national, while he seeks the origin of the dagger outside of the valley of the Yenisei. The bronze scythes, according to the author, were utilized by the nomads to cut the grass necessary for the feeding of cattle during the winter, but in no case were they proper agricultural implements. This stricture, it seems to me, is not necessary, for the ancient and modern nomads of Siberia and Mongolia, as we know from Chinese records and present-day conditions, did to some extent practise agriculture, this being chiefly the task of slaves. The true nature of nomadism still remains to be studied; in fact, however, there is no nomadic tribe in Asia that would ever have subsisted exclusively on its herds and animal products. There is no doubt that the culture of Minusinsk is that of a tribe of horsemen, as evidenced by the numerous horse's bits, stirrups, and other horse equipments, as well as by petroglyphs of mounted archers. I regret that Tallgren has not taken up this important problem as to the identity of this bronze-age people; in view of Radloff's antiquated and unacceptable theories, a restatement of the case becomes imperative. In my opinion, we are here confronted with the culture of a Turkish tribe.

Once more we note the wide distribution of Chinese metal mirrors over Siberia. Those illustrated on Plate VIII, 11, 12, and 13, on the basis of comparative material from China, may with certainty be identified with productions of the T'ang period (A.D 618–906), while No. 10 belongs to the Sung epoch (A.D. 960–1278). As to the mirrors with designs of vine and grapes, Mr. Tallgren adheres to the theory of Hirth that these originated under the Han, and that the motive was introduced into China by General Chang K'ien toward the end of the second century B.C.; this opinion, however, is no longer accepted and rests on a fallacy: these mirrors range from about the fourth to the seventh century A.D., and the motive bears no relation to Greek-Bactrian art, but hails from Sasanian Persia.

The alloy of the bronze objects has not yet been analyzed for economic reasons, but the author, relying on analyses formerly made, emphasizes the curious fact that they contain a remarkable abundance of zinc, which is also found in bronzes of eastern Russia, even in those suspected of North European provenience and showing in type no resemblance to Siberian objects. Another notable fact as yet unexplained is the great variation in the composition of Siberian bronzes, "especially as zinc is unknown in Chinese bronzes according to the canonical books of the Chinese and eventually a Chinese influence in Siberia might be admitted." I believe I can give some information on this point. A careful analysis of a hundred well-defined ancient Chinese bronzes in Field Museum, soon to be published, has revealed the interesting fact that the variability in the composition of the alloys is exceedingly large and that zinc occurs in rather considerable proportions. Hence the affinity of ancient Chinese to ancient Siberian bronze in its purely technical aspect is perfect and will be definitely established,—a fact which for historical reasons had long ago been suspected by me.

The imitation of a cowry-shell (*Cypræa moneta*) in lime-stone (plate XI, 17) is of great interest, as we have from ancient China a large number of such cowries, both real and reproduced in bone and inscribed. Our knowledge of prehistoric Siberian ceramics is still very unsatisfactory; comprehensive collections are stored up in the Museum of Minusinsk, but not yet published. The Tovostin collection contains only five complete pottery jars and a number of sherds. These jars appear to be reproductions of bronze vessels. I am in perfect agreement with the author in regarding the type of the so-called Scythian kettle as of Central-Asiatic origin, particularly emanating from the region of Minusinsk, and I avail myself of this opportunity to state again that, contrary to what has been asserted by superficial observers, this type meets with no counterpart in China, and that China has nothing to do with its development.

I was but able to touch here on some of the problems suggested by the rich and solid publication of Mr. Tallgren, which it is a pleasure to announce. In a letter he has informed me that after the war he is planning to come to America for a study of our museums; he may be sure that his visit will be warmly welcomed.

B. LAUFER

Hindu Achievements in Exact Science. A Study in the History of Scientific Development. BENOY KUMAR SARKAR. Longmans, Green and Co.: New York, 1918. 82 p.

The ethnologist is always gratified at a book in which the achievements of a people outside the pale of our narrow culture-sphere are vividly and forcibly expounded. Professor Sarkar desires to furnish for popular consumption "some of the chronological links and logical affinities between the scientific investigations of the Hindu and those of the Greeks, Chinese, and Arabs," without going into technical details or relating the migration of ideas. He briefly sets forth, without giving new facts, what the Aryan stock of ancient India has accomplished in mathematics, astronomy, physics, chemistry, metallurgy, medicine, and natural history. Owing to its simple and succinct presentation, his book will doubtless find many readers, and I hope that these will not be confined to students interested in the history of science, but that also many ethnologists will imbibe its lessons, for all science has emerged from the domain of folklore. In making this recommendation, however, it is the reviewer's duty to call attention also to the weak points of the book. Mr. Sarkar does not entirely escape from the exaggerations of the specialist, but, what is far worse, writes from the standpoint of the extreme nationalist. The nationalist movement among the highly educated and intelligent Bengali is in itself an interesting phenomenon, yet, whatever the merits and drawbacks of nationalism may be (many of us who have the progress of mankind at heart are absolutely opposed to it), it must never be wedded to science. The history of science can be written only from the universal, broad-minded, and sympathetic viewpoint of humanity, and it makes no difference to the true humanist whether an idea or discovery is due to India or China, to the Greeks or the Arabs, to the Negroes or the Maya. Apodictic and dogmatic assertions, such as "the Hindu were the first to discover gold," and "the Hindu taught the world the art of extracting iron from the ores" (p. 68) cannot be subscribed to by any one; nor is it true that the Hindu discovered zinc during the fourteenth century; at a much earlier date zinc was extracted from the ore in Sasanian Persia and from that quarter became known in China. Caraka and Suçruta were assuredly great physicians, but it is hardly necessary to praise them at the expense of Galen or to belittle Theophrastus or Pliny. Mr. Sarkar's mind is too full of modern scientific facts and terminology and too prone to interpret and to project these without moderation into the thoughts of the Indians. In natural history,

> considerable power of observation was exhibited, as well as remarkable precision in description, and suggestiveness in expression. Their nature study was oriented to the practical needs of socio-economic life. It was minute and comprehensive, and so far as it went, avoided the fallacies of mal-observation and non-observation. Whatever be the value of the results achieved, the investigation was carried on in a genuine scientific spirit (p. 67).

These exuberant remarks are not warranted by the facts. I shall cite but one example: the ancient Indians classified the rhinoceros among the five-toed animals (cf. M. Chakravarti, Animals in the Inscriptions of Piyadasi); until the dawn of our science of zoölogy it was only the Chinese and the great Al-Beruni who knew correctly that the animal was possessed of three hoofs on each foot. In *The Diamond* (p. 65) I have given a good instance of how the modern Indian school proceeds to claim European discoveries as their own simply by reading into their texts what these do not say, and thus to proclaim the phosphorescence of the diamond as an Indian asset centuries before Boyle.

> India was the greatest industrial power of antiquity. It was the manufactures of the Hindu, which, backed up by their commercial enterprise, served as standing advertisements of India in Egypt, Babylonia, Judaea, Persia, etc. To the Romans of the imperial epoch and the Europeans of the middle ages, also, the Hindu were noted chiefly as a nation of industrial experts.

The mere intimation to conceive the Indians as industrialists and advertisers makes me shiver.

The compass as an early invention of India (p. 38) is adopted from Mookerji, but, as shown by me (this journal, 1917, p. 77), this interpretation rests on a fallacy. There is, moreover, no Sanskrit work which mentions the compass. The invention of gun-powder in India is nothing but a learned fable based on the antedating of recent texts, misunderstandings and misinterpretation of terms, seasoned with a strong dose of imagination and uncritical methods. According to the school to which Mr. Sarkar belongs, chemistry, that is, alchemy, is of perfectly indigenous growth in India: the Hindu chemical investigators of the fifth and sixth centuries A.D. even anticipated by one millennium the work of Paracelsus and Libavius, and the physico-chemical theories as to combustion, heat, chemical affinity, were clearer, more rational, and more original than those of van Helmont or Stahl. In my opinion, alchemy is an Egypto-Hellenistic science, first developed in Egypt in the first centuries preceding our era, and was thence transmitted to India and China.

Mr. Sarkar undervalues the expansion and influence of Hellenism, for he states,

> Every attempt on the part of modern scholars to trace the Hellenic or Hellenistic sources of Hindu learning has been practically a failure (p. 5);

and his patriotism culminates in the dogma,

> India's indebtedness to foreign peoples for the main body of her culture is virtually nil.

This is plainly unsound super-Indianism. My enthusiasm for India is no less than that of our author, but I have been inconsiderate enough to demonstrate that burning-lenses and the puppet-play or marionettes were derived by the Indians from the west; and there is no doubt in the minds of all unbiased students that India, especially as to mechanical inventions, owes a large debt, not to the Greeks, but to Alexandrine-Oriental-Hellenistic civilization of western Asia. I also hope to continue my studies in this direction and to furnish exact evidence for the dependence of Indian alchemy. The Indians, in my estimation, cannot be characterized as an inventive nation. There are many points, particularly as to the evaluation of Indian authors, their works and their dates, in which I am at odds with Mr. Sarkar, but discussions of this nature would require many pages and interest only the orientalist. Taken as a whole, his book is a valuable summary and worth reading.

B. LAUFER

Korean Buddhism. History—Condition—Art. Three Lectures by FREDERICK STARR. Boston: Marshall Jones Company, 1918. 104 p. 37 pl.

Korea has always been the step-daughter of Oriental science. There is, of course, the usual number of books, even those which pretend to give a history of the country, a few good papers by specialists, and a mass of worthless printed matter. There is neither a good grammar nor a tolerably satisfactory dictionary of the language. There is but one man, M. Courant at Lyons, France, who has a claim to the title of Korean scholar. Serious research is required for all branches of Korean culture and, above all, for Korean Buddhism. Professor Starr is fortunate enough to have made four journeys to Korea since 1911; thus he has had occasion to see a great deal, to hear and learn much, and to photograph much. He was especially attracted by Buddhism. His lectures make a pleasant causerie, and when the author recites his adventures and impressions, he is always entertaining, but, not having access to original sources, he sometimes treads on unsafe ground as soon as historical questions or Buddhist philosophy come to the fore. In discussing the introduction of Buddhism into Korea, Starr speaks briefly of the first missionaries Syun-to (Chinese Shun-tao), Mārānanda, and Mik-ho-cha (Chinese Mo-hu-tse, anciently Mak-gu-tse) and arrives at the following anthropological theory (p. 16):

> Sundo[1] was a man from Tibet; I suppose he represented the great Mongolian race, that he was a yellow man; Marananda, who brought religion to Pakche was a Hindu; presumably he represented the Caucasic peoples; he may have been dark, but our courts would probably have to call him a white man; Mukocha was called a black man, a negro, and probably really represented the Ethiopian race. Is it not interesting that the peninsula of Korea should have received its first generally spread religion through representatives of the three great races of the world, the yellow, white and black?

Shun-tao was not a Tibetan, but a Chinese monk, who arrived in the kingdom of Kokurye in Korea in A.D. 372 (not 369, as stated on p. 4). Tibet emerges from darkness not earlier than the seventh century A.D.,

[1] Throughout the author transcribes Korean names in their Japanese garb. This procedure is unfortunate, especially with respect to Buddhist nomenclature, and cannot be justified on any rational basis. Korean literature and Buddhism are derived from China, while Japanese Buddhism emanates from Korea (about A.D. 552); accordingly, Korean terms should first be given in Korean, then in Chinese, and finally be identified with their Indian equivalents. Every serious student of Buddhism knows Sanskrit, and all students of Buddhism are familiar with the Indian terminology, and can readily refer to one of the numerous handbooks of Buddhism if in search of explanation.

when Buddhism was first introduced, but in 369 there was no such community as Tibet, nor a Tibetan Buddhist. Mārānanda, who came from China to Korea in 384, may have been an Indian:[1] tradition designates him merely as a Hu, a term which usually refers to the Iranian and other tribes of Central Asia (cf. Courant, *T'oung Pao*, 1900, p. 320, and *Bibliographie coréenne*, III, p. 215). Mik-ho-cha is far from being an African; his name is purely Korean, and all that is known about him is that in the first part of the fifth century he came from Kokurye to the kingdom of Sinra or Silla. The notion of his black skin is purely legendary, as the first element of his name is written with a Chinese character that means "ink."

The highly developed literary cultivation of the Koreans and their achievements in the art of printing are well known. The Chinese translation of the Tripitaka, the sacred canon of the Buddhists, was first printed in A.D. 972. The Koreans followed with the second edition in 1010, which is the oldest and best of all the different editions now in existence, and a copy of which, brought to Japan in the latter part of the fifteenth century, is still preserved in Tokyo (cf. Bunyiu Nanjio, *Catalogue of the Buddhist Tripitaka*, p. xxiv). Starr (p. 26) mentions only a later edition. In many instances the superiority of the Korean text to the Chinese and Japanese versions has been upheld by our scholars.

A few notes are devoted to the curious *miryek* of Korea (p. 23). This word is Korean (not Japanese) and simply means "stone men" (cf. T. de Lacouperie, "The Miryeks or Stone-Men of Corea," *Journ. Roy. As. Soc.*, 1887, with illustration); it has nothing to do with Mi-rok, the Sino-Indian name of Maitreya. I believe that Starr is generally correct in his theory that rude stone figures belonged to the ancient national religion of Korea, and were subsequently adopted by Buddhism and shaped into Buddhistic images. In my opinion, there is some connection here with the stone statues (the *kameniye baby* of the Russians) of Mongolia, southern Siberia, and Russia; but this is a complex problem which remains to be studied at close quarters.

According to Starr (p. 50) "Buddha taught that we end in Nirvana." Buddha taught nothing of the kind. The Nirvana was not for the multitude, but was the highest and ultimate goal and reward of the enlightened one, the Buddha; it meant the extinction of the individual and his absorption in the absolute and infinite.

[1] The Sanskrit name does not allow of an inference as to nationality, as non-Indian monks also usually have an ecclesiastic name in Sanskrit.

A problem not touched upon is the relation of Korean Buddhism to Lamaism. W. W. Rockhill (*China's Intercourse with Korea*, p. 60) has called attention to the fact that the Buddhism of Korea presents many curious analogies with the Tibetan form of Buddhism, and that in the style of church architecture, painting, etc., it has certainly been influenced by it. This coincidence may be explained from the fact that during the seventh century Korean monks were in the habit of making pilgrimages to India, and some of these traveled by way of Tibet and Nepal. The famous Chinese monk and pilgrim, Yi-tsing, has recorded the travels of seven Korean Buddhists (cf. *T'oung Pao*, 1892, p. 462; and Chavannes, *Voyages des pèlerins bouddhistes*, pp. 32–36). Chavannes' work is a complete translation of Yi-tsing's book and merits preference over the rendering of Beal (quoted in the Notes, p. 99), which is incomplete and rather inexact.

The lecture on art does not quite satisfy a student of Buddhist archaeology and iconography. The problem to be pursued would be to study the Korean types and forms in their relation to those of China, Central Asia, and India, and finally to answer the question as to how the Koreans have developed, assimilated, or digested this foreign art and evolved a style of their own. The illustrations form a valuable feature of the book, but no discrimination is made between real art-works, as, for instance, the Bodhisatva in plate IX, who rivals the best Chinese sculptures of the T'ang period, and inane, mechanical modern reproductions, such as the hideous Maharajas on plates XIX–XXI, who are hardly worth the cost of illustration. The paintings on plates XXXI–XXXIII, being reproduced on too small a scale, are unfortunately lost.

It is gratifying to learn that there is a modern movement in the Buddhism of Korea which the author says seems to show that it has real vitality, and he thinks that it may have a political part to play: "if hostile to Japan, when the crisis comes, as it surely will come, when Japan will be tried out again and once for all on Korean soil, Korean Buddhism may be the decisive element in that moment of test." Professor Starr's lectures must be regarded as an *hors d'œuvre;* he has accumulated considerable material on the subject which he should be urged to publish at the earliest possible moment.

B. LAUFER

Quelques considérations sur les jeux en Chine et leur développement synchronique avec celui de l'empire chinois. Captain GEORGE E. MAUGER. (Extrait des Bulletins et Mémoires de la Société d'Anthropologie de Paris. Paris, 1917.) 44 p., 16 text-figures.

This is a highly interesting and suggestive study in which the author endeavors to trace the development of a certain number of Chinese games and to reveal their relations to political and social conditions. The latter idea is novel and merits attention and consideration. The obvious difficulty is that, while we are familiar with the present-day games, their history has been little studied and to a large extent is still very obscure, chiefly owing to the fact that most of the ancient books on games in which a considerable literature still existed under the Sui dynasty (A.D. 590–617) are now lost. Of draught-games, the *wei-k'i* ("game of blockade") is the national game of China traceable to ancient times; it is a war-game for military instruction, a field of tactical problems. Captain Mauger describes it well and concludes,

This game may be regarded as representing the individualistic period of the empire. Then the great chief domineering through his will and his power has but the one object to conquer for his people the largest possible space of territory. Soon, however, the empire develops; the territories grow more considerable and necessitate a more complex organization; feudalism appears, the great chief directs from his palace his vassals who fight for him, and we have chess.

He goes on to explain the ideas underlying chess from the state organization of the Chou dynasty (1122–247 B.C.). I do not deny that this sort of historico-philosophical interpretation of objects is ingenious, but what is ingenious is not necessarily true. The author, of course, accepts as a fact that chess was known in the China of that period, and its origin is even traced to eleven centuries B.C. in a tradition furnished to him by a Chinese friend and taken from a classical schoolbook. The information of our Chinese friends may well serve us as a guide, but must never be accepted without serious and critical examination. This misconception is caused by the verb *yi*, which means "to play at *wei-k'i*," but is erroneously translated also "to play at chess." Granted that it could have the latter meaning, of what help would it be? We have no description of any game approaching chess from that remote period. The present term for chess (*siang-k'i*, "elephant-game"; the word *k'i* denotes any game played on a board with counters) makes its first appearance during the sixth century A.D., but careful examination of subsequent Chinese documents decides in favor of the opinion that this game is not identical with the modern chess, but was one of astronomical lore referring to sun, moon, and constellations; for the word *siang* signifies also "star, constellation." This type, nevertheless, may have also been a war-game, for, in the literature of the Sui, the bibliography relating to this game is placed at the end of military literature (*Sui shu*, ch. 34, p. 9); while the following T'ang dynasty assigns books on games to the department of liberal arts (*T'ang shu*, ch. 59, p. 12). *Wei-k'i* and chess, in the same manner as music and caligraphy, belong to the fundamentals of a liberal education and to the prerequisites of a gentleman. The present chess exhibits some principal differences from the Indian game, but, on the other hand, also very striking coincidences with it, so that, in my opinion, Chinese chess is a cross-breed between a national Chinese game (now extinct) and the *caturaṅga* imported from India. It will thus be seen that the history of chess in China presents a problem of great complexity which can be solved only by minute documentary study. A philosophy of games, as well as of other ideas, can hardly be attempted before their real history is completely and exactly ascertained. The author observes, with reference to chess,

> Pour nous il nous semble que l'idée du jeu aurait pu aussi bien être introduite et adaptée en se simplifiant dans l'Inde venant de Chine, que le contraire si l'un et l'autre n'a pas eu une origine indépendante.

This opinion is not acceptable. Chess is a thoroughly national game of India without a trace of Chinese influence. If Captain Mauger would try to eliminate the Indian features from Chinese chess, he would discover several traits to support his theory; for instance, the king of our chess is replaced by a general in China, because the majesty of the emperor was so exalted that he could not figure in a game of the vulgar, and he himself never went to war, but sent a general to fight his enemies.

The author is perfectly correct in deriving the divination games of China from India, but then it is somewhat surprising that he overlooks the Indian origin of dice; to him the origin of dice in China remains obscure. The Indian (Pāli) word *pāsa* ("die, dice") bears no relation whatever to Chinese *p'ai* ("board, cards, domino"), as he thinks, but the Sanskrit word *prāsaka* (also *pāça*, *pāçaka*) has been adopted by the Chinese in the form *po-lo-sai* (anciently **pa-la-sak*), a very accurate transcription of the Indian model, which appears as early as the fifth century in the Chinese version of the Brahmajālasūtra (§33), translated by Kumārajīva in 406. Subsequently this was abbreviated into *sai* (**sak*); and the modern vernacular names for dice (Peking *šai*, Middle China *sö*, Canton *šik*) are nothing but adaptations of the same Indian word (cf. also Siamese *saka*). Sanskrit *prāsaka* and Chinese *po-lo-sai* denote in particular the Persian game *nard* (our backgammon), introduced into China in the first part of the sixth century, on which I expect to report in the near future. As is well known, dice are of immemorial antiquity in India, being used both for divination and gambling. A standard book on Indian dice was contained in the literature of the Sui. It is positively certain that in the period of Chinese antiquity down to the first centuries A.D. no dice were ever employed. The methods of ancient Chinese divination are perfectly known, being mainly concerned with the consultation of the tortoise and reading the cracks and lines in the burnt shell of this divine animal. These methods could not lead, and in fact did not lead, to the development of any game.

Captain Mauger devotes the greater part of his article to a study of dice games, dominoes, and playing-cards and their interaction. Of cards he describes a number of local variations, also several hitherto unknown, and makes a substantial contribution to the subject. He doubtless possesses a good practical knowledge of Chinese games and others, but should join hands with one in Paris who is posted on historical questions and would lead him more safely through the complex labyrinth of research of this character. Games are hard nuts to crack. It is also somewhat dangerous to write on Chinese subjects without some knowledge of the language and without a clear perception of historical development. It is a rather disturbing *faux pas* to characterize the Chinese of the sixteenth and seventeenth centuries as "un peuple essentiellement féodal" (p. 41), since feudalism was destroyed at the end of the third century B.C. I would finally remark that A. van der Linde, the famed author of the history of chess, was not a Dane (p. 19), but a Hollander, and that the name of another Hollander, quoted twice (p. 29), is Vissering (not Visserung).

B. LAUFER

The Encyclopaedia Sinica. SAMUEL COULING. Shanghai: Kelly and Walsh (or Oxford University Press), 1917. 634 p.

Although I am not a believer in making cyclopædias, as in the present state of science we have better things to do and our knowledge of China is still far from being complete, the work of S. Couling merits a hearty welcome as a pioneer and as the fruit of hard and patient labors. If it does not satisfy in many points the specialist, it will be a useful reference-book to the public at large and to any one in quest of speedy information on a subject connected with China. The author modestly calls his book a beginning and promises greater completeness in future editions; but the beginning he has made is a good one, and he has provided a basis and framework for a larger and finer building to follow.

B. L.

152

对古柯和槟榔咀嚼的见解

both curing and rainmaking functions and organization, we should recall another Zuñi tradition, the tradition that the *shi'wanakwe* is the oldest of the Zuñi fraternities.

ELSIE CLEWS PARSONS

POLYNESIAN TOMBS: A CORRECTION

IN a note published in this journal (vol. XX, no. 4, p. 456) I proposed to amend in some particulars Dr. Rivers' conclusions on "Sun-Cults and Megaliths in Oceania" (vol. XVII. (1915) p. 443). Unfortunately I was on service abroad and having only jotted notes to work upon it was impossible to correct any oversight that might get in. Dr. Rivers points out to me that on p. 460 I have overlooked the strong evidence he brought forward proving the connection between the *areoi* and sun-worship, that it was not a mere inference of his, but a fact vouched for by Maerenhout. This gives the *areoi* a very different aspect from that which I suggested. I must apologize for this oversight; having only extracts to work on I looked to Dr. Rivers' criticisms to prevent any inaccuracy from getting into print; but unfortunately they were attracted by other matter and so missed this unfair statement of his own case.

A. M. HOCART

EXETER COLLEGE,
OXFORD, ENGLAND

COCA AND BETEL CHEWING: A QUERY

IN his work *The American Indian* (p. 30), Dr. Wissler calls attention to the striking coincidence between the method of coca-chewing, as it prevails along the west coast of South America, and the betel-nut consumption in southeastern Asia and Melanesia, in that both narcotics are taken together with pulverized shells or ashes. The analogy is so manifest and complete that the assumption of an historical connection becomes inevitable. The question arises, however, whether the American practice is pre-Columbian or merely the result of circumstances growing out during the period of the *Conquista*. Being engaged for years on the collection of materials for a history of the cultivated plants of this continent, I recently had occasion to read a book by Max Steffen, entitled *Die Landwirtschaft bei den altamerikanischen Kulturvolkern*

ceremonials. In Keresan *shuma* means the dead, the skeleton. The *shumaekoli* masks of both Laguna and Sia were passed over to the Zuñi, but the Zuñi fraternity antedated these gifts. Belonging apparently to the same complex of concepts as the *shi'wanakwe*, the *shuma'kwe* may have been at Zuñi a later institution.

23

(Leipzig, 1883), which seems to be little known in this country. There we are informed as follows (p. 90): "When Oviedo (*Hist. gen., lib.* 26, c. 30) asserts that the Chibcha chewed the *hayo* [-coca] leaf with lime like the Peruvians, he is surely wrong in this point; for Piedrahita (lib. 1, c. 3), a careful and trustworthy writer, reports that they had formerly chewed the plain leaf, and that only since the arrival of the Spaniards they have added the lime of snails introduced by some Spaniards and called *popóro*, as well as another substance, styled *anua*, which intoxicates the senses." The Spanish text of Piedrahita runs as follows:

> De antes usaban mascar esta yerva simple, pero ya la mezclan con cal de caracoles, que han introducido algunos Españoles, y llaman Popóro, y con Anua, que es otro genero de masa que embriaga los sentidos.

On the other hand, however, in speaking of the coca cultivation of ancient Peru, Steffen says (p. 116), "As at present, so also prior to the Conquest, the leaf was rolled up into small globules, usually with unslaked lime, and thus chewed." The authority for this statement is Oviedo (*Hist. gen., lib.* 26, c. 30). Here, Oviedo is upheld, at least not contradicted, by Steffen, while in the case of the Chibcha of Colombia he is wrong. If he should really err in the latter case, is it not equally possible that he may err in the case of Peru? Or if we assume pre-Spanish practice for Peru, why reject it for Colombia, merely on the authority of Piedrahita? Steffen's standpoint seems to me inconsistent. T. A. Joyce (*South American Archaeology*, p. 122) remarks with reference to Peru, "It has been said that lime was not used in times previous to the conquest." Unfortunately he does not tell us by whom it has been said, and on what evidence the statement is based ("one of the Conquerors says" is one of the rubber-stamps gracing his pages).

I appeal to Americanists for help in elucidating this question which is one of importance. If the Spanish importation theory be correct, the historical problem would naturally be much (I even feel like saying, unfortunately, too much) simplified. With all respect for Piedrahita, however, I am not inclined to accept it solely on his testimony. Are there other ancient Spanish sources touching this point? Is there any archaeological evidence? Have remains of coca leaves with or without lime ever been discovered in ancient graves of Colombia and Peru? Any information will be gratefully appreciated.

B. LAUFER

153

古代中国的多胞胎情况

AMERICAN JOURNAL
OF
PHYSICAL ANTHROPOLOGY

EDITED BY

ALEŠ HRDLIČKA

VOLUME III

WASHINGTON. D. C.
1920

MULTIPLE BIRTHS AMONG THE CHINESE

BERTHOLD LAUFER

Field Museum of Natural History, Chicago

INTRODUCTION

The Chinese Annals contain not only records of human events, but also of unusual natural phenomena which left a deep impression upon the minds of the contemporaries. In the early days of historiography, when occurrences were chronicled day by day and year by year, the two categories of human and natural events were noted indiscriminately, merely in the chronological succession as they happened. In the introduction to the Shu king we read, for instance, "The king's uncle, the prince of T'ang, found a head of grain, two stalks in different plots of ground growing into one ear, and presented it to the king." In the Bamboo Annals (*Chu shu ki nien*) this feature is still more conspicuous: solar eclipses, meteoric falls, earthquakes, droughts, extraordinary phenomena in the growth of trees, appearance of a fung-hwang (so-called phœnix), rain of particles of earth, unusual thunderstorms, and other phenomena are there on record, being interspersed with the record of imperial and military affairs. Beginning from the Annals of the Former Han Dynasty (*Ts'ien Han shu*), a novel departure from the old practice was instituted in as much as the natural events were detached from the general narrative to be relegated to a special section, entitled "Records relating to the Five Elements" (*Wu hing chi*). The majority of official annals has adopted this practice. These chapters contain most interesting information, not for the historian, but for the scientist, and therefore merit close study. They give detailed lists, with exact reference to date and place, of great catastrophes, such as famines, droughts, locust-pests, inundations, hail-storms, landslides, earthquakes, conflagrations, excessive cold, electric storms in the winter, etc., abnormal phenomena and monstrosities in domestic animals and human beings, cases of insanity, abnormal customs and practises, etc. It is to this department of records that we owe our principal information on a subject which has not yet been discussed,—the frequency of multiple births among the Chinese.

Amer. Jour. Phys. Anthrop., Vol. III, No. 1.

In ancient times, under the Chou dynasty, the officer presiding over the people (*se min*) was obliged to keep a register of the population. All individuals were recorded from the age when the teeth appear. A separate count was taken of males and females; every year, the number of births was added, while the number of dead was taken off the register (cf. E. Biot, Le Tcheou-li, Vol. II, p. 353). We cannot but regret that documents of this character have not survived. No allusion to twins or other plural births is made at that period.

The chapters Wu hing chi of the two Han Annals contain no records of multiple births. The Wei shu gives a single case of a quadruplet birth. Triplets, but only two cases, are first recorded in the Books of the Tang Dynasty, and there is a long list of them under the following Sung dynasty. There is one case of triplets of early date, not on record in the Annals, but in the *Sou shen ki*, written by Yü Pao in the early part of the fourth century, who reports that "in A.D. 243 there was a woman who gave birth to three sons." I have not embodied this case in my statistical review of the matter, as the work in question is a Taoist book of marvels, and as the extant edition presents merely a retrospective make-up (cf. Wylie, Notes on Chinese Literature, p. 192).

While triplet and quadruplet births are mentioned in the Annals with comparative frequency, they hardly trouble about twins, save a few cases of united twins. This omission may indicate one of two possibilities: either twin-births were too common to attract much attention, or were too rare to be worthy of notice. This alternative cannot be decided without a solid foundation of statistical material, which unfortunately we do not have. At the outset I am not disposed to assume, on a merely empirical basis, a high degree of fecundity of the Chinese woman or a relative frequency of twins; for it is a common experience of our time that personal opinions and impressions along this line are seldom, if ever, upheld by the results of statistical research. Restraint in this case is the more commendable, as in regard to twin-births in Annam we have the following observation of Dr. A.-T. Mondière ("Monographie de la femme annamite," *Mémoires de la Société d'anthropologie*, II, 1875, p. 474): "Les grossesses doubles sont excessivement rares chez la femme annamite. Sur les 153174 naissances que j'ai relevées sur les cahiers des villages de toute la Cochinchine de 1872 à 1877 inclus, je n'ai trouvé que 15 accouchements de jumeaux. Soit 1 sur 10211 naissances. De plus, un arrondissement particulier, celui de Bentré, semble avoir ce privilège, car sur 15 accouchements gémellaires il en a 9 à lui seul, c'est-à-dire 60 pour 100. Les six autres arrondissements (sur 19) qui en ont présenté: Bien-hoà, Chau-Doc, Saigon, Soctrang, Tan-an, Tay-ninh, n'en ont eu chacun qu'un seul cas, en ces six années. D'après ce que les autorités cambodgiennes m'ont déclaré, les jumeaux seraient plus fréquents chez eux, et d'une façon assez sensible, mais ils n'ont pu me fournir de chiffre exact."

A real investigation of the problem in question is impossible for the present, as we lack any vital statistics for the Chinese Republic. Nevertheless I venture to hope that the facts and observations given below will be of some interest to anthropologists. In order to critically balance the data furnished by the Chinese Annals, it would be indispensable to have reliable birth statistics for China, to know the birth-rate for the different provinces, and to depend on good records showing the total number of plural births for at least a decade. In default of such material in the mother-country I anticipated to receive at least some data from those countries outside of China with a large Chinese population, although it must be taken into account that social and economic conditions of the Chinese abroad are different and that, above all, Chinese emigrants hardly ever take their families along, but intermarry, when settled, with women of other nationalities. I have not yet been able to obtain relevant statistics from the British, French and Dutch colonies; but what I have found thus far is not very encouraging. The Birth Statistics for the Registration Area of the United States for 1915 (Washington, 1917) give a total of 74 births (33 males and 41 females) among the Chinese for that year, but nought else.

According to a communication of Dr. William H. Davis, chief statistician in the Bureau of Census, Washington, D. C., there were, in the years 1915–17, 309 births among the Chinese in the registration area for births in the United States (California not being admitted to the registration area is not included), only one pair of twins appearing in this total. The State of California gives in its vital statistics only the number of births and deaths of its Chinese populace, without touching the question of plural births. In 1916 there were 425 births (compared with 727 cases of death); in 1917, 419 births (compared with 818 cases of death) among the Chinese of California (Twenty-Fifth Biennial Report of the State Board of Health of California for the Fiscal Years from July 1, 1916, to June 30, 1918, Sacramento, 1918, pp. 201, 203, 205, 207, 224). The statistics of Mexico contain merely the number of Chinese living in the various provinces, the total, as taken in the third and last census of 1910, being 13,118 men and 85 women = 13,203 (Estados Unidos Mexicanos, *Boletin de la Direccion General de Estadistica*, Num. 5, p. 37, Mexico, 1914), but no tables of births.

A literal translation of all cases of triplet and quadruplet births, as they are chronicled in the Annals, has been prepared by me. In every case, the exact date, the name of the family, the social status of the father, and the place where he lived are given; also the distribution of sex in each birth is indicated. As this material would be unintelligible without the use of Chinese characters, it is here omitted. Readers interested in this phase of the work may be referred to the New China Review of Shanghai, in which the complete article will be published. For some of the bibliographical references mentioned on the following pages I am under obligation to Dr. A. Hrdlička, Curator of Physical Anthropology in the U. S. National Museum of Washington.

UNITED TWINS

The Chinese Annals have preserved a few cases of twins grown together at birth. In this case, the question naturally is of twins produced from a single ovum.

In the fourth year of the period Kien-hing (A.D. 316), under the Emperor Min of the Tsin dynasty, a woman of the family Hu, when she was at the age of twenty-five, the wife of Jen Kiao, a minor official (clerk) in the district of Sin-tsʻai (prefecture of Ju-ning, Honan), gave birth to female twins grown together in the region of the abdomen and the heart, but separated above the breast and beneath the navel.—*Sung shu*, Ch. 34, p. 28.

In A.D. 487 (under the Emperor Wu of the Tsʻi dynasty), the wife of Wu Hiu, one of the people of Tung-tsʻien in Wu-hing (now Hu-chou fu, Che-kiang) gave birth to male twins grown together below the chest down to above the navel.—*Nan Tsʻi shu*, Ch. 19, p. 16b.

In the fourth month of the third year of the period Yi-fung (A.D. 678), King-chou (Kan-su) presented the Court with two infants the hearts of which were connected, but each with a separate body. Formerly it had happened that the wife, *née* Wu, of Hu Wan-nien, a soldier of the guard in the district Shun-ku (Kan-su), gave birth to twins, a male and a female, whose breasts were connected, but who, for the rest, had individual bodies; when separated, both died. At a subsequent birth it was thus again. The twins were boys, and were brought up. In the above mentioned year they had reached the age of four years, and were presented to the Court.—*T'ang shu*, Ch. 36, p. 21.

In A.D. 1610, the wife of Li Yi-ch'en of Fan-ki (in Tai-chou, Shan-si), *née* Niu, brought forth two girls with their heads and faces grown together, but with separate arms and legs.—*Shan-si t'ung chi* ("Gazetteer of Shan-si Province").

Two Chinese twins grown together, born in 1887, were shown by Barnum and Bailey in 1902, and at that time were still unseparated and well. Cf. R. Virchow, Xiphodymie (*Z. Ethn.*, 1891, pp. 366–370).

The modern Gazetteers occasionally record the birth of twins, not, however, on account of any special interest attached to the fact itself, but merely in order to emphasize the interest in the vitality of twins (cf. W. A. Macnaughton, The Longevity of Twins, *Caledon. M. J.*, X, pp. 127–129, Glasgow, 1915). The following examples from the Gazetteer of Hwa-yang (*Hwa-yang hien chi*, Ch. 43, p. 4) will suffice to illustrate this feature.

The wife of Chu Ch'ang-hwa, *née* Lin, had 14 sons, among these 2 pairs of twins, who did not die prematurely, but are still alive.

The wife of Chung Se-kin, *née* Tsou, had 9 sons, among these one pair of twins still alive.

The wife of Chung Chao-k'in, *née* Chang, had 9 sons, among these one pair of twins still alive.

The wife of Li Ch'ao-kung, *née* Lin, had 8 sons, among these one pair of twins still alive.

TRIPLETS

Following is a summary of the Chinese data. For the period of the T'ang dynasty (A.D. 618–906) only two cases of a triplet birth are on record in the Annals. In A.D. 775 a woman of the family Chang gave birth to one male and two females, and in A.D. 905 triplets (males) were born by the wife of P'eng Wen, one of the people of Ju-yin in Ying-chou (Ngan-hwi or An-hwi Province). For the Sung period the data of triplet births are fuller than for any other dynasty. From A.D. 960 down to A.D. 1150 we have a total of 110 cases, listed with exact dates, family and place names, father's social status, and sex distribution in each triplet birth. For the time from A.D. 1023 to 1126 no list of names is given, but merely a statistical record which covers several reign-periods of emperors. It is here reproduced in tabular form.

RECORDS OF MULTIPLE BIRTHS IN CHINA, 1023–1126

Period	Years	Quadruplets (Males)	Quadruplets (3 Males, 1 Female)	Triplets (Males)	Triplets (2 Males, 1 Female)	Total
1023–68.......	46	2	–	44	1	47
1068–83.......	15	1	1	84	–	86
1084–99.......	16	2	–	18	1	21
1100–26.......	27	1	–	19	–	20
Total.....	104	6	1	165	2	174
		Quadruplets total 7		Triplets total 167		

While the preceding cases are not recorded in the way of vital statistics, but solely as unusual events, the above table conveys the impression of embracing a fairly accurate register of all multiple births (save twin births), which took place within the span of a century. The proportion of quadruplet to triplet births in this period is 1: 23.86. The total of triplet births on record during the Sung epoch, accordingly, is 110 + 167 = 277. The total of quadruplet births during the same period is 7 (as shown by the above table) + 7 (recorded in the following section) = 14. The proportion of quadruplet to triplet births for the entire period of the Sung is 1:19.78; while the proportion for the entire period of Chinese history here considered (473–1643) is 1:10.8. This calculation is based on a total of 324 triplets and 30 quadruplets.

There are no multiple births on record in the chapter *Wu hing chi* of the *Kin shi*. The *Yüan shi* (Chs. 50–51), covering the period from 1260 to 1367, contains only 15 cases of triplets (all males), recorded under the years 1261, 1265, 1273, 1285, 1291, 1297, 1300, 1327, 1328, 1335, and 1363. In the years 1273, 1297, 1335, and 1363, two cases are listed for each year; and it is of especial interest that in two instances we have two cases of triplets in the same family, the interval between the two being in either case given as three years. According to Dr. Puech, to whom we owe excellent studies on the causes of multiple births, the more children a woman has had at close intervals, the more she will be inclined toward these physiological anomalies. Three women admitted in the St. Petersburg Midwives' Institute between 1845–59 in their fifteenth pregnancy had triplets, and each had triplets three times in succession (J. M. Duncan, Fecundity, p. 71).

For the period of the Ming dynasty (1368–1643) we lack official records; but the section *jen i* of the *T'u shu tsi ch'eng* gives a list of 30 cases of triplet births, extracted from the provincial and local Gazetteers, and covering a period from 1404 to 1626. In 1413, 1515, and

1520, two cases are recorded in each year. In view of the fact that this material is extracted from a number of scattered books, it cannot lay claim to completeness; the figure 30 is certainly much removed from reality, but even if multiplied by 3 or 4, it is left far behind the total of the Sung period. On the whole, the impression prevails that the number of multiple births has steadily been on the decrease from the days of the Sung. This would agree with an anthropological theory to the effect that the phenomenon of multiple births in man represents a survival of or reversal to his former animal state and that with the advance of civilization the number of such births is liable to decline. There is a correct biological viewpoint in this hypothesis, but it does not account for all facts connected with the phenomenon, and, above all, conflicts with given data and statistics. It is not brought out by the vital statistics of any European country that the frequency of plural births is on the decline; on the contrary, in France, for instance, it is surprisingly high (see below). Further, if that theory were correct, we should naturally anticipate to find the greatest number of multiple births among primitive tribes, which for all we know is not the case. Hardly a century has elapsed that records of plural births have been taken in Europe, and this period is too short to allow us to indulge in much speculation on the subject.

According to the Statutes of the Manchu Dynasty, it was decreed in 1663 that in the case of a triplet birth or a twin birth of a boy and a girl, if it should occur among the people of the Eight Banners, a special report should be submitted to the Board of Rites; if it should occur in the provinces, the governor of such province should report to the Board of Rites, which would have to forward it to the Board of Finance, the latter to grant a premium of five piculs of rice and ten pieces of cloth. In 1674 it was ordered that a special report should be made solely in the case of male triplets, but not in the case of twins or female triplets. In 1684 an edict ordained that in the case of male triplets the Board of Rites and the Board of Finance should submit a joined report to the Throne, and that rewards should be authorized in accordance with law. This benevolent attitude toward the energetic propagators of the race was not an innovation of the Manchu, but a heritage of the Ming; for under the Ming we are frequently informed of special grants of food, cloth, and even paper money, made to these involuntary heroes from public funds.

It may hence be inferred that under the Manchu régime a register of male triplets was kept, and presumably is still preserved in the archives of Peking. If it should ever be published, the fact must be borne in mind that female triplets were not officially reported. Meanwhile we are thrown back for that period on the local and provincial Gazetteers, which in the chapter on untoward or abnormal events sometimes record cases of plural births.

To cite a few instances of this kind in the period of the Manchu dynasty,—the Gazetteer of Ju-chou in Ho-nan (quoted above) enumerates four cases (all males), which occurred in 1770, 1785, 1824, and 1833. In 1797 a triplet birth occurred in Hwa-yang (prefecture of Ch'eng-tu, Se-ch'wan); the case was reported to the throne, and by imperial favor, a picul of rice was granted to the father, Yang Kwo-yü (*Hwa yang hien chi*, Ch. 43, p. 3). The Gazetteer of Mong-chou (prefecture of Hwai-k'ing, Ho-nan) cites only two cases for the years 1682 and 1736. Most Gazetteers which I have looked up are disappointing: thus the Gazetteer of Shen-si Province (*Shen-si t'ung chi*) contains only two cases of triplets, recorded for the years 1470 and 1729.

In the Gazetteer of the Prefecture of Sung-kiang, three cases of triplets are recorded between 1367 and 1640 (according to D. J. Macgowan, Cosmical Phenomena Observed in the Neighborhood of Shanghai, *Journal China Branch R. As. Soc.*, II, 1860, p. 74).

The data of the Chinese certainly are defective, and cannot entirely satisfy the anthropologist. We miss, for instance, data concerning the ages of mother and father and order of birth in triplet deliveries (*rang chronologique de l'accouchement* of the French statisticians). Above all, we should desire information as to the vitality and fecundity of the offspring. What the Chinese may boast of, however, is the fact that they possess lists of plural births for periods of the past when nothing of the kind was ever attempted in any country of Europe. In the vital statistics of France, plural births have been recorded only from 1858; and in no country of Europe did they receive any attention before the nineteenth century (in Berlin from 1825).

The sum of 277 triplet births for the Sung and 324 for the time from the T'ang to the Ming inclusive may seem a high figure to the uninitiated; in fact, however, it is strikingly low. During the four years 1907–1910 there was in France a total of 327 triplet births; 91, 93, 68, 75 in the respective years, making a mean average of 81.75 per year (Statistique général de la France, Statistique du mouvement de la population, Paris, 1912, p. 56). There were, accordingly, more triplet births in France during those four years than in China in the course of many centuries. Or, to cite another example, in the period 1835–47, there were in Bavaria 1,050 triplet, 56,062 twin, and 3,413,763 normal births. The frequency of triplets varies in different years and in different countries. In 1855, triplets were produced in Scotland by 11 mothers out of 92,300 births; that is, one in 8,391. Triplet births in Scotland from 1855 to 1901, a period of 47 years, numbered 644, and averaged 116 per million confinements (C. J. Lewis and J. N. Lewis, Natality and Fecundity, p.62). I do not go any further into the question of the frequency of triplets in Europe and the proportion of triplets to twin and normal births, as the Chinese data are not comparable, and as figures of total births are lacking for the Sung period. Judging from our experience, it must be stated, however, that the Chinese data can hardly be complete; but there is no way of correcting or adjusting the figures, which we are simply compelled to take for what they are worth. The reader should not forget that the material furnished by the Chinese Annals is not intended as statistics, but merely as a record of extraordinary events in human life. In order to give a certain perspective to the number of multiple births, some data concerning the population may follow here. According to the calculations of E. Biot ("Mémoire sur la population de la Chine," *Journal asiatique*, 1836, p. 461), the population of China under the Sung totaled 43,388,380 in the year 1021, and rose to 100,095,250 in 1102; again in 1223, it amounted to only 63,354,005 (in consequence of the loss of northern China to the Kin). These figures, in all probability, are too high; for they are estimated on the number of families given in the Chinese records, the assumption being made that the mean average of the number of individuals in a family is 5, which, in my opinion, is too high a figure.

The total number of triplets recorded for the T'ang and Sung periods is 279. The distribution of sex in this number is as follows: 273 all males, that is, 97.8 per cent; 4 consisting of 2 males and 1 female, that is 1.4 per cent; 1 consisting of 1 male and 2 females, that is, 0.04 per cent; and only one consisting of 3 females (0.04 per cent). Again, the 15 triplet births of the Yüan dynasty and the 30 of the Ming are all males exclusively. The above percentages perhaps give an approximate clew to the actual frequency of sex in triplet births, as far as China is concerned.

C. J. Lewis and J. Norman Lewis (Natality and Fecundity, p. 61, London, 1906), who base their remarkable study on the birth registers of Scotland for the year 1855, during which year there were 11 triplet births in that country (3 males, 5; 3 females, 3; 2 males and 1 female, 3), offer the following conclusion in regard to the distribution of the sexes: "There is a strong probability that in any given occurrence of triplets the children will all be of the same sex, either all males or all females. If the same ratio held in other nations and in other years, it would amount to a law of triplet production that in over 70 per cent of cases the newly-born children are all of the same sex."

In the period from 1858 to 1865, there were in France 1,005 triplet and 4 quadruplet births; among the former, there were 280 entirely males, 218 entirely females, 256 consisting of 1 male and 2 females, and 251 consisting of 2 males and 1 female. The number of twin births during the same period amounted to 83,279; of these 28,056 were two males, 26,310 two females, and 29,363 consisting of one male and one female (A. Puech, *Annales d'hygiène publique*, XLI, 1874).

Of the 277 triplets recorded for the Sung period, the social standing of the fathers is given in only 110 cases, while the remaining 167 cases are merely recorded as chronological-statistical events. Among the 110 cases, the social status of the fathers is distributed as follows:

		Percentage
Rural population	85	76.7
Field-laborers	1	1.1
Workmen	1	1.1
Soldiers	22	20.0
Petty officials	1	1.1
Total	110	100.0

In the Yüan period, 14 common people and 1 soldier share in the 15 cases of triplets placed on record. In the Ming period, 28 common people and 2 soldiers assume responsibility for 30 cases of triplets recorded. It will thus be seen that the bourgeoisie, inclusive of officials, gentry, and merchants, has no share in these records. Peasants and laborers, of course, formed the majority of the populace; but there is no reason why triplet births, if they had occurred in the upper classes, should not have been reported or recorded.

In arranging our data according to families, we arrive at the result that the members of the families Li, Wang, Chang, and Liu, take the uppermost rank. The male Li reach the score with $16^3 + 1^4$, while two female Li figure with 2^4; in the years 986 and 996 respectively we have two male Li participating in triplets. The record of the Wang is $13^3 + 2^4$ (plus one female Wang 1^3); the Chang follow with $9^3 + 2^4$, plus two female Chang (2^3), and the Liu with $9^3 + 1^4$, two members of this family being conspicuous in the same year (1016). This does not mean, of course, that these four families are more prolific than others, but is merely the index of the fact that they are the most numerous and the most widely spread. The share of the members of the Yang family is expressed by the figure 6^3, that of the Chao by 5^3 (plus one female Chao 1^3), that of the Cheng by 4^3 (plus one female Cheng 1^4). The Fung, Sie, and Sü have a 3^3 to their credit; the Wei reach the mark $2^3 + 1^4$, the Kwo $1^3 + 1^4$, while the Chu, Hou, Kao, Mong, and Tung, can only boast of 2^3 each. All other families are represented but once. These figures certainly have a mere relative value, and do not allow of any far-reaching inferences. It is assumed by anthropologists that the tendency to multiple births is frequently hereditary, both in the male and female line, more frequently in the former than in the latter; and there is no doubt that heredity is a potent cause in the perpetuation of plural births. In the case of triplets and to a still higher degree of quadruplets the hereditary tendency is particularly striking. Quadruplets often issue from parents who were multiples themselves. Female twins often give birth to twins.

During the 61 years covering our records 1–109 (= 109^3), the high-water mark is reached in the year 991 with 9^3, and there is only this one year that offers such a record. There are two years (998 and 1015) with 7^3, two years (995 and 996) with 5^3, 4 years (982, 983, 1014, and 1016) with 4^3, 8 years with 3^3, and 11 years with 2^3. In the remaining years there is but 1^3 or 0^3. In the Yüan period we have four years with 2^3.

QUADRUPLETS

There is a total of 30 on record, the first in A.D. 473, the last in A.D. 1608, a span of 1,136 years.

In this total of 30, 4 quadruplets fall to the lot of a single woman. Twenty-five out of the number of 30, that is 5/6 or 83.33 per cent, consist of males exclusively. The remaining 5 are distributed as follows: 3 cases consisting of 3 males and 1 female (10 per cent), 1 case being 2 males and 2 females (3.33 per cent) and 1 case being 4 females (3.33 per cent).

In 1907 two quadruplet births in France produced 5 males and 3 females; in 1908 there was one quadruplet birth of 4 boys; in 1909 three quadruplet births produced 10 boys and 2 girls; and in 1910, there was one quadruplet birth of 2 males and 2 females (Statistique du mouvement de la population, p. 56).

For 7 cases no personal data are on record; in a single case of the Ming period the father's social status is not indicated. In the remaining 22 cases we find 2 soldiers, 1 falconer, and 19 common people, in all probability, farmers. Again, we accordingly meet here with the same social status of the parents as in the case of triplets.

As to the relative proportion of quadruplet to triplet births, see above, p. 50.

Pliny (VII, 3, § 33) records the example of a quadruplet birth of two males and two females toward the end of the reign of Augustus and ascribed to Fausta, a Plebeian woman of Ostia (Fausta quaedam e plebe Ostiæ).

QUINTUPLETS

It is striking and worthy of especial mention that the Chinese Annals do not record a single example of a quintuplet birth; at least I have failed in tracing any. Both Aristotle and Pliny were convinced of such an occurrence. Aristotle (Historia animalium, transl. of D'Arcy W. Thompson, p. 584b) states: "Some animals produce one and some produce many at a birth, but the human species does sometimes the one and sometimes the other. As a general rule and among most nations the women bear one child at a birth; but frequently and in many lands they bear twins, as for instance in Egypt especially. Sometimes women bring forth three and even four children, and especially in certain parts of the world. The largest number ever brought forth is five, and such an occurrence has been witnessed on several occasions. There was once upon a time a certain woman who had twenty children at four births; each time she had five, and most of them grew up." Pliny (VII, 3, § 33) has it that in the Peloponnesus a woman was delivered of five children at a birth four successive times, and that the greater part of these survived (Reperitur et in Peloponneso quinos quater enixa, maioremque partem ex omni eius vixisse partu),—perhaps the same event alluded to by Aristotle. Nijhoff, in his interesting study "Vijflinggeboorten" (Groningen, 1904, 4°) has fully described and figured a case which came under his notice. He further reviews from literary records 29 more cases of quintuplet births, one of which only seems to be of doubtful authenticity. Cf. also S. Shishido, The Birth of Five Infants at One Parturition (*Iji Shinbun*, Tokyo, 1901, pp. 433–438).

SEXTUPLETS

In regard to sextuplet birth, I have found only two cases on record. According to the Gazetteer of Chi-li Province (*Ki fu t'ung chi*), it was in 1574 that a woman of the people of Fei-hiang (in Kwang-p'ing fu, Chi-li) brought forth six children at one birth. The name of the woman and the husband is not given. The other case is reported in the *Sü K'ien shu* (Ch. 5, p. 8), a record of Kwei-chou Province, written by Chang Chu in 1805. In a certain village of western Kwei-chou a woman, *née* Wang, gave birth to six sons at one time, both children and mother being well. The author, however, had this merely from hearsay.

Nijhoff (p. 66) reports the case of a sextuplet birth in a peasant family at Castagnola near Lugano (Italy) in 1888 (4 males, 2 females, who were alive at the time of birth, but died in a few seconds) and another from Alburi on the Gold Coast in Africa after Dr. H. Vortisch of the Basle Mission. In the latter case a Negro woman is said to have been delivered of 5 boys and 1 girl, who for lack of care died shortly; the woman stated that it was her fifth deliverance, at the second she had twins, at the third quadruplets, and at the fourth triplets. An Italian woman, who in the fifth month of pregnancy miscarried, expelled six fœtuses; the truthfulness of this report is generally conceded (J. Parvin, Science and Art of Obstetrics, 3d ed., p. 161). Other cases of sextuplet delivery are described in the *Boston Medical and Surgical Journal*, XXXV, 2, 1847, and by J. W. Kerr and H. Cookman (*Med. Pres. and Circ.*, LXXV, p. 537, London, 1903: five boys and one girl). Cf. also Shishido, Examination of the Records of More than Five Infants at a Birth (*Iji Shinbun*, Tokyo, 1901, pp. 1897–1901).

SEPTUPLETS

The Gazetteer of Chi-li Province (*Ki fu t'ung chi*) has it on record that in 1527 a woman, *née* Ch'en, of Ho-kien (Chi-li), was delivered of seven girls at one birth, but that none of them survived. According to Trogus, there was a case of seven children at one birth in Egypt (Et in Aegypto septenos uno utero simul gigni auctor est Trogus, Pliny, VII, 3, § 33). Cases of septuplet birth are mentioned by Roy (Couches avec 7 fœtus, in *Revue médicale française et étrangère*, Paris, 1877, I, p. 225); cf. also R. C., Multiple Pregnancy with a Vengeance (*Med. Rec.*, LXIII, p. 267, New York, 1903).

It is reported in verse on a tombstone of Hameln that on January 9, 1600, two boys and five girls were born to Thiele Roemer and Anna Breyers. The tombstone is adorned with a sculptured scene which shows six babes in swaddling-clothes on a pillow, while the lucky (?) father holds the seventh on his arm toward the Savior. Nijhoff

(p. 71–72), who has reproduced the tombstone, holds that the question is here of a veritable fact, as at that time no mockery was made of religious subjects. I am far from sharing this opinion. It is well known that the Germans have displayed a great deal of fun and humor on their epitaphs, collections of which have been made. The tombstone itself does not suffice to bear out the historicity of the case. It would be necessary to trace it in the parish-register, death-lists, or any other documents in the archives of the town of Hameln; but this evidence, as far as I know, has not yet come forward.

A case of eight children at a birth, as far as I know, is nowhere on record, but the following curious passage occurs in the book "Cosmographie de Levant" (p. 114, Lyon, 1554) by F. André Thevet d'Angoulesme: "Non seulement ce païs abonde en fruits, et herbes: mais aussi en fleurs odoriferentes, Les femmes y sont tant fecondes, qu'elles engendrent communement trois, et quatre, et bien souvent huit enfans: et encores qu'ils naissent au huitième mois, ils vivent: ce qu'aucuns attribuent à la bonté du Nil: Outreplus elles surengendroient, (comme Pline raconte d'une femme d'Alexandrie) ce qui n'est pas tant signe de merveille, que argument de fecondité." Thevet is doubtless influenced by the passage of Pliny, and is somewhat inclined toward exaggerations. I doubt very much that a case of octoplets has ever come under his actual experience: his statement is generalized, but no reference is made to a specific case.

The preceding article discloses the fact that a department of vital statistics, either in connection with the Government or as a private enterprise, is urgently required for China. The motive which principally guided me in writing this notice was to demonstrate by a concrete example the necessity of founding such an institution. It goes without saying that a statistical research into the population of China would mean a considerable advance of our knowledge, from which the economists and sociologists all the world over might learn and benefit, and that the Chinese would yield the most fruitful material for all problems of heredity and eugenics. It is particularly genealogical research that could be carried on in China with most promising results. Another problem which is much on my heart is that of longevity and the average duration of a generation among the Chinese and Japanese; and if nothing interferes, I hope to make a small contribution to this question in the near future.

Vol. II. April, 1920 No. 2.

The New China Review

EDITED BY SAMUEL COULING, M.A.

CONTENTS

OFFICE OF

THE NEW CHINA REVIEW

73 CHAOUFOONG ROAD

SHANGHA

THE NEW CHINA REVIEW

VOL. II APRIL, 1920 No. 2

CONTENTS

The Editor will be pleased to receive suitable articles, notes, etc., and will return MSS. not accepted for publication.

When accepted articles are printed, twenty-five copies will be sent to the writer.

Subscriptions are payable in advance and are as follows for the six numbers of 1920: Great Britain, 25/–, *post free;* United States, G. $6.50, *post free;* China, Japan, etc., Mex. $7.50, *post free.*

Subscribers in Europe and U. S. A. should send orders and subscriptions (crossed cheques or P. O. O.) direct to the Editor, and not through agents.

Books for review should be sent to the Editor as early as possible.

Editor: *Samuel Couling,*

73 Chaoufoong Road, Shanghai, China.

THE NEW

CHINA REVIEW

VOL. II April 1920 No. 2

MULTIPLE BIRTHS AMONG THE CHINESE

BY

BERTHOLD LAUFER

The Chinese Annals contain records not only of human events, but also of unusual natural phenomena which left a deep impression upon the minds of the contemporaries. In the early days of historiography, when occurrences were chronicled day by day and year by year, the two categories of human and natural events were noted indiscriminately, merely in the chronological succession as they happened. In the introduction to the *Shu king* we read, for instance: "The king's uncle, the prince of T'ang, found a head of grain, two stalks in different plots of ground growing into one ear, and presented it to the king." In the Bamboo Annals (*Chu shu ki nien*) this feature is still more conspicuous: solar eclipses, meteoric falls, earthquakes, droughts, extraordinary phenomena

in the growth of trees, appearance of a *fung-hwang* (so-called phœnix), rain of particles of earth, unusual thunder-storms, and other phenomena are there on record, being interspersed with the traditions of imperial and military affairs. Beginning from the Annals of the Former Han Dynasty (*Ts'ien Han shu*), a novel departure from the old practice was instituted inasmuch as the natural events were detached from the general narrative to be relegated to a special section, entitled "Records Relating to the Five Elements" (*Wu hing chi*, 五行志). The majority of official annals has adopted this practice. These chapters contain most interesting information, not for the historian, but for the scientist, and therefore merit close study. They give detailed lists, with exact reference to date and place, of great catastrophes, such as famines, droughts, locust-pests, inundations, hail-storms, landslides, earthquakes, conflagrations, excessive cold, electric storms in the winter, etc., abnormal phenomena and monstrosities in domestic animals and human beings, cases of insanity, abnormal customs and practices, etc. It is to this department of records that we owe our principal information on a subject which has not yet been discussed—the frequency of multiple births among the Chinese.

In ancient times, under the Chou dynasty, the officer presiding over the people (*se min*, 司民) was obliged to keep a register of the population. All individuals were recorded from the age when the teeth appear. A separate count was taken of males and females; every year, the number of births was added, while the number of dead was taken off the register (cf. E. Biot, *Le Tcheou-li*, Vol. II, p. 353). We cannot but regret that documents of this character have not survived. No allusion to twins or other plural births is made at that period.

The chapters *Wu hing chi* of the two Han Annals contain no records of multiple births. The *Wei shu* gives a single case of a quadruplet birth. Triplets, but only two cases, are first recorded in the Books of the T'ang Dynasty, and there is a long list of them under the following Sung dynasty. There is one case of a triplet birth of early date, not on record in

the Annals, but in the *Sou shen ki* (搜神記), written by Yü Pao (于寶) in the early part of the fourth century, who reports that "in A.D. 243 there was a woman who gave birth to three sons." I have not embodied this case in my statistical review of the matter, as the work in question is a Taoist book of marvels, and as the extant edition presents merely a retrospective make-up (cf. Wylie, *Notes on Chinese Literature*, p. 192).

While triplet and quadruplet births are mentioned in the Annals with comparative frequency, they hardly trouble about twins, save a few cases of united twins. This omission may indicate one of two possibilities: either twin births were too common to attract much attention, or were too scarce to be worthy of notice. This alternative cannot be decided without a solid foundation of statistical material, which unfortunately we do not have. At the outset I am not disposed to assume, on a merely empirical basis, a high degree of fecundity of the Chinese woman or a relative frequency of twins; for it is a common experience of our time that personal opinions and impressions along this line are seldom, if ever, upheld by the results of statistical research. Restraint in this case is the more commendable, as in regard to twin births in Annam we have the following observation of Dr. A.-T. Mondière (Monographie de la femme annamite, *Mémoires de la société d'anthropologie*, II, 1875, p. 474): "Les grossesses doubles sont excessivement rares chez la femme annamite. Sur les 153,174 naissances que j'ai relevées sur les cahiers des villages de toute la Cochinchine de 1872 à 1877 inclus, je n'ai trouvé que 15 accouchements de jumeaux. Soit 1 sur 10,211 naissances. De plus, un arrondissement particulier, celui de Bentré, semble avoir ce privilège, car sur 15 accouchements gémellaires il en a 9 à lui seul, c'est-à-dire 60 pour 100. Les six autres arrondissements (sur 19) qui en ont présenté: Bien-hoà, Chau-Doc, Saigon, Soctrang, Tan-an, Tay-ninh, n'en ont eu chacun qu'un seul cas, en ces six années. D'après ce que les autorités cambodgiennes m'ont déclaré, les jumeaux seraient plus fréquents chez eux, et d'une façon assez sensible, mais ils n'ont pu me fournir de chiffre exact."

A real investigation of the problem in question is impossible for the present, as we lack any vital statistics for the Chinese Republic. Nevertheless I venture to hope that the facts and observations given below will be of some interest to anthropologists. In order critically to balance the data furnished by the Chinese Annals, it would be indispensable to have reliable birth statistics for China, to know the birth-rate for the different provinces, and to depend on good records showing the total number of plural births for at least a decade. In default of such material in the mother-country I anticipated to receive at least some data from those countries outside of China with a large Chinese population, although it must be taken into account that social and economic conditions of the Chinese abroad are different and that, above all, Chinese emigrants hardly ever take their families along, but intermarry, when settled, with women of other nationalities. I have not yet been able to obtain relevant statistics from the British, the French, and the Dutch colonies; but what I have found thus far is not very encouraging. The Birth Statistics for the Registration Area of the United States for 1915 (Washington, 1917) give a total of 74 births (33 males and 41 females) among the Chinese for that year, but nought else.

According to a communication of Dr. William H. Davis, chief statistician in the Bureau of Census, Washington, D. C., there were, in the years 1915–17, 309 births among the Chinese in the registration area for births in the United States (California, not being admitted to the registration area, is not included), only one pair of twins appearing in this total. The state of California gives in its vital statistics only the number of births and deaths of its Chinese populace, without touching the question of plural births. In 1916 there were 425 births (compared with 727 cases of death); in 1917, 419 births (compared with 818 cases of death) among the Chinese of California (Twenty-Fifth Biennial Report of the State Board of Health of California for the Fiscal Years from July 1, 1916, to June 30, 1918, Sacramento, 1918, pp. 201, 203, 205, 207, 224). The statistics of Mexico

contain merely the number of Chinese living in the various provinces, the total, as taken in the third and last census of 1910, being 13,118 men and 85 women = 13,203 (Estados Unidos Mexicanos, *Boletin de la Dirección General de Estadistica*, Num. 5, p. 37, Mexico, 1914), but no tables of births.

The data are arranged under the headings—Triplets, Quadruplets, Quintuplets, Sextuplets, and Septuplets, and finally United Twins. The reader should keep in mind that these data are not co-ordinated in the Annals, as they appear here, but that they are scattered there and intermixed with other items; they were hence extracted from the text. These data naturally are monotonous, as birth-certificates and statistics in general are bound to be; but in the same manner as it is the statistician's duty to publish his tables of dry figures before offering his conclusions, I deem it my duty to present the Chinese data literally and unchanged in their sober chronicle style in order to enable the reader to formulate conclusions for himself and to check up my own.

TRIPLETS

For the period of the T'ang dynasty (A.D. 618–906) only two cases of a triplet birth are on record in the Annals.

1. In A.D. 775 a woman of the family Chang (張), of Chao-ying (昭應), gave birth to one male and two females.
2. In A.D. 905 triplets (males) were borne by the wife of P'eng Wen (彭文), one of the people of Ju-yin (汝陰), in Ying-chou (潁州) (Ngan-hwi).—*T'ang shu*, ch. 36, p. 21b.

For the Sung period (A.D. 960–1278) the data of triplet bi-ths are fuller than for any other dynasty. The following are placed on record:

1. In A.D. 960 triplets (males) were born to Liu Tsin (劉進), a soldier of the Kwei-yi army (歸義軍民), in Hiung-chou (雄州) (prefecture of Pao-ting, Chi-li).—*Sung shi*, ch. 62, p. 19b.

2, 3. In A.D. 961 triplets (males) were born to Mong Fu (孟福), one of the people of Mong-chou (孟州), (prefecture of Hwai-k'ing, Ho-nan); and to Mong Kung-li (孟公禮), one of the people of Ting-chou (定州) (Chi-li). The former case is also on record in the *Mong hien chi* (ch. 10B, p. 2).

4. In A.D. 962 triplets (males) were born in the family of a soldier, Yi Ch'ao (宜超), of the Lung-tsie army in the capital (京師龍捷軍卒).

5. In A.D. 965 triplets (males) were borne by the wife of Liu Hwi (劉暉), one of the people in the prefecture of Kiang-ling (江陵) (Hu-pei).

6. In A.D. 966 triplets (males) were borne by the wife of Chao Yüan (趙遠), a soldier of the Hiao-kien army (驍健軍卒), of Ngan-chou (安州) (prefecture of Pao-ting, Chi-li).

7–9. In A.D. 967 triplets (males) were born to Kao Yü (高輿), one of the people of Kwang-chou (光州) (Ho-nan); to Chao Se (趙嗣), one of the people of Te-chou (德州) (prefecture of Tsi-nan, Shan-tung), and to Wang Tsin (王進), a soldier of the K'ien-ning army (乾寧軍卒).

10–12. In A.D. 968 triplets (males) were born to Wang Cheng (王政), one of the people of Yi-chou (沂州) (Shan-tung); to Sie Hing (謝興), one of the people of Shan-chou (澶州) (now K'ai-chou (開州), in the prefecture Ta-ming, Chi-li); and to Wang Ngan (王安), one of the people in the district of Ch'ang-chou (長洲縣), in the prefecture of Su-chou (蘇州) (Kiang-su).

13, 14. In A.D. 969 triplets (males) were born to Sun Yen-kwang (孫延廣), one of the people of Lang-chou (閬州) (Se-ch'wan); and to Tung Yüan (董遠), one of the people of K'ai-chou (開州) (prefecture of Ta-ming, Chi-li).

15, 16. In A.D. 973 triplets (males) were born in the family of Wang Yu (王宥), in the district Ts'ing-ch'eng (青城) (prefecture of Wu-ting, Shan-tung);

and to Liu Yüan (劉元), one of the people in the prefecture of Ho-nan.

17–20. In A.D. 977 triplets (males) were born to Li Yü (李遇), a soldier of the Chao-shou army (招收軍卒), of Hing-chou (邢州) (now Shun-te fu, Chi-li); to Yü Pa (魚霸), a soldier of the Kwei-hwa army (歸化軍卒), of Ju-chou (汝州) (Ho-nan); to Sie Tsu (謝祚), one of the people of Ch'ang-chou (常州) (now Ch'ang-chou fu, Kiang-su); and to Yang Wan (楊萬), one of the people in the district Tsin-yüan (晉原縣) (Se-ch'wan).

21–24. In A.D. 982 triplets were born to Kin Hing (靳興), a soldier of the Lung-wei army (龍衛軍卒), of Shan-chou (澶州) (see above, 10–12); to Cheng Yen-fu (鄭彥福), one of the people of Tsin-chou (晉州) (prefecture of Cheng-ting, Chi-li); three girls were born to Cheng Sün (鄭訓), one of the people of Fen-chou (汾州) (Shan-si); triplets, two males, and one female, were born to Ngan Wang (安旺), a soldier of the Kwei-hwa army (as above), of Hwa-chou (滑州) (prefecture of Wei-hwi, Ho-nan).

25–27. In A.D. 983 triplets (males) were born to Yü Chao (俞釗), a soldier of the Shun-hwa army (順化軍卒), of Yang-chou (揚州) (Kiang-su); to Li Yü (李遇), one of the people of Wen-chou (溫州) (prefecture of Hwai-k'ing, Ho-nan); and to Li Tsu (李祚), one of the people of Yung-chou (雍州).

28. In A.D. 985 triplets (males) were born to Ho Tsing (何靖), one of the people in the district of Fung-sin (奉新縣) (prefecture of Nan-ch'ang, Kiang-si).

29. In A.D. 986 triplets (males) were born to Chang Mei (張美), one of the people in the district Lu-shan (魯山縣) (Ho-nan); and to Chang K'in (張欽), one of the people in the district of Lin-lü (林慮縣) (now Lin (林) hien, prefecture of Chang-te, Ho-nan).

30, 31. In A.D. 987 triplets (males) were born to Chou Ch'eng-hwi (周承暉), one of the people in the district Tsin-yüan (晉原縣) (Se-ch'wan); and to Yang Sheng (楊昇), one of the people in the district of Ku-shi (固始縣) (Ho-nan).

32. In A.D. 988 triplets (males) were born to Fung Yü (馮遇), one of the people of K'i-chou (祁州) (prefecture of Pao-ting, Chi-li).

33–35. In A.D. 989 triplets (males) were born to Sü Mei (徐美), one of the people of Ts'i-chou (齊州) (Shan-tung); to Hou Yüan (侯遠), one of the people of Ping-chou (并州) (Se-ch'wan); and to Sü Liu (徐流), a soldier of Ch'ang-chou (常州) (Kiang-su).

36, 37. In A.D. 990 triplets (males) were born to Wang Pin (王斌), one of the people in the district Ho-yang (河陽縣) (now Mong (孟) hien, Hwai-k'ing fu, Ho-nan); and to Li Kwei (李珪), one of the people in the district Sin-si (新息縣). The former case is also on record in the *Mong hien chi* (ch. 10B, p. 2b).

38–46. In A.D. 991 triplets (males) were born to Hwang Chao (黃釗), one of the people in the district Tsin-ling (晉陵) (Ch'ang-chou fu, Kiang-su); to P'eng Kung-pa (彭公霸), one of the people in the district Nan-ch'ung (南充縣) (prefecture of Shun-k'ing, Se-ch'wan); to Chou Sin-wang (周信王), one of the people in the district Lung-yang (龍陽) (prefecture of Chang-te, Ho-nan); to Li Ts'ing (李清), one of the people in the district Wu (屋縣); to Kwo Chung (郭忠), one of the people in the district Lin-ts'ing (臨清) (Shan-tung); to Sie Yüan-sheng (謝元昇), *li* (吏) of the district Lin-shwi (鄰水) (prefecture of Shun-k'ing, Se-ch'wan); to Chu Wang (朱旺), a soldier in the district Fung-hwa (奉化) (prefecture of Ning-po, Che-kiang); to Hu Li (胡立), one of the people of Ying-chou (瀛州) (Ch'ao-chou fu, Kwang-tung);

and to Kao Te (高德), one of the people of Hing-chou (邢州) (now Shun-te fu, Chi-li).

47–50. In A.D. 993 triplets (males) were born to Cheng Ngan (鄭安), one of the people in the district Han-tan (邯鄲) (prefecture of Kwang-p'ing, Chi-li); to Wang Hi-nien (王希年), one of the people in the district Ho-kien (河間) (Chi-li); and to Sung Ho (宋和), one of the people of Ngan-chou (安州) (prefecture of Pao-ting, Chi-li).

51. In A.D. 994 triplets (males) were born to Sheng T'ai (盛泰), a soldier in the district Yung-k'iu (雍丘).

52–56. In A.D. 995 triplets (males) were born to Li Shen (李深), a sergeant (*ti kün kiao* 商軍校) in Pao-chou (保州) (now Pao-ting fu, Chi-li); to Wang Hia (王洽), one of the people in the district Sung-ch'eng (宋城); to Ho Yung (賀用), one of the people in the district Lin-hwai (臨淮) (Ngan-hwi); to Tung Mei (董美), one of the people in the district Yung-ts'ing (永清) (prefecture of Shun-t'ien, Chi-li); and to Ma Fang (馬方), one of the people in the district Chüan-ch'eng (鄄城) (now Pu-chou (濮州), in Ts'ao-chou fu, Shan-tung).

57–61. In A.D. 996 triplets (males) were born to Chao Yen (趙演), one of the people of Fen-chou (汾州) (Shan-si); to Li Se (鍘), one of the people of Yi-chou (沂州) (Shan-tung); to Liu Siang (劉相), one of the people of Nan-kien-chou (南劍州) (now Yen-p'ing fu, Fu-kien); to Lu Lwan (陸鸞), one of the people in the district Ngan (安) (in Mien-chou, Se-ch'wan); and to Li Yün (李筠), a soldier of Wei-chou (衛州) (now Wei-hwi fu, Ho-nan).

62–68. In A.D. 998 triplets (males) were born to Wang Wang (王旺) in the district Yung-ngan (永安), in T'ai-chou (台州) (Che-kiang), to Cheng Hwi (鄭穗), a soldier of the Tsing-jung army (靜戎軍卒), of Shan-chou (澶州) (prefecture of

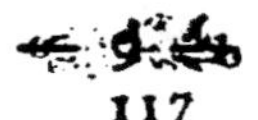

Ta-ming, Chi-li); to Hwai Liang (懷梁), one of the people in the district Shen (莘) (prefecture of Tung-ch'ang, Shan-tung); to Wang Kwei (王貴), one of the people in the district Hwo-kia (獲嘉) (prefecture of Wei-hwi, Ho-nan); to Lo Yen-t'ao (羅彥瑫), one of the people in the district Yung-k'ang (永康) (prefecture of Kin-hwa, Che-kiang); to Yang Jung-p'i (楊榮毗), one of the people in the district Wen (溫) (prefecture of Hwai-k'ing, Ho-nan); and to Wei Ki (魏吉), one of the people in the district Ling (陵) (prefecture of Tsi-nan, Shan-tung).

69–71. In A.D. 1000 triplets (males) were born to Chu Tsin (朱進), one of the people in the district Sui (睢) (Kwei-te fu, Ho-nan); to Sü Jao (徐遶), a soldier of the Wu-wei army (武威軍卒), of Yün chou (鄆州) (now Yün-ch'eng in the prefecture Ts'ing-chou, Shan-tung); and to P'eng Yüan (彭遠), one of the people of Shen-chou (深州) (Chi-li).

72, 73. In A.D. 1001 triplets (males) were born to Kwo Jung (郭瑩), one of the people in the district Wang-tu (望都) (prefecture of Pao-ting, Chi-li); and to Liang Tsi (梁濟), a soldier of the Teng-hai army (澄海軍卒), of Yung-chou (邕州) (prefecture of Nan-ning, Kwang-si).

74. In A.D. 1002 triplets (males) were born to Chao Nien (趙輦), one of the people in the district Hia-tsin (夏津) (Shan-tung).

75, 76. In A.D. 1003 triplets (males) were born to Liu Sien (劉銑), one of the people in the district Shi-ch'eng (石城) (Kiang-si, another of the same name in Kwang-tung); and to Tai Hing (戴興), one of the people in the district T'ang-yi (堂邑) (Kiang-ning fu, Kiang-su).

77. In A.D. 1004 triplets (males) were born to Li Tsung (李總), one of the people in the district Nan-ch'ang (南昌) (prefecture of Nan-ch'ang, Kiang-si).

78. In A.D. 1005 triplets (males) were born to Wei Yung (魏勇), one of the people in the district Fung-sin (奉新) (Nan-ch'ang fu, Kiang-si).

79–81. In A.D. 1007 triplets (males) were born to Chao Jung (趙榮), a workman (匠), of Pa-tso se (八作司); to Jen Teng-lao (任登老), one of the people in the district Na-tun (南頓); and to Chang Sü (張緒), one of the people in the district Tsao-k'iang (棗強) (Chi-li).

82, 83. In A.D. 1009 triplets (males) were born to Chang Liu (張留), one of the people in the district Kwo (崞) (in Tai-chou, Shan-si); and to Yang Ts'üan (楊泉), who served in the army of Ts'ing-p'ing (清平) (Tung-ch'ang fu, Shan-tung).

84, 85. In A.D. 1010 triplets (males) were born to Feng K'o (馮可), one of the people in the district Hwo-kia (獲嘉) (prefecture of Wei-hwi, Ho-nan). Triplets (two males and one female) were born to Li Hwi (李crypt), one of the people in the district Sung-ch'eng (宋城).

86. In A.D. 1011 triplets (males) were born to Feng Shou-k'in (馮守欽), one of the people in the district Ho-ch'i (河池) (K'ing-yüan fu, Kwang-si).

87, 88. In A.D. 1012 triplets (males) were born to Sü Lin (徐璘), a soldier of the Süan-yung army (宣勇軍卒) in the prefecture Ta-ming (大名) (Chi-li); and to Li Chao (李釗), one of the people in the district Tsan-hwang (贊皇) (Chêng-ting fu, Chi-li).

89–92. In A.D. 1014 triplets (males) were born to Li Kien (李謙), one of the people in the district T'ung-ti (銅鞮) (Shan-si); to Pai Te (白德), one of the people in the district Sung-ch'eng; to Chu Lin (朱璘), one of the people in the district Ho-k'iu (霍丘) (prefecture of Ying-chou, Ngan-hwi); and to Tsiao Se-shun (焦思順), one of the people in the district P'ing-liang (平涼) (Kan-su).

93–99. In A.D. 1015 triplets (males) were born to Sung Tsai-hing (宋再興), one of the people of Ho-nan fu; to Chou Yüan (周元), one of the people in the district Chen-yang (眞陽) (now Cheng-yang (正陽) prefecture of Ju-ning, Ho-nan); to Hou Yen (侯言), a field-laborer of the district Li-t'ing (歷亭) (now Ku-ch'eng (故城), in Ho-kien fu, Chi-li); to Wang Chung (王忠), one of the people of the district Ho-k'iu (霍丘) (Ngan-hwi); to Wei Chi-ts'ung (衛志聰), one of the people of the district Mung-yang (濛陽); to Chang Ki (張吉), a soldier of the Hiao-wu army (驍武軍), of Ting-chou (定州), Chi-li); and to Hwang Tsin (黃進), a soldier of the Hwai-yung army (懷勇軍), of the district Yung-k'iu (雍丘).

100-103. In A.D. 1016 triplets (males) were born to Nie Te (聶德), a soldier of the Hiung-yung army (雄勇軍), of Ts'ao-chou (曹州) (Shan-tung); to Liu Yüan (劉元), one of the people of Ying-chou (瀛州) (Ch'ao-chou fu, Kwang-tung); to Chang Kwei (張貴), one of the people of Li-chou (澧州) (Hu-nan); and to Liu Ki (劉吉), one of the people of Kwang-chou (廣州) (Kwang-tung).

104. In A.D. 1017 triplets (males) were born to Ch'en Pa (陳霸), one of the people of the district Lien-kiang (連江) (Fu-chou fu, Fu-kien).

105, 106. In A.D. 1019 triplets (males) were born to Wen Sin (文信), one of the people of the district Ts'ien-t'ang (錢塘) (Hang-chou fu, Che-kiang); and to Li Ch'eng-yü (李承遇), one of the people of the district of Swi-ngan (遂安) (Yen-chou fu, Che-kiang).

107-109. In A.D. 1020 triplets (males) were born to Tu Ming (杜明), one of the people of the district Hiao-kan (孝感) (Han-yang fu, Hu-pei); to Liu Shun (劉順), one of the people of the district P'ing-ngen (平恩) (Shan-tung); in the seventh month, to Chang Chung (張中), one of the people of the district Lei-yang (耒陽) (Heng-chou fu, Hu-nan). On their foreheads there were white moles, over an inch square, and covered with white hair.

110. In A.D. 1150 triplets (males) were born to a family of the people [name not given] in the district Chen-fu (眞符) (Shen-si).

For the time from A.D. 1023 to 1126 no list of names is given, but merely a statistical record covering several reign-periods. It is here reproduced in tabular form (*Sung shi*, ch. 62, p. 22b).

Period	Years	Quadruplets (males)	Quadruplets (3 males, 1 female)	Triplets (males)	Triplets (2 males, 1 female)	Total
1023–68	46	2	—	44	1	47
1068–83	15	1	1	84	—	86
1084–99	16	2	—	18	1	21
1100–26	27	1	—	19	—	20
Total	104	6	1	165	2	174
		Quadruplets total 7		Triplets total 167		

While the preceding cases are not recorded in the way of vital statistics, but solely as unusual events, the above table conveys the impression of embracing a fairly accurate register of all multiple births (save twin births) which took place within the span of a century. The proportion of quadruplet to triplet births in this period is 1:23.86. The total of triplet births on record during the Sung epoch, accordingly, is 110+167=277. The total of quadruplet births during the same period is 7 (as shown by the above table)+7 (recorded in the following section)=14. The proportion of quadruplet to triplet births for the entire period of the Sung is 1:19.78; while the proportion for the entire period of Chinese history here considered (A.D. 473–1643) is 1:10.8. This calculation is based on a total of 324 triplets and 30 quadruplets.

There are no multiple births on record in the chapter *Wu hing chi* of the *Kin shi*. The *Yüan shi* (chs. 50, 51), covering the period from 1260 to 1367, contains only 15 cases of triplets (all males), recorded under the years 1261, 1265, 1273, 1285, 1291, 1297, 1300, 1327, 1328, 1335, and

1363. In the years 1273, 1297, 1335, and 1363, two cases are listed for each year; and it is of especial interest that in two instances we have two cases of triplets in the same family, the interval between the two being in either case given as three years. According to Dr. Puech, to whom we owe excellent studies on the causes of multiple births, the more children a woman has had at close intervals, the more she will be inclined toward these physiological anomalies. Three women admitted in the St. Petersburg Midwives' Institute between 1845–59 in their fifteenth pregnancy had triplets, and each had triplets three times in succession (J. M. Duncan, *Fecundity*, p. 71).

For the period of the Ming dynasty (1368–1643) we lack official records; but the section *jen i* (人異) of the *T'u shu tsi ch'eng* gives a list of 30 cases of triplet births, extracted from the provincial and local Gazetteers, and covering a period from 1404 to 1626. In 1413, 1515, and 1520, two cases are recorded in each year. In view of the fact that this material is extracted from a number of scattered books, it cannot lay claim to completeness; the figure 30 is certainly much removed from reality, but even if multiplied by 3 or 4, it is left far behind the total of the Sung period. On the whole, the impression prevails that the number of multiple births has steadily been on the decrease from the days of the Sung. This would agree with an anthropological theory to the effect that the phenomenon of multiple births in man represents a survival of or reversal to his former animal state and that with the advance of civilization the number of such births is liable to decline. There may be a correct biological viewpoint in this hypothesis, but it does not account for all facts connected with the phenomenon, and, above all, conflicts with given data and statistics. It is not brought out by the vital statistics of any European country that the frequency of plural births is on the decline; on the contrary, in France, for instance, it is surprisingly high (see below). Further, if that theory were correct, we should naturally anticipate to find the greatest number of multiple births among primitive tribes, which for all we

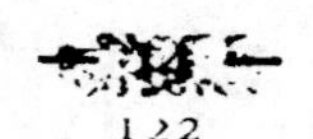

know is not the case. Hardly a century has elapsed that records of plural births have been taken in Europe, and this period is too short to allow us to indulge in much speculation on the subject.

According to the Statutes of the Manchu Dynasty (大清會典), it was decreed in 1663 that in the case of a triplet birth or a twin birth of a boy and a girl, if it should occur among the people of the Eight Banners, a special report should be submitted to the Board of Rites (禮部); if it should occur in the provinces, the governer of such province should report to the Board of Rites, which would have to forward it to the Board of Finance (戶部), the latter to grant a premium of five piculs of rice and ten pieces of cloth. In 1674 it was ordered that a special report should be made solely in the case of male triplets, but not in the case of twins or female triplets. In 1684 an edict ordained that in the case of male triplets the Board of Rites and the Board of Finance should submit a joined report to the throne, and that rewards should be authorized in accordance with law. This benevolent attitude toward the energetic propagators of the race was not an innovation of the Manchu, but a heritage of the Ming; for under the Ming we are frequently informed of special grants of food, cloth, and even paper money, made to these involuntary heroes from public funds.

It may hence be inferred that under the Manchu régime a register of male triplets was kept, and presumably is still preserved in the archives of Peking. If it should ever be published, the fact must be borne in mind that female triplets were not officially reported. Meanwhile we are thrown back for that period on the local and provincial Gazetteers, which in the chapter on untoward or abnormal events sometimes record cases of plural births.

To cite a few instances of this kind in the period of the Manchu dynasty—the Gazetteer of Ju-chou in Ho-nan (quoted above) enumerates four cases (all males), which occurred in 1770, 1785, 1824, and 1833. In 1797 a triplet birth occurred in Hwa-yang (prefecture of Ch'eng-tu, Se-ch'wan); the case was reported to the throne, and by imperial favor, a picul of rice was granted to the father, Yang Kwo-yü (楊國玉) (*Hwa yang hien chi*, ch. 43, p. 3). The Gazetteer of Mong-chou (prefecture of Hwai-k'ing, Ho-nan) cites only two cases for the years 1682 and 1736. Most Gazetteers which I have looked up are disappointing: thus the Gazetteer of Shen-si Province (*Shen-si t'ung chi*) contains only two cases of triplets, recorded for the years 1470 and 1729.

In the Gazetteer of the Prefecture of Sung-kiang, three cases of triplets are recorded between 1367 and 1640 (according to D. J. Macgowan, Cosmical Phenomena Observed in the Neighborhood of Shanghai, *Journal China Branch R. As. Soc.*, II, 1860, p. 74).

The cases of triplets referring to the Sung period and listed in the Gazetteers are usually copied from the *Sung shi*. A few, however, seem to be derived from independent local sources; thus the *Ju-chou ts'üan chi* (汝州全志), published in 1840, gives the case of Chang Mei from Lu-shan under the year 986 in accordance with the Sung Annals, but has a further case concerning a certain Chang Shou (張壽), from Yi-yang (伊陽) (in Ju-chou, Ho-nan), whose wife had male triplets in A.D. 991—not recorded in the *Sung shi*.

The data of the Chinese certainly are defective, and cannot entirely satisfy the anthropologist. We miss, for instance, data concerning the ages of mother and father and order of birth in triplet deliveries (*rang chronologique de l'accouchement* of the French statisticians). Above all, we should desire information as to the vitality and fecundity of the off-spring. What the Chinese may boast of, however, is the fact that they possess lists of plural births for periods of the past when nothing of the kind was ever attempted in any country of Europe. In the vital statistics of France, plural births have been recorded only from 1858; and in no country of Europe did they receive any attention before the nineteenth century (in Berlin from 1825).

The sum of 277 triplet births for the Sung and 324 for the time from the T'ang to the Ming inclusive may seem a

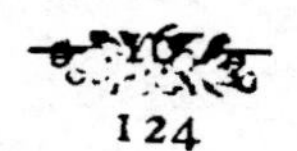

high figure to the uninitiated; in fact, however, it is strikingly low. During the four years 1907–10 there was in France a total of 327 triplet births; 91, 93, 68, and 75 in the respective years, making a mean average of 81.75 per year (Statistique général de la France, Statistique du mouvement de la population, Paris, 1912, p. 56). There were, accordingly, more triplet births in France during those four years than in China in the course of many centuries. Or, to cite another example, in the period 1835–47, there were in Bavaria 1,050 triplet, 56,062 twin, and 3,413,763 normal births. The frequency of triplets varies in different years and in different countries. In 1855, triplets were produced in Scotland by 11 mothers out of 92,300 births; that is, one in 8,391. Triplet births in Scotland from 1855 to 1901, a period of 47 years, numbered 644, and averaged 116 per million confinements (C. J. Lewis and J. N. Lewis, *Natality and Fecundity*, p. 62). I do not go any further into the question of the frequency of triplets in Europe and the proportion of triplets to twin and normal births, as the Chinese data are not comparable, and as figures of total births are lacking for the Sung period. Judging from our experience, it must be stated, however, that the Chinese data can hardly be complete; but there is no way of correcting or adjusting the figures, which we are simply compelled to take for what they are worth. The reader should not forget that the material furnished by the Chinese Annals is not intended as statistics, but merely as a record of extraordinary events in human life. In order to give a certain perspective to the number of multiple births, some data concerning the population may follow here. According to the calculations of E. Biot (Mémoire sur la population de la Chine, *Journal asiatique*, 1836, p. 461), the population of China under the Sung totaled 43,388,380 in the year 1021, and rose to 100,095,250 in 1102; again in 1223, it amounted to only 63,354,005 (in consequence of the loss of northern China to the Kin). These figures, in all probability, are too high; for they are estimated on the number of families given in the Chinese records, the assumption being made that the mean average of the number

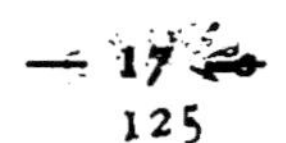

of individuals in a family is 5, which, in my opinion, is too high a figure.

The total number of triplets recorded for the T'ang and Sung periods is 279. The distribution of sex in this number is as follows: 273 all males, that is, 97.8%; 4 consisting of 2 males aud 1 female, that is, 1.4%; 1 consisting of 1 male and 2 females, that is, 0.04%; and only one consisting of 3 females (0.04%). Again, the 15 triplet births of the Yüen dynasty and the 30 of the Ming are all males exclusively. The above percentages perhaps give an approximate clew to the actual frequency of sex in triplet births, as far as China is concerned.

C. J. Lewis and J. Norman Lewis (*Natality and Fecundity*, p. 61, London, 1906), who base their remarkable study on the birth registers of Scotland for the year 1855, during which year there were 11 triplet births in that country (3 males 5, 3 females 3, 2 males and 1 female 3), offer the following conclusion in regard to the distribution of the sexes: "There is a strong probability that in any given occurrence of triplets the children will all be of the same sex, either all males or all females. If the same ratio held in other nations and in other years, it would amount to a law of triplet production that in over 70 per cent. of cases the newly-born children are all of the same sex."

In the period from 1858 to 1865, there were in France 1,005 triplet and 4 quadruplet births; among the former, there were 280 entirely males, 218 entirely females, 256 consisting of 1 male and 2 females, and 251 consisting of 2 males and 1 female. The number of twin births during the same period amounted to 83,279; of these 28,056 were two males, 26,310 two females, and 29,363 consisting of one male and one female (A. Puech, *Annales d'hygiène publique*, XLI, 1874).

Of the 277 triplets recorded for the Sung period, the social standing of the fathers is given in only 110 cases, while the remaining 167 cases are merely recorded as chronologic-statistical events. Among the 110 cases, the social status of the fathers is distributed as follows:

			Percentage
Rural population (民)	..	85	76.7
Field-laborers (田 用)	..	1	1.1
Workmen (匠)	.. ..	1	1.1
Soldiers (兵)	.. ..	22	20.0
Petty officials (吏)	.. ..	1	1.1
Total	.. ..	110	100.0

In the Yüan period, 14 common people and 1 soldier share in the 15 cases of triplets placed on record. In the Ming period, 28 common people and 2 soldiers assume responsibility for 30 cases of triplets recorded. It will thus be seen that the bourgeoisie, inclusive of officials, gentry, and merchants, has no share in these records. Peasants and laborers, of course, formed the majority of the populace; but there is no reason why triplet births, if they had occurred in the upper classes, should not have been reported or recorded.

In arranging our data according to families, we arrive at the result that the members of the families Li (李), Wang (王), Chang (張), and Liu (劉), take the uppermost rank. The male Lis reach the score with 16^3+1^4, while two female Lis figure with 2^4; in the years 986 and 996 respectively we have two male Lis participating in triplets. The record of the Wangs is 13^3+2^4 (plus one female Wang 1^3); the Changs follow with 9^3+2^4, plus two female Changs (2^3), and the Lius with 9^3+1^4, two members of this family being conspicuous in the same year (1016). This does not mean, of course, that these four families are more prolific than others, but is merely the index of the fact that they are the most numerous and the most widely spread. The share of the members of the Yang (楊) family is expressed by the figure 6^3, that of the Chao (趙) by 5^3 (plus one female Chao 1^3), that of the Cheng (鄭) by 4^3 (plus one female Cheng 1^4). The Fung (馮), Sie (謝), and Sü (徐) have a 3^3 to their credit; the Wei (魏) reach the mark 2^3+1^4; the Kwo (郭) 1^3+1^4, while the Chu (朱), Hou (侯), Kao (高), Mong (孟), and Tung (董), can only boast of 2^3 each. All other families are represented but once. These figures certainly have a mere relative value, and do not allow of any far-reaching inferences. It is assumed by anthropologists that the tendency to multiple births is frequently hereditary, both in the male and female line, more frequently in the former than in the latter; and there is no doubt that heredity is a potent cause in the perpetuation of plural births. In the case of triplets and to a still higher degree of quadruplets, the hereditary tendency is particularly striking. Quadruplets often issue from parents who were multiples themselves. Female twins often give birth to twins.

During the 61 years covering our records 1–109($=109^3$), the high-water mark is reached in the year 991 with 9^3, and there is only this one year that offers such a record. There are two years (998 and 1015) with 7^3, two years (995 and 996) with 5^3, 4 years (982, 983, 1014, and 1016) with 4^3, 8 years with 3^3, and 11 years with 2^3. In the remaining years there is but 1^3 or 0^3. In the Yüan period we have four years with 2^3.

QUADRUPLETS

Wei Dynasty, A.D. 386–532.

1–4. In A.D. 473 a woman in Siu-jung (秀容) (now Ting-siang, Shan-si) gave birth to four sons four times and thus had 16 sons.—*Wei shu.*

T'ang Dynasty, A.D. 618–906.

5, 6. In A.D. 655 quadruplets (all males) were borne by the wife of Wu Wei (吳威), one of the people of Kao Yüan (高苑) in Tse-chou (淄州) (Shan-tung); and to Hing Tao-hwo (辛道讓), one of the people of Kia-chou (嘉州) (now Kia-ting fu, Se-ch'wan). If creation turns away from the regular norm, supernatural phenomena are bound to come forth; if the female element is abundant, the mother's way is fertile.—*T'ang shu*, ch. 36, p. 21.

7. In A.D. 826 quadruplets (all males) were borne by the wife of Ho Wen (賀文), a man of Yen-chou (延州) (now Yen-ngan fu, Shen-si).

Sung Dynasty, A.D. 960–1278.

8. In A.D. 982 quadruplets (all males) were born to Liu Si (劉習), one of the people in the district of Yen-men (鴈門縣) (Shan-si).

9. In A.D. 1003 the wife of Kwo Jang (郭讓), one of the people in the district P'ing-hiang (萍鄉) (prefecture of Shun-te, Chi-li), gave birth to four sons.

10. In A.D. 1008 the wife of Wang Yen (王言), one of the people in the army of Kao-yu (高郵軍民), brought forth four sons.

11. In A.D. 1015 the wife of Chang Pao (張保), one of the people in the district Yung-kia (永嘉) (Wen-chou fu, Che-kiang), gave birth to four sons.

12, 13. In the period from 1023 to 1068 there were two cases of quadruplet births (all males).

14, 15. In the period from 1068 to 1083 there was one case of a quadruplet birth (males) and another of 3 males and 1 female.

16, 17. In the period from 1084 to 1099 there were two cases of quadruplet births (all males).

18. In the period from 1100 to 1126 there was one case of a quadruplet birth (males).

19. In A.D. 1097 a woman of the people of Süan-ch'eng (宣城) (Ning-kwo fu, Ngan-hwi), brought forth four sons.—宋史哲宗本紀.

20. In A.D. 1099 a woman of the people of Yi-shi (猗氏) in Ho-chung (河中) (Shan-si) brought forth four sons.—*Ibidem.*

21. In A.D. 1118 a woman of the people of Hwang-yen (黃巖) (Tai-chou fu, Che-kiang) brought forth four sons.—宋史徽宗本紀.

Yüan Dynasty, A.D. 1260–1367.

22. In A.D. 1283 the wife of Chang Chou (張丑) of Kao-chou (高州), *née* Li (李), brought forth four children at one birth—three boys and one girl.

23. In A.D. 1297 four males were brought forth at one birth by Ho-li-mi (和里迷), wife of Na Hwai (那

惕) (Mongols), a falconer (*po-lan-ki*, 孛蘭奚) of Liao-yang (遼陽) (Sheng-king, Manchuria). Regarding the term *po-lan-ki*, see *Yüan shi*, ch. 101, p. 16b.

24. In A.D. 1306 the wife of Fang Ping (方丙) in the district Hu-k'ou (湖口) in Kiang-chou (江州) (Kiang-si), gave birth to four sons.

Ming Dynasty, A.D. 1368–1643.

25. In 1471 the wife of Wei Süan (魏宣), one of the people of the town Ho-yang (郃陽) (T'ung-chou fu, Shen-si), had four sons at one birth.—*Shen-si t'ung chi* (Gazetteer of Shen-si Province).

26. In 1543 the wife, *née* Li (李), of Wang Shi (王世), one of the people of K'i (祁) hien (Pao-ting fu, Chi-li), brought forth 3 boys and 1 girl.—*Shan-si t'ung chi* (Gazetteer of Shan-si Province).

27. In 1543 the wife of Lu Ngan (陸安), of Heng-chou (橫州) (Nan-ming fu, Kwang-si), brought forth four boys.—*Kwang-si t'ung chi* (Gazetteer of Kwang-si Province).

28. In 1557 quadruplets (males) are recorded in K'ing-yang fu (慶陽府) (Shen-si).—*Ibidem* (no further data are given).

29. In 1583 four girls were born to Li Hwang (李鍠) of Hing-hwa (興化) (Fu-kien).—*Hing-hwa hien chi* (Gazetteer of Hing-hwa).

30. In 1608 two boys and two girls were born to Su Kiu-lang (蘇九郎), a soldier; his wife was *née* Cheng (鄭).—*Fu-kien t'ung chi* (Gazetteer of Fu-kien Province).

The 7 cases numbered 12–18 are identical with those given in the statistical table of the Sung period (above, p. 5). It is noteworthy that the three cases under Nos. 19–21 are recorded in the biographies of the emperors, not in the section *Wu hing chi*; it seems that particular importance was attached to them.

In this total of 30, 4 quadruplets fall to the lot of a single woman. Twenty-five out of the number of 30, that is,

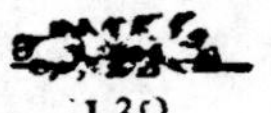

$\frac{5}{6}$ or 83.33 per cent, consist of males exclusively. The remaining 5 are distributed as follows: 3 cases consisting of 3 males and 1 female (10 per cent), 1 case being 2 males and 2 females (3.33 per cent) and 1 case being 4 females (3.33 per cent).

In 1907 two quadruplet births in France produced 5 males and 3 females; in 1908 there was one quadruplet birth of 4 boys; in 1909 three quadruplet births produced 10 boys and 2 girls; and in 1910, there was one quadruplet birth of 2 males and 2 females (*Statistique du mouvement de la population*, p. 56).

The period covered by the above 30 cases ranges from A.D. 473 to 1608; that is, a span of 1,136 years.

For the 7 cases Nos. 12–18 no personal data are on record; in a single case of the Ming period the father's social status is not indicated. In the remaining 22 cases we find 2 soldiers, 1 falconer, and 19 common people, in all probability, farmers. Again, we accordingly meet here with the same social status of the parents as in the case of triplets.

As to the relative proportion of quadruplet to triplet births, see above, p. 5.

Pliny (VII, 3, §33) records the example of a quadruplet birth of two males and two females toward the end of the reign of Augustus and ascribed to Fausta, a Plebeian woman of Ostia (*Fausta quaedam e plebe Ostiae*).

Quintuplets

It is striking and worthy of especial mention that the Chinese Annals do not record a single example of a quintuplet birth; at least I have failed in tracing any. Both Aristotle and Pliny were convinced of such an occurrence. Aristotle (*Historia animalium*, transl. of D'Arcy W. Thompson, p. 584b) states: "Some animals produce one and some produce many at a birth, but the human species does sometimes the one and sometimes the other. As a general rule and among most nations the women bear one child at a birth; but frequently and in many lands they bear twins, as for instance in Egypt especially. Sometimes women bring forth three and even four children, and especially in certain parts of the world. The largest number ever brought forth is five, and such an occurrence has been witnessed on several occasions. There was once upon a time a certain woman who had twenty children at four births; each time she had five, and most of them grew up." Pliny (VII, 3, § 33) has it that in the Peloponnesus a woman was delivered of five children at a birth four successive times, and that the greater part of these survived *(Reperitur et in Peloponneso quinos quater enixa, maioremque partem ex omni eius vixisse partu)*—perhaps the same event alluded to by Aristotle.

Nijhoff, in his interesting study "Vijflinggeboorten" (Groningen, 1904, 4°) has fully described and figured a case which came under his notice. He further reviews from literary records 29 more cases of quintuplet births, one of which only seems to be of doubtful authenticity. Cf. also S. Shishido, "The Birth of Five Infants at One Parturition" *(Iji Shinbun*, Tokyō, 1901, pp. 433–438).

SEXTUPLETS

In regard to sextuplets, I have found only two cases on record. According to the Gazetteer of Chi-li Province *(Ki fu t'ung chi)*, it was in 1574 that a woman of the people of Fei-hiang (肥鄉) (in Kwang-p'ing fu, Chi-li) brought forth six children at one birth (一產六子). The name of the woman and the husband is not given. The other case is reported in the *Sü K'ien shu* (續黔書) (ch. 5, p. 8), a record of Kwei-chou Province, written by Chang Chu (張澍), in 1805. In a certain village of western Kwei-chou a woman, *née* Wang (王氏), gave birth to six sons at one time, both children and mother being well. The author, however, had this merely from hearsay.

Nijhoff (p. 66) reports the case of a sextuplet birth in a peasant family at Castagnola near Lugano (Italy) in 1888 (4 males, 2 females, who were alive at the time of birth, but died in a few seconds) and another from Alburi on the Gold Coast

in Africa after Dr. H. Vortisch of the Basle Mission. In the latter case a Negro woman is said to have been delivered of 5 boys and 1 girl, who for lack of care died shortly; the woman stated that it was her fifth deliverance, at the second she had twins, at the third quadruplets, and at the fourth triplets. An Italian woman, who in the fifth month of pregnancy miscarried, expelled six fœtuses; the truthfulness of this report is generally conceded (J. Parvin, *Science and Art of Obstetrics*, 3d ed., p. 161). Other cases of sextuplet delivery are described in the *Boston Medical and Surgical Journal*, XXXV, 2, 1847; and by J. W. Kerr and H. Cookman, *Med. Pres. and Circ.*, LXXV, p. 537, London, 1903: five boys and one girl. Cf. also Shishido, Examination of the Records of More than Five Infants at a Birth (*Iji Shinbun*, Tokyo, 1901, pp. 1890–1901).

SEPTUPLETS

The Gazetteer of Chi-li Province (*Ki-fu t'ung chi*) has it on record that in 1527 a woman, *née* Ch'en (陳氏), of Ho-kien (Chi-li) was delivered of seven girls at one birth (一孕七女), but that none of them survived. According to Trogus, there was a case of seven children at one birth in Egypt (*Et in Aegypto septenos uno utero simul gigni auctor est Trogus.*—Pliny, VII, 3, §33). Cases of septuplet birth are mentioned by Roy (Couches avec 7 foetus, in *Revue médicale française et étrangère*, Paris, 1877, I, p. 225); cf. also R. C., Multiple Pregnancy with a Vengeance (*Med. Rec.*, LXIII, p. 267, New York, 1903).

It is reported in verse on a tombstone of Hameln that on January 9, 1600, two boys and five girls were born to Thiele Roemer and Anna Breyers. The tombstone is adorned with a sculptured scene which shows six babes in swaddling-clothes on a pillow, while the lucky(?) father holds the seventh on his arm toward the Savior. Nijhoff (pp. 71, 72), who has reproduced the tombstone, holds that the question is here of a veritable fact, as at that time no mockery was made of religious subjects. I am far from sharing this opinion. It is well known that the Germans have displayed a great deal of

fun and humor on their epitaphs, collections of which have been made. The tombstone itself does not suffice to bear out the historicity of the case. It would be necessary to trace it in the parish-register, death-lists, or any other documents in the archives of the town of Hameln; but this evidence, as far as I know, has not yet come forward.

A case of eight children at a birth, as far as I know, is nowhere on record, but the following curious passage occurs in the book *Cosmographie de Levant* (p. 114, Lyon, 1554) by F. André Thevet d'Angoulesme: "Non seulement ce païs abonde en fruits, et herbes: mais aussi en fleurs odoriferentes, Les femmes y sont tant fecondes, qu'elles engendrent communement trois, et quatre, et bien souvent huit enfans: et encores qu'ils naissent au huitieme mois, ils vivent: ce qu'aucuns attribuent à la bonté du Nil: Outreplus elles surengendroient, (comme Pline raconte d'une femme d'Alexandrie) ce qui nest pas tant signe de merveille, que argument de fecondité." Thevet is doubtless influenced by the passage of Pliny, and is somewhat inclined toward exaggerations. I doubt very much that a case of octoplets has ever come under his actual experience: his statement is generalized, but no reference is made to a specific case.

UNITED TWINS

The Chinese Annals have preserved a few cases of twins grown together at birth. In this case, the question naturally is of twins produced from a single ovum.

In the fourth year of the period Kien-hing (建興) (A.D. 316), under the Emperor Min (愍帝), of the Tsin dynasty, a woman of the family Hu (胡), when she was at the age of twenty-five, the wife of Jen Kiao (任僑), a minor official (clerk) in the district of Sin-ts'ai (新蔡) (prefecture of Ju-ning, Ho-nan), gave birth to female twins grown together in the region of the abdomen and the heart, but separated above the breast and beneath the navel.—*Sung shu*, ch. 34, p. 28.

In A.D. 487 (under the Emperor Wu of the Ts'i dynasty) the wife of Wu Hiu (吳休), one of the people of Tung-ts'ien (東遷), in Wu-hing (吳興) (now Hu-chou fu, Che-kiang), gave birth to male twins grown together below the chest down to above the navel.—*Nan Ts'i shu*, ch. 19, p. 16b.

In the fourth month of the third year of the period Yi-fung (儀鳳) (A.D. 678) King-chou (涇州) (Kan-su) presented the Court with two infants the hearts of which were connected, but each with a separate body (二小兒連心異體). Formerly it had happened that the wife, *née* Wu (吳), of Hu Wan-nien (胡萬年), a soldier of the guard in the district Shun-ku (鶉觚) (Kan-su), gave birth to twins, a male and a female, whose breasts were connected, but who, for the rest, had individual bodies; when separated, both died. At a subsequent birth it was thus again. The twins were boys, and were brought up. In the above-mentioned year they had reached the age of four years, and were presented to the Court.—*T'ang shu*, ch. 36, p. 21.

In A.D. 1610, the wife of Li Yi-ch'en (李宜臣), of Fan-ki (繁峙) (in Tai-chou, Shan-si), *née* Niu (牛氏), brought forth two girls with their heads and faces grown together, but with separate arms and legs.—*Shan-si t'ung chi* (Gazetteer of Shan-si Province).

Two Chinese twins grown together, born in 1887, were shown by Barnum and Bailey in 1902, and at that time were still unseparated and well. Cf. R. Virchow, Xiphodymie (*Z. Ethn.*, 1891, pp. 366–370).

The modern Gazetteers occasionally record the birth of twins, not, however, on account of any special interest attached to the fact itself, but merely in order to emphasize the interest in the vitality of twins (cf. W. A. Macnaughton, The Longevity of Twins, *Caledon. M.J.*, X, pp. 127–129, Glasgow, 1915). The following examples from the Gazetteer of Hwa-yang (*Hwa-yang hien chi*, ch. 43, p. 4) will suffice to illustrate this feature.

The wife of Chu Ch'ang-hwa (朱昌華), *née* Lin (林氏), had 14 sons, among these 2 pairs of twins, who did not die prematurely, but are still alive (俱存無殀).

The wife of Chung Se-kin (鍾思謹), *née* Tsou (鄒氏), had 9 sons, among these one pair of twins still alive.

The wife of Chung Chao-kin (鍾超錦), *née* Chang (張氏), had 9 sons, among these one pair of twins still alive.

The wife of Li Ch'ao-kung (李朝恭), *née* Lin (林氏), had 8 sons, among these one pair of twins still alive.

The preceding article discloses the fact that a department of vital statistics, either in connection with the Government or as a private enterprise, is urgently required for China. The motive which principally guided me in writing this notice was to demonstrate by a concrete example the necessity of founding such an institution. It goes without saying that a statistical research into the population of China would mean a considerable advance of our knowledge from which the economists and sociologists all the world over might learn and benefit, and that the Chinese would yield the most fruitful material for all problems of heredity and eugenics. It is particularly genealogical research that could be carried on in China with most promising results. Another problem which is much on my heart is that of longevity and the average duration of a generation among the Chinese and the Japanese; and if nothing interferes, I hope to make a small contribution to this question in the near future.

154

中国的变性人和两性人

AMERICAN JOURNAL
OF
PHYSICAL ANTHROPOLOGY

EDITED BY

ALEŠ HRDLIČKA

VOLUME III

WASHINGTON, D. C.
1920

SEX TRANSFORMATION AND HERMAPHRODITES IN CHINA

BERTHOLD LAUFER

Field Museum of Natural History, Chicago

The following data are excerpted from Chinese records, chiefly the official historical annals, and may be of interest to some anthropologists and students of sexual psychology. As they are merely intended for the inner circle, I have deemed it superfluous to add a commentary. Those who are familiar with the writings of E. Laurent in particular will have no difficulty in understanding these phenomena. There is no doubt that under the heading of sex transformation different phenomena are involved, but the Chinese data are too succinct to admit a complete and positive interpretation in each and every case.

There are in the Chinese Annals numerous records of female chickens being transformed into males (cf., for instance, *Sung shu*, Ch. 30, pp. 14*b*, 15). Unfortunately we do not receive any details, and the Chinese hardly discriminate between primary and secondary sexual characters. "It is a singular fact," says Darwin,[1] in discussing the sexual differences in fowl, "that the males in certain sub-breeds have lost some of their secondary masculine characters, and from their close resemblance in plumage to the females, are often called hennies. There is much diversity of opinion whether these males are in any degree sterile; that they sometimes are partially sterile seems clear, but this may have been caused by too close interbreeding. That they are not quite sterile, and that the whole case is widely different from that of old females assuming masculine characters, is evident from several of these hen-like sub-breeds having been long propagated. . . . There is also a breed of Game-fowls, in which the males and females resemble each other so closely that the cocks have often mistaken their hen-feathered opponents in the cock-pit for real hens, and by the mistake have lost their lives." Note also his observation, "I may add that at the first exhibition of Poultry at the Zoölogical

[1] Variation of Animals and Plants under Domestication, I, p. 306 (Murray's ed., 1905).

Gardens, in May, 1845, I saw some fowls, called Friesland fowls, of which the hens were crested, and the cocks furnished with a comb."

"The transformation of women into males is not a fable (ex feminis mutari in mares not est fabulosum). We find in the *Annals* that under the consuls P. Licinius Crassus and C. Cassius Longinus (171 B.C.) it happened in Casinum that a virgin was transformed into a boy under the eyes of the parents and by order of the Haruspices was deported to a desert island. Licinius Mucianus relates that he saw at Argos a certain Arescon, formerly named Arescusa and even married to a man; soon after she grew a beard and produced other marks of virility, so that she then took a wife. The same author asserts that he saw a boy of the same condition at Smyrna. I myself saw in Africa a citizen of Thysdris, L. Consitius, who on the very day of her wedding with a husband had been changed into a man." Thus Pliny (VII, 4, § 36) reports in his Natural History.

Sex 'transformations' in human beings must have been observed by the Chinese at an early time. King Fang, a philosopher of the first century B.C., who made a special study of the ancient book of divination, the *Yi king*, indulged in some philosophical speculations on the subject and regarded it as a foreboding of evil; thus the transformation of a woman into a man augurs that worthless creatures will become kings. The following eleven cases are on record.

In the thirteenth year of King Siang of Wei (306 B.C.), a woman became transformed into a man. During the period Kien-p'ing (6–2 B.C.) of the Emperor Ai of the Han dynasty, there was a man at Yü-chang (Kiang-si Province), who changed into a woman, was married to a man, and gave birth to a son. In the seventh year of the period Kien-ngan (A.D. 202), there was a man in Yüe-swi (Se-ch'wan Province), who became transformed into a woman.—*Ts'ien Han shu,* Ch. 27 B, p. 23; *Hou Han shu,* Ch. 27, p. 4 ("Annals of the Former and Posterior Han Dynasties" respectively).

In the beginning of the period Ning-k'ang (A.D. 373–376) of the Emperor Hiao Wu of the Tsin dynasty, a woman, *née* T'ang, of Chou-ling (Hu-pei Province), was gradually transformed into a man.—*Sung shu,* Ch. 34, p. 29*b* ("Annals of the Liu Sung Dynasty").

In A.D. 886, a girl who had not yet her teeth, in the district of Mei, prefecture of Fung-siang (Shen-si Province), was transformed into a male. She died after ten days.—*T'ang shu,* Ch. 36, p. 22*b*.

In 1512, there was in a village of Shen-si Province a woman, who was transformed into a man and grew a beard. Subsequently she brought forth two sons.—*Shen-si t'ung chi* ("Gazetteer of Shen-si Province").

In 1547, there was in the prefecture of Ta-t'ung (Shan-si Province) a woman who became transformed into a man.—*Shan-si t'ung chi* ("Gazetteer of Shan-si Province").

In 1620, a man of Kwang chou (in Ho-nan Province), Wu Lo by name, married a woman of the family Ch'en; after several days she changed into a man and grew a little moustache.—*Ju-ning fu chi* ("Gazetteer of the Prefecture of Ju-ning").

In 1625, a woman, *née* Ma, of T'ung Ch'eng (Ngan-hwi Province) changed into a man at the age of seventy.—*Kiang nan t'ung chi* ("Gazetteer of Kiang-nan").

In 1631, a woman, *née* Li, of Hwa-t'ing (prefecture of Sung-kiang, Kiang-su) changed into a man.—*Ibidem.*

In 1638, a poor woman, *née* Sun, in the Home for the Old at Lo-yang (Ho-nan Province), at the age of over seventy, grew a moustache.

In regard to hermaphrodites, let us begin with Pliny again. "Men of double sex are also born, and these we call hermaphrodites; formerly they were called Androgyni ('men-women'), and were taken as prodigies; now they serve sensual purposes" (Pliny, VII, 3, § 34). In regard to the Androgyni, a tribe living in what is now Tripolis, Pliny (VII, 2, § 15) informs us after Calliphanes that they have a double sex and alternately cohabit with one another (androgynos esse utriusque naturæ, inter se vicibus coeuntes).

In Chinese sources I have found the following four cases on record.

In A.D. 306 a son was born to Sie Chen of Kwei-ki (prefecture of Shao-hing, Che-kiang). He had a large head covered with hair, and the two soles were turned up. In the upper part of the body he was shaped like a male and a female. At the hour of his birth he possessed the voice of a male. At the end of a day he died.—*Tsin shu* ("Annals of the Tsin Dynasty").

At the time of the Emperors Hwi and Hwai (A.D. 290–312) of the Tsin dynasty, there was in the capital Lo (in Ho-nan Province) an hermaphrodite (literally, "one in whom the body of a man and a woman is united"). He was capable of having sexual intercourse as a man or a woman, and was inclined to excesses. This is the result of a disturbance of the vital forces.—*Sung shu*, Ch. 34, p. 27*b* ("Annals of the Liu Sung Dynasty").

In the first year of the period Yüan-hi (A.D. 304), under the Emperor Kung of the Tsin dynasty, a man of Kien-ngan (Fu-kien) had no glans on his penis; it was straight and flat. From below his trunk he had the shape and body of a woman.—*Ibid.*, Ch. 34, p. 29*b*.

In A.D. 1436 there were in T'ai-ts'ang chou (Kiang-su Province) two hermaphrodites; people called them "the double-shaped."—*T'ai-ts'ang chou chi* ("Gazetteer of T'ai-ts'ang chou").

In the modern written language, the term for hermaphrodism is *yin-yang* ("combination of female and male elements"); also *tse hiung t'ung t'i* ("body in which male and female traits are united"), and *liang sing kü yu* ("being possessed of a double nature"). The hermaphrodite is accordingly styled *yin yang jen* or *yu yin yu yang* ("man with female and male qualities"), and *pan nan pan nü jen* ("an individual half man, half woman"). In the colloquial language of northern China we hear *er wei-tse* ("one with two tails"); in Amoy (in southern China), people speak of a *liang hing dzin* ("man of two shapes"), *poan-ts'i-hiong* ("half-female-male"), or *tsu ui ts'i hiong tsi dz'in* ("individual who can pose as a female or male"); the Cantonese say *pun nam nü ke yan* ("half-man-woman-individual") or *yam yeung ping yau tik* ("one who unites in himself female and male qualities").

H. Ramsay (Western Tibet, p. 61) states that hermaphrodites are not known in Ladākh. Ladākhis think ill luck is caused by human monstrosities, and it is therefore probable that these are killed as soon as they are born.

Hermaphrodites, as far as I know, have never found expression in Chinese art as in Greek sculpture (cf., for instance, S. Reinach, Hermaphrodite, in his Cultes, Mythes et Religions, II, pp. 319–337; M. Houel, Pièces d'hermaphrodites conservées au Musée Dupuytren, *Bull. Soc. d'Anthr.*, XIV, 1881, pp. 554–556). Chinese art is asexual and anti-sexual.

In the European literature on China I have not been able to find any allusion to the two subjects here treated. In India hermaphrodites seem to be better known. Thevenot (Travels, part 3, containing the Relation of Indostan, p. 23, London, 1687) writes that for the first time he saw hermaphrodites at Surat in India. "It was easy to distinguish them, for seeing there is a great number in that town, and all over the Indies, I was enformed before hand, that for a mark to know them by, they were obliged under pain of correction, to wear upon their heads a turban like men, though they go in the habit of women."

155

再论驯鹿

AMERICAN ANTHROPOLOGIST

NEW SERIES

ORGAN OF THE AMERICAN ANTHROPOLOGICAL ASSOCIATION
THE ANTHROPOLOGICAL SOCIETY OF WASHINGTON,
AND THE AMERICAN ETHNOLOGICAL
SOCIETY OF NEW YORK

PLINY E. GODDARD, *Editor*, New York City
JOHN R. SWANTON and ROBERT H. LOWIE, *Associate Editors*

VOLUME 22

LANCASTER, PA., U. S. A.
PUBLISHED FOR
THE AMERICAN ANTHROPOLOGICAL ASSOCIATION

1920

DISCUSSION AND CORRESPONDENCE

The Reindeer Once More

There is as yet no exhaustive or real history of any animal domestication or plant cultivation, and such a task will still be impossible for a long time to come. Naturalists, biologists, geographers, historians, ethnographers, and orientalists, have made numerous contributions to these subjects, every one from the particular angle of his field; and, as is well known, their results are widely divergent and cannot yet be harmonized. Whoever has had occasion to work on these problems feels only too well that he is merely able to make a contribution to a problem, and makes no pretense of solving the problem in its entire complexity. Dr. G. Hatt has recently published an article on Reindeer Nomadism (*Memoirs, American Anthropological Association*, vol. VI, no. 2), which is partially devoted to a criticism of my former contribution to these Memoirs on the same subject. Dr. Hatt's paper doubtless contains many interesting references and notes, especially as far as his own field, the Lapp, is concerned; but I find it necessary to point out a number of misunderstandings and to discuss briefly some of his conclusions which are unacceptable to me.

Dr. Hatt claims that my disregard of the biology of the reindeer "seriously impairs the value of my theories about the origin of reindeer-domestication." This criticism is hardly fair, for I have not given any theories in regard to such origins, nor do I believe that in general origins can be explained satisfactorily. I hold that facts mean everything and that theories are of no account, and have plainly enough indicated (p. 129 of my article) that we are ignorant of how the initial domestication of the reindeer was brought into effect. I have then arrayed a number of available data which might give us a clew as to how this process came about, leaving it to whoever so desired to reconstruct this process according to his own liking. I did not attempt "to trace the evolutionary history of reindeer nomadism," as Dr. Hatt wishes me to do; for like Boas, Lowie, and others, I have always opposed the evolutionary method in its application to anthropological problems (cf. this journal, 1917, p. 299, with reference to Dr. Hatt's theory of the evolution of moccasins).

The essential points discussed by Hatt are all contained in my notice of the reindeer. It seems to me that the points made in this paper, regarding the relative age of the cultural elements of reindeer nomadism, can hardly be maintained. The criteria made out for earlier and later phenomena are purely subjective and a matter of debatable opinion. The vagueness of his chronology is not helpful in historical investigation. History must be based on historical data and documentary evidence, not on speculation. The account of Rubruck of the thirteenth century, Dr. Hatt quotes as proving the early use of ox or horse carts by the nomads, is of little value in view of the ancient accounts of the carts used by the nomadic Scythians in Hippocrates (*cf.* Minns, *Scythians and Greeks*, p. 50) and by the Turkish tribes in the Chinese Annals many centuries before that time. I fully maintain my point that the domestication of the reindeer presents a secondary and imitative process leaning toward horse and cattle, and as regards driving, toward the dog. Hatt denies the former influence, but then he hastens to explain that "reindeer milking certainly must be due to influence from cow or horse culture," and again, "the use of the reindeer for riding and carrying and as a milk-giving animal must have come into the Tungusian-Soyotian area as a result of contact with horse and cow culture;" and finally, "it is not to be denied that some reindeer nomads have taken over certain things from horse and cattle breeding."

I do not see that the data relating to the milking of reindeer which Dr. Hatt quotes alter the views expressed by me. Pekarski states plainly in regard to the Tungus of Ayan that butter is made to a small extent only, and this isolated case of modern origin is an exception which confirms the rule that butter was formerly not made by the Tungus in general. The fact that the Soyot consume reindeer milk in the shape of butter or cheese was stated by myself (p. 127). The chapter "Beginning of Reindeer Nomadism" is based on unproved premises and hypothetical and arbitrary speculations. Olsen (p. 113), according to Hatt, is wrong in his observations among the Soyot and must have misunderstood what he saw, because it so happens that his data contradict a theory of Hatt. This procedure seems to me entirely inadmissible, because it is based on the desire of the speculative theorist who combats the facts which disturb or shatter his dreams.

I strictly maintain my interpretation of Ohthere's account. Nowhere have I entertained any doubt as to the nationality of the Finn, as Hatt supposes. His two objections to my interpretation (p. 120) are not valid. How do we know that a Norseman in the ninth century would never think of keeping deer in a park? The supposition that "the reindeer is a migratory animal which cannot be kept in parks or enclosures" is unproved. For at least twelve years I have observed a couple of reindeer in a zoological garden, and they were perfectly happy and content there. The reindeer of the Soyot is not at all migratory, but during the summer the herds constantly remain in the forest in the proximity of human habitations (after Olsen, in my article, p. 127). When in the summer of 1898 I resided in the settlement Wal among the Ewunki Tungus on the northeast coast of Saghalin island, the reindeer herds of this tribe were kept in confinement on a small isle hardly two miles square, which they were unable to leave; they were held there in a perfect enclosure formed by water. Any park or enclosure may certainly be large enough to allow an animal to yield to its migratory habit. Giles Fletcher (*Of the Russe Common Wealth*, London, 1591, ed. of E. A. Bond, 1856, p. 101), in his description of the life of the Lapp, states, "Their travaile to and fro is upon sleds, drawen by the Olen deer; which they use to turne a grasing all the sommer time in an iland called Kilden (of a very good soile compared with other partes of that countrie), and towards the winter time, when the snow beginneth to fall, they fetch them home again for the use of their sledde."

We do not read more from or into our documents than is warranted by their contents, and Ohthere does not say a word about the Lapp tending his herds. There is as yet no proof for the allegation that the Lapp of the ninth century were reindeer nomads. Frijs says advisedly, "The Lapp in the north of Scandinavia during the ninth century were still fishermen and hunters, and were only acquainted with reindeer as game, while they did not yet possess tame animals" (C. Keller, *Naturgeschichte der Haustiere*, p. 200). Dr. Hatt objects to this statement that our forefathers "were not ethnographers"; but this is no argument. The interpretation of reindeer into the harts put to the cart of Hotherus (p. 125) is not safe: the tale of Saxo is legendary, not historical. Also the Romans and Chinese harnessed stags to carriages (see my article, pp. 132, 133), and no one would think of claiming that these were tamed reindeer. In the opinion of the best philologists of our time, particularly those of France, no historical facts should be deduced from the status of loan-words and other linguistic phenomena (against Hatt, p. 128); if this is done, however, the conclusion will always remain an hypothesis, but will never rise into a fact. That Hatt, after offering not a single piece of tangible evidence, should advance the assertion, "That reindeer nomadism existed in Scandinavia in the ninth century, and even somewhat earlier, may accordingly be regarded not as a mere hypothesis, but as a solid fact," is beyond my comprehension. I apprehend that a deep gulf separates us as to what constitutes a solid fact. There is not even room here for an hypothesis.

Dr. Hatt's discussion of the Kalewala is based more on an attempt to sustain his theory than on objective evidence. I had occasion myself to read this work repeatedly and at different times, and considerable literature about it. The book of Comparetti on which I chiefly relied is justly regarded as a classic throughout the civilized world, and it may be expected that a man who devoted a lifelong and serious study to this vast and complex subject knows at least as much about it as Hatt. Naturally there is much controversial matter and divergence of opinion with respect to the Kalewala, in the same manner as in the case of the Homeric poems, the Rigveda, or the Avesta. Dr. Hatt passes off his own ideas as "the truth about Kalewala" (p. 127) and denies categorically that it represents a true and perfect picture of the Finn prior to their christianization. May it not be that his judgment is influenced by the fact that the Kalewala runs counter to his theories, for it does not contain the faintest allusion to domesticated reindeer, while the wild reindeer was an object of the hunt, while sledge-driving is most frequently mentioned, but the sledges are always drawn by horses (my article, p. 191).[1]

Little troubled Lemminkäinen,
And he spoke the words which follow:
"Make a snowshoe left to run with,
And a right one to put forward!
I must chase the elk on snowshoes,
In the distant field of Hiisi."

XIII, 59–64.

"Let the men who live in Lapland,
Help me all to bring the elk home;
And let all the Lapland women
Set to work to wash the kettles;
And let all the Lapland children
Hasten forth to gather splinters;
And let all the Lapland kettles
Help to cook the elk when captured." etc.
But the third time he rushed onward,
Then he reached the elk of Hiisi.
Then he took a pole of maple,
And he made a birchen collar;
Hiisi's elk he tethered with it,

[1] I select several passages (translation of W. F. Kirby) in support of my above statement. If all this is not realism of cultural conditions, the Kalewala assuredly is pretty well consistent in its madness.

In a pen of oak he placed it.
"Stand thou there, O elk of Hiisi,
Here remain, O nimble reindeer!"

XIII, 203–210, 217–224.

Glide throughout the land of Hiisi,
And across the heaths of Pohja,
There to chase the elk of Hiisi,
And to catch the nimble reindeer.

XIV, 19–22.

Thereupon the colt he harnessed,
In the front she yoked the bay one,
And she placed old Väinämöinen
In the sledge behind the stallion.

VII, 349–352.

Väinämöinen, old and steadfast,
Took his horse of chestnut color,
And between the shafts he yoked him,
Yoked before the sledge the chestnut,
On the sledge himself he mounted,
And upon the seat he sat him.

X, 1–6.

Thus the smith, e'en Illmarinen,
Clothed himself, and made him ready,
Robed himself, and made him handsome,
And his servant he commanded:
"Yoke me now a rapid courser,
In the sledge adorned so finely,
That I start upon my journey,
And to Pohjola may travel."
Thereupon the servant answered,
"Horses six are in the stable,
Horses six, on oats that fatten;
Which among them shall I yoke you?"

XVIII, 379–390.

I cannot see what gives us the right to say that "To regard the descriptions of Lapland and the Lapp, contained in Kalewala, as realism, would be perfectly ridiculous (p. 127)." Had it so happened that the Kalewala furnishes the opposite data which would support Hatt's presumptions, he would probably have accepted them without hesitation.

While I have to disagree with Dr. Hatt on many points, and am compelled to reject his claims, there is one point, however, on which I am heartily in accord with him, and this is his plea for collecting more material, before we shall know all about reindeer nomadism. What we need are facts and research based on serious information. We live to learn and to work.

Finally I may be allowed to quote a passage from a letter of the late Dr. Herman K. Haeberlin, in memory of a friend who was always dear to me. On November 1, 1917, our regretted friend wrote me from Columbia University, New York, as follows: "I was very much interested in your paper on the reindeer. Aside from its value as the investigation of a concrete cultural trait, its methodology I think is highly instructive for us anthropologists. It shows what can be attained by a scholarly coöperation of direct historical reconstruction and indirect ethnological inference. Furthermore, an important methodological point is that you trace the origin of reindeer domestication to a definite geographical area rather than to a certain tribe. This methodological distinction ought to be borne in mind more clearly than we have thus far done. I shall attempt to make the same point when I discuss the center of distribution of imbricated basketry in North America."

B. Laufer

156

维纳《非洲与发现美洲的关系》书评

Africa or America

"AFRICA AND THE DISCOVERY OF AMERICA." Volume I. By LEO WIENER. Philadelphia: Innes Sons.

Reviewed by BERTHOLD LAUFER.

Curator of Anthropology, Field Museum of Natural History, Chicago.

THERE is something highly pathetic about this book: a formidable arsenal of erudition, both historical and philological, but unfortunately wasted on a whimsical, impossible theory, bred in the dim atmosphere of the study, but doomed to evaporate in the light of the day. Professor Wiener endeavors to prove that tobacco, the sweet potato, peanut, and cassava, which are typical cultivated plants of American citizenship, were introduced into this continent by negroes from Africa, in post-Columbian times. The idea itself can hardly lay claim to originality, for there were in the past many anti-American heretics, who were inclined to ascribe an African origin to those cultivations. What is original about Wiener's work is the peculiar method of dealing with these complex problems from a purely philological viewpoint. A history of cultivated plants cannot be based on a more or less arbitrary interpretation of vague and usually faulty vernacular nomenclature, but, above all, must be firmly grounded on botanical research, which in Wiener's book is simply ruled out of court. Even such a well-known handbook as A. de Candolle's "Origin of Cultivated Plants," in which the American origin of the plants in question is forcibly vindicated, is not utilized, not to speak of numerous monographs contributed by Gray, Trumbull, Wittmack, Safford, Engler, Stuhlmann, Ficalho, and many others. Wiener devotes a long discussion to yams; but what does "yams" mean? It is a general popular designation for the genus *Dioscorea*, which comprises about two hundred species, distributed over the tropical regions of both the New and Old World and the Far East; in view of this situation it takes more than the mere will-o'-the-wisp of seeming phonetic similarity of words to make out the history of a plant migration.

It has been demonstrated over and over again that the linguistic method applied to historical problems is a fallacy, and that historical facts cannot be deduced from the status of loan-words and etymologies, which necessarily remain problematic and speculative. It is quite natural that the first conquerors and colonists of America transferred familiar names to the novel animals and plants encountered in the New World. Thus the Spaniards called the puma a lion, the jaguar a tiger, the alpaca a sheep, the peccary and tapir a hog, the turkey a peacock, etc. In harmony with Wiener's ideas, the Spanish notices pertaining to such animals would either be forgeries or the animals in question must have been imported from Africa. A further source of confusion was given by the fact that in the sixteenth century a great part of America was called the Indies, that simultaneously the sea route to India was discovered by the Portuguese, that a lively interchange of plants and products between America, Africa, and Asia soon sprang up with amazing rapidity, and that the attribute "Indian" did refer to the East and the West Indies as well. The early Spaniards and Portuguese may very well have applied ancient European, Arabic, and African words to American productions, but this question of terminology and confusion of names has no bearing whatever on the objective botanical and historical conditions. One might argue as well that because the natives of this continent were universally called Indians they must have emigrated from India. The cardinal deficiency of this book rests on the author's incapability of discriminating between objective truth and subjective error, and his tendency to base his speculations on the latter, instead of proceeding from the well-ascertained facts of botanical science.

The Manihot or cassava, at the time of the conquest, was raised from the Isthmus of Tehuantepec, to the Paraná, being the staple food in the West Indies and in the Eastern part of South America; it is clearly mentioned in Columbus's Diary under the name *cabi*. Wiener has an almost Talmudic way of treating the documents relative to the discovery of America, either rejecting their statements as falsehoods or twisting their meaning so as to fit into the fabric of his speculations; he constantly battles with windmills and hugs shadows for realities. Thus he indulges in many fanciful assertions, as for instance: "Bread was never made of sweet potatoes, nor can it be made, because they contain too much sugar" (p. 193). I refer solely to G. Hughes, who, in his classical work, "Natural History of the Island of Barbados" (London: 1750, p. 228), describes sweet-potato bread as follows: "The potatoes being first grated, and the juice pressed out, the flowery or mealy part is mix'd with sugar and spice, and made into paste, which being baked in the oven, in the form of a plum-cake, its taste is far from being disagreeable; this they call Pone."

The peanut (*Arachis Hypogæa*) has no claim to African origin. Its wild congeners, six species, occur in Brazil only, and there the cultivated form was developed. This fact was recently demonstrated again in a very thorough manner by R. A. Waldron. The Chinese received the plant from the Philippines, between 1575 and 1580; the Japanese only in the beginning of the eighteenth century. Wiener's theory that Arabic *habaziz* might be derived from Chinese *lo hwa sheng* ("Born from flowers sinking into the ground") is impossible for chronological reasons, as the Chinese term it first used only in 1587 by Wang Shi-mou with reference to the peanut; the Arabic word, moreover, does not relate to the peanut, but to the tubers of *Cyperus esculentus*. The erroneous notion that the peanut may be a native of Africa was prompted by the fact that in Africa we have a nut, or rather a pea-like fruit maturing underground like a peanut, but of an entirely different genus—*Voandzeia subterranea*. This plant was introduced into Brazil, where it is known as "peanut of Angola" *(mandubi di Angola.)*

The chapter on tobacco is a veritable performance of jazz music: by stress of imagination the author endeavors to show that the word *tabaco* is of Arabic-African origin, and this again suggests to him that Arabs and Africans were familiar with tobacco smoking prior to the discovery of America. The outstanding facts in the history of tobacco are briefly the following: All European botanists and pharmacists of the sixteenth and seventeenth centuries are agreed on the point that the herb came from America and was previously unknown; in Turkey, Egypt, Persia, and India it appears in the beginning of the seventeenth century; in China and Japan a few decades earlier, as the Far East received the plant and its use directly from the Philippines, where the Spaniards had introduced it. We are well informed as to the cultivated plants of Africa in the successive ages by mediæval Arabic travellers and the early Portuguese navigators; but none of these makes any reference to tobacco or the habit of smoking. The first authentic allusion to tobacco smoking among the inhabitants of Guinea is made by Capt. Stephan van der Hagen, who visited the country in 1598, and whose *Descriptio navigationum versus Guineam* was published by the famous botanist Clusius in his *Curæ posteriores* (1611). The Dutch explorer states that "the people of those regions acquired the habit of inhaling the fumes of tobacco or Nicotiana from the Hollanders only a short while ago, and that previously they were plainly ignorant of this matter, being content with chewing betel." The Cape Colony was actually purchased by the Hollanders from the poor Hottentot for tobacco leaves. Both cultivation and preparation of the plant are still but imperfectly understood by the negroes, and African tobacco does not suit a European or American palate. America, the home of the plant, still produces the finest qualities. He who is able to believe in the African origin of tobacco may fancy as well that the sun moves around this globe or that man lives on the inner surface of the earth.

The author is rather prone to charge others with blunders, without taking the trouble to verify the original texts. Thus he upbraids Barbosa for saying that on Sam Lourenço (Madagascar) "their principal food is roots, which they sow, and it is called yname, and in the Indies it is called maize." He relies upon Stanley's faulty English version, made after a corrupted Spanish translation of Barbosa's Portuguese text. In the Portuguese original Barbosa writes merely: "Yams are their principal food." The wrong definition is solely an interpolation due to the Spanish translator (see M. L. Dames, "The Book of Duarte Barbosa," Vol. I, p. 25).

While I am still willing to forgive the Harvard professor, in a spirit of conciliation, his errings and vagaries in rebuilding the migration of cultivated plants, on the plea that this field is familiar ground to him, it is difficult to find an apology for his intemperate and morbid supercriticism of Columbus, and his attempt at discrediting the whole Spanish literature relative to the discovery and the work of American archæologists, for no other apparent reason than that his own theory of the discovery of America in Africa might triumph. Columbus is branded as a "liar." "He lied, every time he said that the Indians he had with him told him so and so" (p. 9), and "his philological madness surpasses all belief" (p. 22). We hold that language unbecoming a gentleman is not worthy of a professor either, and that a professor should not lecture to a genius of the mettle of Columbus. We subscribe to every word that Markham has said with reference to Columbus's journal: "Even in the mutilated condition in which it has come down to us, it is a document of immense value. . . . It impresses us with his knowledge and genius as a leader, with his watchful care of his people, and with the richness of his imagination. Few will read the journal without a feeling of admiration for the marvellous ability and simple faith of the great genius whose mission it was to reveal the mighty secret of ages. The journal is the most important document in the whole range of the history of geographical discovery because it is a record of the enterprise which changed the whole face, not only of that history, but of the history of mankind."

On the other hand, while Wiener's main results are unacceptable, it must be candidly admitted that his book is not altogether devoid of merits. His accumulation of data and source material, although far from being exhaustive, is of great service, and he has confronted us with the interesting problem of early negro influence on the populations of America. This subject merits a more profound study from the hands of trained specialists. There is no doubt also that many African plants were at an early date introduced into our continent, possibly by negro slaves, or even more probably, through Portuguese agency. Thus the oil-palm (*Elæis guineensis*), cultivated in Mexico, the Antilles, Guiana, and Brazil, hails from the west coast of Africa. Old Piso, writing in 1658, enumerates four African plants transmitted by negroes to Brazil. According to Soarez de Sousa (*Noticia do Brasil,* 1587), the coconut-palm was transplanted to Bahia, Brazil, from Cape Verde. There are, further, sorghum, okra (*Hibiscus esculentus*), sesame (*Sesamum orientale*), the pigeon-pea (*Cajanus indicus*), Guinea grass (*Panicum maximum*), the watermelon, and others—really African plants and naturalized in America; but none of these is mentioned by Wiener. As he promises to continue his Afro-American studies in a second volume, this matter may suitably be pointed out here as a fruitful field of research.

The New Women

"FOUNDATIONS OF FEMINISM." By AVROM BARNETT. McBride.

THE feminist movement has glutted itself with arguments; many of them contradictory and most of them founded on doubtful assumptions. It has become sluggish through the attempt to digest its own propaganda, and has reached the stage, says Mr. Barnett, "where a strong and vigorous catharsis is the only thing that will energize it." He intends his own volume as the necessary medicine; it is an unpleasant dose, perhaps, but the radical feminists would think and act more clearly if they swallowed it.

The arguments adduced in favor of feminism may be divided into three categories: the biological, the psychological-physiological, and the sociological. Mr. Barnett's critique is similarly divided. His best writing—and his most vindictive—is in the first chapter, in which he discusses the biological foundation. Blanche Shoemaker Wagstaff, Charlotte Perkins Gilman, Scott Nearing, and other writers with similar beliefs have gone back to the original amœba—and beyond—to prove the superiority of the female over the male. In three or four incisive pages Mr. Barnett disposes of their claims to scientific authority. He quotes authorities on biology; he discusses the theory of chromosomes; most effective of all, he quotes the more rabid feminists themselves in a sort of encircling attack on their position. He is extremely capable at this destructive warfare; when he is finished he leaves the corpses of his opponents strewn about the field of battle.

One expects him to exhaust his polemic vigor in this chapter: on the contrary, he returns immediately to the attack. This time it is the psychological feminists whom he ridicules, picking out W. L. George for special attention. In a final chapter he discusses the sociological implications of the New Woman, this time attacking the writers who claim all labor as the proper possession of woman. When he has finished one is willing to subscribe to the declaration of Professor Beard: that science, logic, ethics, sociology have no conclusions on the question of feminism which have the validity of physical law.

And yet Mr. Barnett might be spoken of as a feminist himself. He has no wish that we should return to the conditions of the last century, when women were confined to sewing and sweeping and taking care of innumerable children. No more does he believe that the conditions of the present are desirable; the modern factory, with its monotonous and exhausting labor, is no place for women who hope some day to become the mothers of the race. Yet the intrusion of woman into commerce and industry has been a fortunate education for her sex. Through the broadened outlook it has given, a new woman may be developed who will possess a stronger intellect than the woman of the past. "She may or may not retain the so-called graces," Mr. Barnett says in conclusion, "but she will be a better mate, a more efficient mother, and a true, living, breathing, inspired, and aspiring individual."

J. G. COTTA announces "Goethe, Geschichte eines Menschen," by Emil Ludwig. The work is in three volumes. This makes the third elaborate life of Goethe that has appeared in Germany since 1917, aside from Brandes's in Danish and Brown's in English. Klara Hofer, who has made a specialty of works of fiction with poets as heroes, has also brought out with Cotta "Goethes Ehe."

At Headquarters

"G. H. Q." By "G. S. O." E. P. Dutton & Co.

Reviewed by COL. FITZHUGH LEE MINNIGERODE

"G. S. O." is, of course, a General Staff Officer, on duty at Field Marshal Haig's headquarters. He is more than a Staff Officer. He is an extremely interesting and entertaining writer, but why should the author, avoiding all argument on anything that since the armistice has become a controversial matter between high authorities, withhold his name?

"G. H. Q." takes the reader into the atmosphere of General Headquarters of the British Expeditionary Force situated in the little but historic village of Montreuil. You share in the life and labor and subconsciously absorb the optimism or pessimism of the officers charged with sending Britain's armies to victory or defeat. You mess at the club, and after dinner, with a high heart in success or a firmer determination to "carry on" in defeat, watch "Hannibal Napoleon" move the colored pins, which mark the lines, forward or back, to denote the day's advance or retirement, on the large map over which he presides. You walk the ramparts of the old walls in sunshine or rain for your daily exercise.

"G. S. O."—like many of his countrymen—cannot refrain from taking his little fling at American naiveté. Harmless as this discredited pastime has become, it leaves a note of discord in the ear. For instance,

> There is a Forest of Josse just near Montreuil, and I regret to say that some American officers were persuaded to believe that it got its name from being the site of a Chinese labor Joss house, to the lessening of the glory of St. Josse.

This may be entirely true, but why make your American guest at G. H. Q. squirm because of his brother's gullibility?

And again in the chapter devoted to Americans:

> American rank marks were puzzling to British officers at first. An American liaison officer obliged me with a mnemonic aid to their understanding.
>
> "You just reckon that you are out to rob a henroost. Right. You climb up one bar; that's a lieutenant. You climb up two bars; that's a captain. When you get up to the chickens, that's the colonel" (the colonel's badge was an eagle on the shoulder straps). "Above the chicken there's the stars" (a star was the badge of a general).

I cannot picture an American liaison officer on duty at Sir Douglas Haig's headquarters being of the stamp indicated by "G. S. O.'s" report of this conversation. Perhaps this was the exception that proved the rule.

But "G. S. O." does not withhold praise of us. Writing of the various methods pursued by the German Government with a view to belittling the results that might follow from America's entry into the war he says:

> Unhappy German people to have been fed by their leaders with such delusions! The United States a "quitter"! Had any German read the history of the eighteenth and nineteenth centuries, heard of Washington, of Hamilton, of Lincoln? If the German had searched back only so far as 1861 he would have found that the nation which he was told, might throw up the sponge at the first hint of hardship and danger faced a war which probably, for nerve strain and grim resolution, surpassed even this great war. The United States had then to fight not a foreign foe but domestic discord. It had to set its teeth through a series of great military disasters. It had to hold firmly to a forlorn hope, while it was faced by the ever present prospect of foreign interference. No nation in modern times has been put to a harsher test of courage and resolution than the United States in 1861 and the following year. No nation in history showed a more indomitable courage. And this was the nation that the German leaders would fain persuade their people was likely to prove a "quitter"! I ventured to say at the time that before the German military despotism was through with the war it would recognize that the reluctance of the United States to enter the war would be matched by the reluctance of the United States to go out of the war until its purpose was finally accomplished.

"G. S. O." does not say that England or France or any other country won the war. He does say, however, that the entry of the United States made victory a certainty.

Delightful bits of humor are scattered here and there throughout the book. There is much color, much human interest, and I think a little fiction in "G. H. Q." Beyond question, it is a very interesting story of a tremendously important and vital cog in the mechanism of the greatest army ever assembled by England.

UNDER the title "Nach Paris," the Swiss writer Louis Dumur published last year a novel composed of a rehash of the most bestial atrocities reported from Belgium during the German invasion. The success of this effort was such that the author has written an analogous *réchauffé* of the war's horrors called "Le Boucher de Verdun," which is running serially in the *Mercure de France*.

157

考狄《玉尔〈马可·波罗之书〉补注》书评

THE
AMERICAN HISTORICAL REVIEW

VOLUME XXVI
OCTOBER 1920 TO JULY 1921

THE MACMILLAN COMPANY
LONDON: MACMILLAN AND CO., LTD.
1921

Crusade, and with his long imprisonment after the disaster at Tinchebray; these, though of some interest and necessary to the completeness of the work as a biographical study, are of minor importance. The narrative closes with a chapter on "Robert Curthose in legend", in which the author follows his subject into the field of romance and shows how within a single generation the story of the duke's military achievements in the Orient had become overlaid with legendary growth.

Dr. David has added several useful appendixes, most of them dealing with problems relating to Duke Robert's participation in the First Crusade. Appendix A is devoted to a critical discussion of the sources, which, though somewhat brief, will be found of real value. Students of military history will be interested in Appendix F, in which the author reviews the controversy as to the tactics employed at Tinchebray; the conclusion is that Oman, though he exaggerates the importance of the infantry in this fight, is more nearly correct than most of his critics who have generally held that the battle of Tinchebray was chiefly a matter of cavalry warfare.

While Dr. David has not presented any new conclusions of startling importance, he has produced a volume which students of English and Norman history will find exceedingly useful. His researches have cleared up a number of controversies as to biographical and political details, and he has been able to correct the conclusions of earlier writers, like Freeman and Gaston Le Hardy, on many significant points. The result is that our knowledge of Norman affairs during the period covered is far more accurate and specific than it formerly was. The volume is carefully indexed and is provided with a map showing the principal places in England and Normandy referred to in the narrative. The reviewer is pleased to add that the work of the printer and the proof-reader seems to have been done with unusual care.

LAURENCE M. LARSON.

Ser Marco Polo: Notes and Addenda to Sir Henry Yule's Edition, containing the Results of Recent Research and Discovery. By HENRI CORDIER, D.Litt., Professor at the École des Langues Orientales Vivantes, Paris. (New York: Charles Scribner's Sons. 1920. Pp. x, 161. $4.00.)

THE names of Henry Yule and Marco Polo will always remain inseparable in the minds of those who have the medieval geography of Asia and the study of the Mongol period on their hearts. Hardly any other medieval traveller has exerted such a profound influence on modern research, whether it be geography or cultural history, nor could he have been more lucky in finding so competent and sympathetic an interpreter as Yule. His edition of Polo, first published in 1870 (second ed., 1874; third ed., 1903, by H. Cordier), has become a classic and household book in the hands of all students interested in Asia; and during twenty-five years of activity I do not know of any work that I have consulted and quoted more frequently than Yule's *Book of Ser Marco Polo,* which is an inexhaustible mine of information on almost all questions bearing on the history, geography, ethnography, and folk-lore of medieval Asia. Professor Henri Cordier, to whom we are indebted for a revised and largely increased edition of Yule's *Cathay and the Way Thither,* has collected in this small volume of 162 pages additional information apt to shed light on Polo's observations or on Yule's comments, and either published in print after 1903 or contributed to him by his correspondents and collaborators directly. The volume thus presents a harvest mostly of brief notes and essays with reference to the third edition and arranged according to Polo's chapters. This book is easily readable only for those who know their Polo by heart, or who have closely followed the discussion of pending problems, and this small band of readers will doubtless peruse the volume with great pleasure and profit. Others will have to refer constantly to Yule's edition, and must first read up in order to appreciate the fresh evidence. There are no new contributions from the hand of the editor; the most valuable notes are from the pen of Sir Aurel Stein, chiefly concerning the route and topography of the Venetian. Sir Richard C. Temple has supplied a very interesting notice of the Andaman and Nicobar Islands, which gives a good summary of the present knowledge of the inhabitants. The reviewer has contributed twelve short articles, also the only illustration, which serves as the frontispiece, and which represents a Lo-han out of a Chinese series of Five Hundred Lo-han. Most readers will be at a loss to grasp the *raison d'être* of this illustration in the book, no explanation to this effect or a page-reference being added to the plate. It refers to the article on the alleged Marco Polo Lo-han of Canton (pp. 8–11).

The editor has given a few additions to Yule's bibliography, to the manuscripts of Polo's work, and to Polo literature. The proof-reading is not carefully done, and even whole words have occasionally dropped from a sentence. Nor does the editor discuss or decide contradictory opinions of his collaborators. Thus on pp. 69–70 two conflicting interpretations of the Mongol word *chinuchi* peacefully follow each other. In my opinion, that given by Pelliot is far-fetched and wrong; but how is the unsophisticated reader to decide for himself? The same difficulty is prominent in Cordier's third edition: the method adopted is simply to quote authorities in full and *verbatim,* and in many cases one statement flatly contradicts another. What will the editor of the fourth edition do? New materials will doubtless come to light during the next years; and if this method of mere citation, without an intelligent discussion of the problems, should be kept up, Yule's head will finally be buried under a mass of débris, and the commentary will no longer be intelligible or useful. It seems to me that the new editor should break away from the past, fling the superfluous ballast overboard, retain only what is good, and present a co-ordinated essay in the place of a massed attack of bewildering notes.

B. LAUFER.

Thought and Expression in the Sixteenth Century. By HENRY OSBORN TAYLOR. In two volumes. (New York: Macmillan Company. 1920. Pp. xiv, 427; viii, 432. $9.00.)

IT is now almost a decade ago that Henry Osborn Taylor gave us *The Medieval Mind,* a work which, in a masterly manner, traced for us the gradual formation of the medieval spirit until it found the end of its development and proper issue of its genius, at the close of the thirteenth century, in the immortal *Divina Commedia.* There were many who looked forward to another book from the same pen that should have to do with the Renaissance; and so, when at last a new work by the same author was announced, and we learned that its chief purpose was to give an exposition of thought and expression in the sixteenth century, some of us wondered why the two intervening centuries had been ignored. Slighted they are, but not ignored. It would have been impossible to have overlooked them altogether even in a book that has for its purpose the presenting of a survey of only the sixteenth century. Even in the preface, the fifteenth century assumes its rightful place side by side with its immediate successor. "We shall treat the fifteenth and sixteenth centuries," says our author, "as a final and objective present." And what has he done with the fourteenth? "All that went before," he tells us, "will be regarded as a past which entered into them." Thus, evidently, he would date a new era from the close of the fourteenth century. But the attempt fails. The century of Petrarch and Boccaccio and Giotto refuses to be regarded as medieval. Its place as the first modern century quickly becomes evident. In the first pages of the book we find our author telling us that "Petrarch was a great inaugurator", that Boccaccio, in "looking to life" and "drawing from life", was not medieval, that "no man is medieval who goes straight to the life about him", and that the work of Giotto, "summing up the past's attainment" and "incorporating riches of its own", was "altogether a prefigurement of Italian painting in the *Cinquecento*".

The truth of the matter is that a new era began towards the close of the thirteenth century. More than once our author finds himself obliged to repeat that "emotionally as well as intellectually, the final *summa,* and a supreme expression, of the Middle Ages was the *Divina Commedia*". There is, of course, much that is medieval in Dante; but to summarize a period is to end it. Dante could not have been "the voice of ten silent centuries", as Carlyle said he was, had not the time permitted him to view the work of those centuries as being essentially completed. And so to the present writer it seems that it would have

158

女真语和蒙古语中的数字

I. KÖTET. 2. SZÁM. 1921 DEC. 10.

KŐRÖSI CSOMA-ARCHIVUM

A KŐRÖSI CSOMA-TÁRSASÁG FOLYÓIRATA

A TÁRSASÁG MEGBÍZÁSÁBÓL

CHOLNOKY JENŐ, GYŐRFFY ISTVÁN, F. TAKÁCS ZOLTÁN, gr. TELEKI PÁL ÉS gr. ZICHY ISTVÁN KÖZREMŰKÖDÉSÉVEL

SZERKESZTI

NÉMETH GYULA

TARTALOM:

SZERKESZTŐSÉG ÉS KIADÓHIVATAL:
BUDAPEST, I., BERCSÉNYI-UTCA 10. III. 3.

Jurči and Mongol Numerals

— By BERTHOLD LAUFER —

Jurči	Manchu	Golde	Tungus (Castrén)	Oročon	Mongol
1 *omu*	*emu* (*umudu*, ("isolated")	*emu*	*umun*	*omu*	*nigän*
2 *jo*	*juwe* (*jue*, *juo*)	*jur* (*jul*)	*jur*	*jur*	*xuyar* / *xuyur*
3 *ilan*, *ilai*	*ilan*	*elá* (*elán*)	*ilan*	*ela*	*gurban*
4 *duyin*, *duin*	*duin*	*duin*, *duyí*	*digin*	*dyi*	*dürbän* / *dörbön*
5 *šunja*	*sunja*	*tęŋgá*	*toŋa*	*tuŋa*	*tabun*
6 *niŋju* (*ningu?*)	*niŋgun*	*n'uŋgú*	*nuŋun*	*nuŋu*	*jirgugan* / *jirgan* (Kalmuk *jurgan*)
7 *nadan*	*nadan*	*nadá*	*nadan*	*nada*	*dulugan* / *dolōn*
8 *jakun* (10—2)	*jakôn*	*japkún*	*japkun*	*jaxpu*	*naiman*
9 *uyen* (*uyewen*) (10—1)	*uyun*	*xuiyú*	*jägin*	*xuyu*	*jisun*
10 *jua*, *jwa*	*juwan*	*jua*, *juan*	*jan*	*ja*	*arban*
11 *anšo* (*an* = 10)	*juwan emu* *omšon biya* (*b'a*) 11th month)	*jua emu*	*jan umun*		*arban nigän*, etc.
12 *jirxoan* (2 + 10)	*jorxon* (12th month)				
13 *gorxoan* (3 + 10)					
14 *durxoan* (4 + 10)					
15 *tobuxoan* (5 + 10)	*tofoxon* (5 + 10)				
16 *nixun* (*n'üruxun*) (6 + 10)					
17 *darxoan*, *darxwan* (7 + 10)					

Jurči	Manchu	Golde	Tungus (Castrén)	Oročon	Mongol
18 *n'üxun* (8+10)					
19 *won'i-xoan, on'üxoan*					
20 *worin, orin*	*orin*	*xorē*	*orin*	*oi*	*xorin*
30 *gušin, guši*	*gôsin* (**gursin*)	*gočē*	*gutin, gučin, ilaŋi,*	*kudty*	*gučin* (**gurčin*)
40 *texi*	*dexi* (**derxi*)	*derxi*	*digiŋi, dučin*	*dyšja*	*döčin*
50 *susai*	*susai*	*sosai*	*toŋaŋi*	*tommonja*	*tabin*
60 *niŋju*, old Jurči *inju*	*ninju*	*n'uŋgu-iŋ-gú*	*nuŋuni*	*nunkuja*	*jiran*
70 *nadanju*	*nadanju*	*nada-iŋ-gú*	*nadaŋi*	*nadanja*	*dalan*
80 *jakunju*	*jakônju*	*japk-ŋ-gú*	*japkuŋi*	*jaxpunja*	*nayan*
90 *uyenju*	*uyunju*	*xuiyuŋgu*	*jägiŋi*	*xuyunja*	*järän* / *jirän*
100 *taŋgu*	*taŋgô*	*em-taŋgú*	*namaji-(d'i)*	*tanku*	*dzagun*
1000 *miŋgan meŋan*	*miŋgan*	*emeŋgá* (= *emu* 1)	*miŋan*	*minka*	*miŋxan*
10000 *tuman*	*tumen*		*tuman*	*ja minka*	*tümän*

The Jurči numerals from 1 to 10 all show the common Manchurian type. *Omu* 1 agrees with Oročon *omu* and Tungusian *umu-n,* Manchu *umu-du* ("isolated"). In *jo* 2 the final *r* is eliminated as in Manchu, but is still preserved in *jir-xoan* 12. *Šunja* 5 coincides with Manchu *sunja,* while Golde and the Tungusian languages have initial *t*. *Jwa* 10 is identical with Golde *jua*, Oročon *ja*. The formation of the numerals from 11—19 are based on the principle 1 + 10, 2 + 10, etc. The second element *-xoan*, *-xun* is met in *-xon* of Manchu *tofo-xon* 15, and means 10. It may be related to Turkish *on* 10 and the endings *-an*, *-än*, *-in* in the Mongol decimal numerals, two of which, those for 20 and 30, coincide with Manchu-Tungusian. The only exceptional formation in Jurči is *anšo* 11, in which I take *an* = 10. The most striking fact in the

Jurči numerals from 12—18 is that the units exhibit a decidedly Mongol affinity.

The Mongol numeral for six has been explained by Schott,[1] and, I believe correctly, as 2 × 3 : *jirgugan* = *jir-gu-gan* (*jir* = 2; *gu* = 3), — an analysis confirmed by W. Radloff.[2] Now we meet the form *jir* in Jurči *jirxoan*, "twelve" (2 + 10), and the variant *jor* in Manchu *jorxon* ("twelfth month"); cf. further Manchu *jur*, *juru* ("a pair"), and *jurjun* ("the game of backgammon", Chinese *šwan-liu*, 雙六 or *-lu-* 陸 where the element *jur* doubtless means "two"). Certainly these *jir* and *jor* are identical with *jur* of Golde and Tungusian, so that there is a type of the numeral 2 common to Mongol-Manchu-Tungusian. In Jurči *gor-xoan* (3 + 10), the element *gor* is the equivalent of Mongol *gur-ban*; with the final *r* eliminated we find it in the word for 30 (*gušin*). It is likewise not difficult to recognize in *durxoan* (14) Mongol *dür-bän* (*dur-bän*) and in *tobuxoan* (15) Mongol *tabu-n* (cf. Manchu *tofo-xon*). For 16 we have two forms, — an older one handed down by the *Kin ši*, *n'üruxun*, and the more recent one of the Glossary *nixun*. I regard *n'üru-* as related to Mongol *jirgu-gan*, *jur-gan*, *n'* and *j* being interchangeable. The element *dar-* of *darxoan* (17) agrees with Mongol *dul-u-gan*, *dol-ōn* (7); and *n'ü-* of *n'üxun* (18), with Mongol *naiman* (8). The form for 19, *on'üxoan*, is irregular: in my opinion it is framed on the basis of the formula 1 + 8 + 10 : *o* 1 (= *o-mu*) + *n'üxun* 18. 20 *orin* = *or-in* 2 × 10; or from *xor* (cf. Golde *xorḗ* and Mongol *xorin*), related to Mongol *xuyar*, *xuyur* 2.

30 *guši*, *gušin*, from **gur-šin* 3 × 10 (cf. *gor-xoan* 13).

40 *texi*, from **ter-xi* (cf. Golde *d̦erxi*); *ter*, *der* = Mongol *dür-*, *dör-* 4; Jurči *dur-xoan* 14.

There is thus no doubt that, contrary to what has previously been supposed, there is virtually a connection between the Tungusian and Mongol numerals, and that the Mongol stratum is preserved in Jurči, while merely faint survivals of it are retained in Manchu, and a complete extinction of it characterizes the living Tungusian forms of speech. It is reasonable to conclude that in a former stage of development two series of numerals have existed

[1] Das Zahlwort in der tschudischen Sprachenklasse, p. 11.
[2] Phonetik der nördlichen Türksprachen, p. 183.

in Tungusian, — one which still survives in modern Tungusian, and another which still prevails in Mongol. Unfortunately, for lack of materials, we are not in a position to trace the primeval condition of affairs: this would only be possible if we knew the numerals of the extinct languages of the Hiuń-nu and Kitan, the vocabulary of both of which, as far as we know it from Chinese records, displays many affinities with Mongol.

We know but a few numeral of the Kitan language; that for 5 is transcribed in Chinese *t'ao* 討; that is, Kitan *taw*, *tab*, identical with Mongol *tab-un*. Possibly also Kitan *nai* ("the first day"), as proposed by Shiratori (p. 43), is related to Mongol *nigän* 1.

Dömsödi török oklevelek.

— Közli: ZSINKA FERENC. —

Azok az oklevelek, melyeket itt hasonmásban, átiratban és fordításban közlünk, a magyar történetkutatás előtt nem ismeretlenek. Egy részüket Repiczky János fordította le és kéziratban hagyta hátra 1852-ben. (Nemz. Múz. Kézirattára. Quart. Hung. 448.) Az egészet pedig Szilády Áron szólaltatta meg először magyar nyelven még 1863-ban, amikor a Török-magyarkori Történelmi Emlékek II. kötetében fordításukat közölte. Dömsöd, pestmegyei község multjához szolgáltatnak adalékot s természetesen Dömsödön keresztül az egész magyarországi hódoltság történetéhez is egy morzsányit. Dömsöd nem volt jelentékeny helység; a hódoltság korában még csak kerületi székhely sem volt s így a róla szóló oklevelek történeti anyagot nem nyujtanak elsőrendűen fontosat. Mindamellett találunk okot, mely kívánatossá teszi, hogy e levelek így egy csomóban még egyszer napvilágot lássanak. Egy magyar falu küzdelmes életét mutatják a hódoltság alatt. Már ez magában elég beszédes ok volna megjelenésük mellett. Azonkívül, bármily nagyra becsüljük is Szilády Áron tudományos jelentőségét a török oklevelek fordítása körül, nem hallgathatjuk el, hogy neki a hatvanas években könnyebb dolga volt: kevesebb szem ellenőrizte s így nem kellett törekednie filológiailag teljes fordításra s ha már a levelek magyarul szólaltak meg, eleget tett a legszigorúbb kritika követelésének is. Emiatt több kisebb pontatlanság maradt a lefordított oklevelekben. Mindamellett Szilády for-

159

梵语中的猫眼石

MÉMOIRES

DE

LA SOCIÉTÉ DE LINGUISTIQUE

DE PARIS

TOME VINGT-DEUXIÈME

PARIS

LIBRAIRIE ANCIENNE HONORÉ CHAMPION, ÉDITEUR

ÉDOUARD CHAMPION

QUAI MALAQUAIS, 5 (VIe)

1922

SANSKRIT *KARKETANA*.

La traduction arménienne de la Bible emploie le mot *karkehan* comme l'appellation d'une pierre précieuse rouge. Dans le lapidaire arménien du XVIIe siècle (cf. l'édition avec traduction russe de Patkanov, p. 16), le même mot se présente comme un synonyme de *seilan* (le zircon trouvé presque exclusivement dans l'île de Ceylan), et Patkanov propose à juste titre d'identifier le *karkehan* avec l'hyacinthe, qui en effet n'est qu'une variété du zircon (cf. G. Ferrand, *Textes relatifs à l'Extrême-Orient*, p. 387). Cette identification est beaucoup plus probable que l'améthyste, suggérée par Clément-Mullet (*Essai sur la minéralogie arabe*, *Journal asiatique*, 1868, p. 55).

Il est évident que la forme arménienne ne procède ni du sémitique, ni, comme nous verrons, du grec, mais présuppose un iranien arsacide ou parthe **karkaδan*, comme l'a déjà reconnu Hübschmann (*Arm. Gram.*, p. 167, 512 et *Pers. Studien*, p. 199); mais, autant que je sache, on n'en a pas encore rapproché le parallèle sanskrit *karketana*, *karketaṇa* ou *karketila*, ordinairement traduit par chrysobéryl, avec une forme prākrite *kakkeraa*, qui se trouve dans le drame *Mṛcchakaṭikā* (acte IV) et qui a abouti à une forme apabhraṃça **kekeru*, adoptée par le tibétain (cf. *Loan-Words in Tibetan*, n° 73) et le mongol (c'est par inadvertance que Kovalevski, dans son *Dictionnaire mongol*, p. 2498, donne à ce mot l'équivalent «nom d'arbre, *Pandanus odoratissimus*»). Le terme sanskrit est expliqué dans la *Ratnaparīkṣā* de Buddhabhaṭṭa (antérieur au VIe siècle) et dans la *Bṛhatsaṁhitā* de Varāhamihira (505-587); il est aussi contenu dans l'*Amarakoṣa* et dans le *Mahāvyutpatti* (sect. 235). Il est dit dans le lapidaire de Buddhabhaṭṭa : «Vāyu saisit les ongles du roi des Daityas et, joyeux, les jeta juste dans le pays des Yavanas. Le chrysobéryl (*karketana*) en naquit, produit du pays des Yavanas (*javanopapannaṁ*), très estimé sur la terre» (L. Finot, *Les Lapidaires in-*

diens, p. 49). Ici le mot Yavana fait allusion, sinon aux Grecs, du moins à un pays étranger, et s'il en est ainsi, il serait assez séduisant au premier abord de tirer la conclusion que skr. *karketana* serait dérivé de l'iranien parthe **karkaδan*. Il n'y a pas de doute que des noms sanskrits de pierres précieuses soient empruntés de l'iranien; par exemple skr. *lājavarta* ou *rājāvarta* («lapis lazuli») provient du persan *lājvard* ou *lāžvard* (arabe *lāzvard*, arménien *lazvart'*). Mais, dans ce cas, le mot indien est assez tardif (Finot, *loc. cit.*, p. XVIII, 191; Garbe, *Die indischen Mineralien*, p. 191; le *Rājanighaṇṭu* de Narahari traduit par Garbe n'est pas antérieur au commencement du XVe siècle). La forme *rājāvarta* repose sur une assimilation populaire à *rāja* («roi»); mais bien que traduite «propre au front d'un roi» (Garbe), ce n'est qu'une reproduction du mot persan, comme il est bien démontré par le parallèle *lājavarta*. Le lapis lazuli est une pierre caractéristique de l'Iran, les mines principales étant situées au Badaxšān. De même, skr. *peroja*, *perojā* ou *pīroja* est sans doute dérivé du persan *fīrūza* فيروزه; la turquoise ne fut introduite dans l'Inde par les Mahométans que dans la dernière moitié du Xe siècle. Mais le cas du *karketana* est un peu différent. Comme l'indique le dictionnaire sanskrit de Boehtlingk, le mot se trouve dans Caraka (6, 23). Or la tradition bouddhiste rend Caraka contemporain du roi indoscythe Kaniṣka. La date de Kaniṣka est un des problèmes les plus complexes et les plus discutés, et le texte de Caraka est très embrouillé. Mais tout cela rappelé, il se peut que Caraka ait été contemporain de la dynastie arsacide et que le mot *karketana* lui ait été connu; en tout cas il n'y a rien d'incroyable à cela. En outre nous savons que l'œuvre de Caraka fut traduite dans les langues persane et arabe (Sachau, *Alberuni's India*, I, p. XXXV, 159; Jolly, *Ind. Med.*, p. 11). Voilà une indication de la route que le mot *karketana* peut avoir prise dans sa migration de l'Inde à l'Ouest. La source principale du chrysobéryl, c'est l'île de Ceylan; c'est aussi le cas pour le zircon. Ainsi il y a assez de raisons pour supposer que skr. *karketana* soit le point de départ pour la formation du mot iranien.

Hübschmann a aussi ajouté aux mots arménien et iranien le syriaque *qarkeδnā*, le latin *calchedonius* de Pline et le grec χαλκηδών. Je ne crois pas que ces rapprochements soient justes. D'abord il y a ici une différence sémantique : le mot gréco-latin désigne la chalcédoine, un groupe nettement distinct de pierres semi-précieuses. Le mot syriaque est sans doute tiré du grec καρχηδόνιος, mais non de χαλκηδών; peut-être aussi le néo-hébreu *karkdōn*. Naturellement, les deux mots grecs n'ont rien de commun : χαλκηδών se rapporte à la ville Chalkedon, dans la Bithynie, vis-à-vis de Byzance, tandis que καρχηδόνιος est dérivé du nom de Carthage, où la pierre en question était fréquemment trouvée. Dans cette note philologique, je ne pousserai pas la question minéralogique. Il est évident que le mot arménien, s'il est emprunté à l'iranien parthe et si le mot iranien est tiré du sanskrit, ne peut pas être mis en rapport avec un type grec. Mais il est possible de préciser le problème un peu plus près.

Il y a un mot hébreu ancien כדכוד *kadkōd*, qui semble désigner une espèce de rubis et qui est traduit χόρχος et κρύσταλλος par la Septante, *chodchod* et *jaspis* par la Vulgate. Quelques manuscrits hébreux et Symmaque écrivent *karkōd* ou *karkōr*, et E. Levesque (*Dictionnaire de la Bible* de Vigouroux, V, col. 1263) préfère ces variantes, qui, dit-il, rappellent le grec χαρχηδων, un *carbunculus*; en outre, selon lui, l'arabe *karkand* rappelle le nom spécifique du *carbunculus carchedonius* et le *karkōd* hébreu. Je ne partage pas cette opinion : *kadkōd* est bien la forme hébraïque ancienne et seulement correcte, et n'a rien à voir avec le grec ou l'arabe; les variantes *karkōd* ou *karkōr* sont assez tardives et le résultat de plusieurs confusions, ainsi qu'il est bien démontré par la traduction hébraïque du lapidaire d'Aristote, où nous trouvons les formes *kadkōd*, *kadkōr*, *karkōr* et *karkahin* comme équivalents de l'arabe *karkand* et *karkahan* (la traduction latine offre *cortond*, *cortaud* et *corheen*; cf. Ruska, *Steinbuch des Aristoteles*, p. 136, n. 2). Il est assez curieux que nous avons deux termes minéralogiques sanskrits encore inexplicables, *karkoda* et *karkoṭika*, d'origine médiévale et trouvés par Finot (*Lapidaires indiens*, p. 135, 138) seulement dans l'*Agastimata*, traité sur les gemmes par Agasti, qui paraît avoir écrit entre le VIe et le XIIIe siècle. Mais que faire avec les types arabes *karkand* et *karkahan*?

L'arabe *karkand* كركند est ainsi défini par Ibn al-Baiṭār, selon al-Jāfikī (al-Ghafeky), qui mourut en 1164 : «On dit que c'est une pierre qui ressemble au rubis, mais elle n'en a ni l'éclat ni la valeur. Si elle est exposée au feu, elle se brise en morceaux. La lime n'agit sur cette pierre que faiblement» (L. Leclerc, *Traité des simples*, III, p. 168). R. Dozy (*Supplément aux dictionnaires arabes*, II, p. 459) ne cite que Baiṭār sous ce mot, mais donne كركهن *karkahan* et كركهان *karkahān* comme «améthyste». Cependant il y en a une mention du IXe siècle dans le lapidaire arabe du Pseudo-Aristote, qui dit, dans sa notice sur le *yāqūt* : «Le *karkand* ressemble au *yāqūt* rouge, mais il ne soutient pas comme lui l'action du feu; aussi le *karkahan* ressemble-t-il au *yāqūt*, mais de même celui-ci n'appartient pas au genre du *yāqūt*» (Clément-Mullet, *loc. cit.*, p. 54; J. Ruska, *Steinbuch*, p. 136). Cette définition est identique à celle de al-Jāfikī, mais il s'ensuit que le Pseudo-Aristote établit quelque divergence entre le *karkand* et le *karkahan*. Tout d'abord nous n'avons pas affaire à des emprunts classiques :

l'œuvre d'Ibn al-Baiṭār est décisive à cet égard, car elle est basée sur le livre de Dioscoride, dont la traduction est entourée de commentaires arabes. Dans tous les cas, Ibn al-Baiṭār se rapporte aux termes grecs si le mot arabe en question en est issu; mais il ne dérive pas *karkand* du grec, il ne cite pas Dioscoride à ce sujet. D'autre part, *karkand* et *karkahan* ne sont pas des mots arabes. L'un et l'autre sont d'origine iranienne. *Karkahan* est au même niveau que l'arménien *karkehan*; comment le mot a pénétré de l'arménien à l'arabe, je ne prétends pas le savoir. *Karkand*, il me semble, se réduit à une forme iranienne sassanide ou pehlevie **karkandan* ou **karkadan*, transmise aux Arabes après la conquête de la Perse.

Cette pierre n'a pas fait grande fortune et n'est jamais devenue l'objet d'un commerce animé. Elle est inconnue aux Chinois et aux peuples malais; du mot *karketana*, nous n'avons pas encore découvert de trace dans la littérature chinoise bouddhiste ou minéralogique. En tibétain et mongol, le mot *ke-ke-ru* n'a qu'une existence littéraire et est étranger au langage familier. Même dans l'Inde, les parlers aryens modernes paraissent l'avoir perdu. Le persan n'a pas perpétué la tradition parthe et sassanide. Abu Mansur ne mentionne pas la pierre dans ses *Principes pharmacologiques*. L'arménien et l'arabe l'ont seuls retenue.

B. Laufer.

160

密勒日巴——藏文文本选译

Schriften-Reihe

KULTUREN DER ERDE

MATERIAL ZUR KULTUR-
UND KUNSTGESCHICHTE
ALLER VÖLKER

Abteilung: Textwerke

TIBET I

1922

FOLKWANG-VERLAG G. M. B. H.
HAGEN i. W. UND DARMSTADT

MILARASPA

Tibetische Texte

in Auswahl übertragen von

BERTHOLD LAUFER

1922

FOLKWANG-VERLAG G. M. B. H.
HAGEN i. W. UND DARMSTADT

DIE AUSFUHR UNSERER BÜCHER INS AUSLAND UNTERSAGEN WIR UND JEDER KÄUFER EINES BUCHES IN DEUTSCHLAND VERPFLICHTET SICH DURCH DEN ERFOLGTEN ABSCHLUSS DES KAUFES, DAS BUCH NICHT OHNE UNSERE ERLAUBNIS INS AUSLAND WEITER ZU VERKAUFEN. ZUWIDERHANDELNDE MÜSSEN WIR GESETZLICH VERFOLGEN, DA DIE VERLUSTE, DIE FÜR UNS AUS DER DAUERNDEN UNERLAUBTEN AUSFUHR VON BÜCHERN ENTSTEHEN, AUSSERORDENTLICH BEDEUTEND SIND. WIR MÜSSEN JEDEN EINZELNEN, DER DEM VERBOT DES VERKAUFES INS AUSLAND ENTGEGENHANDELT, FÜR DIE GESAMTEN SCHÄDLICHEN FOLGEN VERANTWORTLICH HALTEN. VERKÄUFE INS AUSLAND GENEHMIGEN WIR NUR BEI BEZAHLUNG ENTSPRECHEND DER JEWEILIGEN VALUTA ABZÜGLICH EINER PROVISION FÜR DIE BEMÜHUNGEN DER AUSFUHR.

FOLKWANG-VERLAG.

PRINTED IN GERMANY

DRUCK VON BALD & KRÜGER, HAGEN i.W.

VORWORT.

Nachdem unser Verlag es zu seiner Aufgabe gemacht hatte, die geistigen Leistungen aller Zeiten und Völker einem größeren Publikum zugänglich zu machen, stießen wir vor einiger Zeit in den Denkschriften der Kaiserlichen Akademie der Wissenschaften in Wien auf die Uebertragungen des Herrn Dr. Berthold Laufer, die unter dem Titel „Aus den Geschichten und Liedern des Milaraspa" im Jahre 1902 veröffentlicht wurden. Auf unsere Anfrage teilte uns Herr Laufer aus Chicago mit, daß bereits im Jahre 1901 im „Archiv für Religionswissenschaft" zwei Legenden des Milaraspa zum Abdruck gekommen waren. Beide Zeitschriften sind weder im Buchhandel noch auch sonst für den Laien leicht erreichbar. Wir waren über diese tibetischen religiösen Dichtungen so sehr erstaunt, daß wir glaubten, diese Dichtungen könnten die Psyche des Landes und Volkes, das wie kein anderes unbekannt geblieben ist, am besten den Lesern näher bringen.

Herr Laufer teilte uns mit, daß er auf Grund seiner weiteren Forschungen natürlich in der Lage wäre, manche Verbesserungen in der Uebertragung zu machen. Da wir nun einerseits eine so umfangreiche Arbeit einem Verfasser, der in Amerika lebt, nicht zumuten mochten, und andererseits nicht glauben, daß diese Verbesserungen an dem Inhalt und Geist der Dichtungen Wesentliches ändern würden, sondern wiederum nur wissenschaftlichen Kreisen von Interesse wären, haben wir Herrn Laufer gebeten, die Uebertragungen so gut wie ohne Veränderungen zum Abdruck für ein größeres Publikum bringen zu dürfen. Das Werk enthält so außerordentlich viel wertvolle Aufschlüsse über die Vorstellungen, die in Tibet mit dem Leben und der Macht der Heiligen verbunden sind, daß wir glauben, daß diese Schriften ein außerordentliches Interesse finden werden bei einem Publikum, das bestrebt ist, seine Begriffe vom Geist vergangener Zeiten zu erweitern.

AUS DER EINLEITUNG DES ÜBERSETZERS.

Milaraspa oder kurz Mila genannt, sagt Jäschke von ihm, ist der Name eines buddhistischen Asketen des elften Jahrhunderts, der zwischen seinen Meditationsperioden als Bettelmönch im südlichen Teile Mitteltibets umherwandernd, durch seine stets in gebundener Rede und Gesangweise vorgetragenen Improvisationen Lehrbegierige unterrichtet, Weltlichgesinnte zum Glauben bringt, Ketzer niederdisputiert und bekehrt und mannigfache Wunder (Rdzu-p̗rul) verrichtet, und dessen nicht ohne Witz und Poesie geschriebene Legenden das beliebteste und verbreitetste Volksbuch in Tibet sind. Ob nun Milaraspa selbst der Verfasser des Mgur ˳bum*) ist, steht einstweilen dahin. In dem Buche selbst läßt sich keine darauf oder überhaupt auf die Autorschaft desselben bezügliche Angabe nachweisen. Ebensowenig ist etwas über die Zeit der Abfassung gesagt. Nach der chronologischen Tafel Reu mig hat Milaraspa von 1038 bis 1122 gelebt. Zunächst ist sicher, daß das vorliegende Werk keine posthume Erfindung vorstellen kann, denn es führt die Ereignisse aus Milaraspa's Zeit mit solch lebendiger Frische und unmittelbarer Treue der Darstellung, mit einer Schärfe der Charakteristik des Trägers der Handlung und der dabei beteiligten Personen, mit solch feiner Wiedergabe der Orts- und Zeitfärbung vor, daß man nicht anders als dem Gedanken Raum geben kann, der Inhalt dieser Blätter müsse einer gleichzeitigen Aufzeichnung entstammen. Das gilt in erster Reihe von den zahlreichen Liedern, welche die erzählenden Prosapartien so stark überwuchern, daß das Werk mit Recht nach ihnen benannt worden ist. Sie rühren zweifellos von dem Dichter selbst her und sind sicher von ihm auch niedergeschrieben worden; sie sind zu sehr von der Subjektivität der Persönlichkeit erfüllt, als daß man mit gutem Grunde annehmen könnte, ein Schüler des Meisters habe dieselben, wie er sie aus seinem Munde gehört, aufgezeichnet und nachträglich bearbeitet. Gleichwohl kann nicht das ganze Werk in dem Zustande, wie es uns jetzt vorliegt, als Erzeugnis des Milaraspa angesprochen werden. Dagegen streiten verschiedene in den Prosateilen vorkommende Stellen, welche deutlich auf ein späteres Zeitalter als das des Milaraspa anspielen. So werden wiederholt Stätten der Verehrung, die durch die Erinnerung an den Aufenthalt des verehrten Lama geweiht sind, mit ihren späteren Benennungen aufgeführt; solche Plätze konnten ja naturgemäß auch erst nach dem Tode des Meisters ihre rechte Bedeutung erlangen, als der Ruf seiner Tätigkeit und seiner Lieder sich weiter

*) d. h. „Die Hunderttausend Gesänge". Der vollständige Titel heißt: Rje btsun Mi la ras pai rnam thar rgyas par phye ba mgur˳bum „die des ehrwürdigen Milaraspa Lebensgeschichte ausführlich darlegenden Hunderttausend Gesänge". Die auf Seite 8 gegebenen Zitate beziehen sich auf einen im Besitze des Herrn Dr. Laufer befindlichen, aus 263 fol. bestehenden Holzdruck des Werkes.

und weiter im Volke verbreitete. So heißt es in dem Prosaabschnitt nach dem ersten Liede der folgenden Proben (S.19), daß der Felsblock, auf dem Milaraspa einen hüpfenden Tanz aufführte und Spuren seiner Füße und Schneeschuhe hinterließ, in früherer Zeit „der flache weiße Felsblock", später dagegen „Spuren der Schneeschuhe" genannt wurde. Diese Bemerkung kann wohl nur der Zusatz eines späteren Bearbeiters sein. Eine ähnliche Stelle begegnet auf fol. 6 b 3: Nach einem durch Dämonenspuk erregten Unwetter läßt sich Milaraspa auf einem Steine nieder, wo ebenfalls seine Fußspur zurückbleibt; dann geht er einige Schritte weiter, der Himmel heitert sich auf, in freudig gehobener Stimmung setzt er sich auf einen Hügelvorsprung, wo er sich einer Betrachtung der Liebe zu allen Wesen hingibt. Dieser Ort nun, so lautet ein folgender Zusatz, ist bekannt als Byams sgang „der Hügelvorsprung der Liebe". In derselben Erzählung, fol. 9 b 5, erhält ein anderer, durch eine phantastische Vision Milaraspa's berühmt gewordener Ort den Beinamen La dgu lung dgu „die neun Pässe, die neun Täler". Solche Namen werden im Volke erst in späterer Zeit, als die Geschichten des beliebten Lehrers populär wurden, zum Andenken an seine Wanderungen entstanden sein.

Die Beliebtheit des Mgur ₒbum wird einmal durch die Aussage von Jäschke bezeugt, daß das bekannteste und am meisten gelesene Werk in Bezug auf Milaraspa eben diese Hunderttausend Gesänge sind, welche in verschiedenen Holzdruckausgaben in ganz Tibet verbreitet sind, sodann durch den Historiker ₒJigs med nam mkha, der an zwei Stellen seines Werkes sowohl das Rnam thar als Mgur ₒbum zusammen mit philosophischen Abhandlungen als Vortrags- und Studienschriften in der Mongolei für die zweite Hälfte des 18. Jahrhunderts erwähnt. An zwei anderen Stellen zitiert derselbe Autor einmal drei und das andere Mal vier Verse aus Mila.

Wenn wir auch vorläufig noch nicht imstande sind, das Datum der Abfassung des vorliegenden Werkes zu bestimmen, so gibt uns doch dieses selbst Hilfsmittel in die Hand, um die Zeit der Ereignisse festzusetzen, welche den Rahmen der vorliegenden Geschichten bilden. Auf fol. 6 b 3 findet sich nämlich die Angabe einer bestimmten Jahreszahl. Als Milaraspa zu dem Flüßchen Chu bzang („treffliches Wasser") gelangte und sich in die Betrachtung des fließenden Wassers vertiefte, heißt es: me pho stag gi lo ston zla ra bai tśes bcui nub „da ging der 10. Tag des ersten Herbstmonats des Feuer-Tiger-Jahres zur Rüste". Das Feuer-Tiger-Jahr ist das 60. Jahr des Cyklus, in diesem Falle des 1. Prabhava, also das Jahr 1085. Der erste Herbstmonat wäre nach der indischen Einteilung der Jahreszeiten der September, nach tibetischer Anschauung der August. Dies ist die Zeit, zu der Milaraspa die Dämonen auf dem La phyi bekehrt. Darauf bringt er einen ganzen Monat in Gnas ₒthil zu, worauf er sich in das Dorf Gña nam begibt (fol. 9 b 7). Wie lange er dort weilt, ist nicht mit Bestimmtheit gesagt, doch läßt sich die Zeit ungefähr aus der Angabe in unserer ersten Erzählung bemessen, daß unmittelbar nach seiner zweiten Besteigung des La phyi ein neun Tage und Nächte andauernder heftiger Schneefall, d. h. also der Winter eintritt. Dies Ereignis muß demnach im Oktober stattgefunden haben, was durch zwei weitere Zeitangaben bestätigt wird. Im Pferd-Monat (Rtai zla ba), d. i. im 3. Monat des Jahres 1086, und zwar in der zweiten Hälfte desselben, bringen dem Meister seine Verehrer in dem Glauben, daß er während des strengen Winters auf dem Schneeberge umgekommen sei, ein Totenopfer dar; im folgenden Monat Saga machen sie sich auf, ihn zu suchen, treffen ihn lebend an und kehren mit ihm nach Rtsar ma zurück. Seite 21, Vers 16, gibt Milaraspa die Zeit seines Winteraufenthaltes auf sechs Monate an, was also mit der Annahme, daß derselbe im Oktober 1085 begonnen, übereinstimmt. Seite 17, Vers 25, wird der Neujahrstag 1086 erwähnt. Der Zeitraum, den die Handlung unserer ersten Erzählung einnimmt, dehnt sich also von September 1085 bis April 1086 aus. Dem gegenüber ist die auf Seite 16, Vers 8—10, gemachte Zeitangabe befremdend, daß Milaraspa am Jahresende des Tigerjahres, im Jahresanfang des Hasenjahres, am sechsten Tage des Monats Wa-rgyal, der Weise des Kreislaufes überdrüssig, auf den La phyi gegangen sei. Unter dem Tigerjahr kann nur das Jahr 1085 (me stag) und unter dem Hasenjahr nur das Jahr 1086 (me-yos) verstanden werden. Der Monat Wa-rgyal wird in den Wörterbüchern nicht erklärt; wa ist aber die Bezeichnung des 7. oder 8. Naksatra (= Skr. açlesa) und rgyal Name des 6. Naksatra (= Skr. pusya), und das Sternbild beider ist der Krebs. Darnach könnte es sich möglicherweise um den Monat Juli handeln. Nach der vorausgegangenen Darlegung ist es jedoch unverständlich, wie hier dieser Monat in Betracht kommen könnte, und ferner, in welcher Beziehung er zu den beiden vorhergehenden Jahreszahlen stehen sollte, aus denen man nur schließen kann, daß es sich um den Winter von 1085 auf 1086 handelt. Man darf nicht übersehen, daß die Zeitangabe in jenen drei Versen eine stark poetische Färbung hat, wie aus den ungewöhnlichen Ausdrücken yong stag für stag, yos bu für yos, aus der Wiederholung von lo in V. 8 und 9 und dem sonst nicht belegten Wa-rgyal hervorgeht; ebenso ist ña (V. 10) ganz ungewöhnlich statt tshes. Schon aus diesem Grunde kann diese Stelle nicht die gleiche Bedeutung beanspruchen wie die übrigen in dem gewöhnlichen Stile der Prosa gegebenen Zeitbestimmungen, deren Richtigkeit dadurch in keiner Weise angefochten wird. Die zweite, dritte und vierte Geschichte schließen sich zeitlich eng an die erste an und spielen gleichfalls im Jahre 1086. Aus den Naturschilderungen des Liedes Seite 48 u. 49 in der dritten Erzählung geht hervor, daß die Zeit derselben der Spätfrühling oder der Sommer sein muß. Beachtenswert ist, daß die von Sum pa mkhan po verfaßte chronologische Tafel Reu-mig außer anderen Jahreszahlen aus Milaraspa's Leben die Vollziehung seiner Bußübungen zur Erlangung der Heiligkeit dem Jahre 1083 zuweist.

Alle diese Zeitbestimmungen weisen deutlich auf die historische Grundlage des vorliegenden Werkes hin. Einen weiteren Fingerzeig für dieselbe geben die Namen und Schilderungen der Oertlichkeiten, die sich selbst bei unseren spärlichen Hilfsmitteln zum Teile noch im heutigen Tibet nachweisen lassen. Der Schauplatz von Milaraspa's Erzählungen ist das nordwestliche Tibet in den nördlichen Teilen des Himalaya zwischen der Grenze Nepals und dem Oberlaufe des Brahmaputra (Ya ru gtsang po). Die Schönheiten der Gebirgswelt und der Natur überhaupt sind wiederholt Gegenstand der Lieder. Die in den Geschichten genannten Dörfer liegen alle nicht weit von Gung thang, Milaraspa's Geburtsort (vergl. Seite 54, Vers 10), der sich gleichzeitig rühmt, den Uebersetzer R v a l o zu seinen Söhnen zu zählen. Es ist eine anziehende Erscheinung, daß das Volk in diesem Distrikte die Erinnerung an Milaraspa noch bewahrt hat; denn in der tibetischen Geographie des M i n c h u l C h u t u k t u (gest. 1839) wird bei der Erwähnung von Gung thang ausdrücklich der Grotte gedacht, in der Mila die Vollendung erlangte, und einer Insel (Chu bar), wo er predigte.

Außer diesen äußeren Momenten der Zeit und des Ortes sprechen für den geschichtlichen Hintergrund des Buches der historische Charakter der Persönlichkeit des Milaraspa selbst, sein Zusammentreffen mit anderen historisch beglaubigten Personen und die innere Wahrscheinlichkeit der in chronologischer Reihenfolge vorgetragenen Ereignisse, soweit sie nicht durch Wunder und andere rein legendenhafte Zutaten ausgeschmückt sind. Ueber Milaraspa's Person ist uns freilich bis jetzt nicht viel bekannt geworden, und eine vollständige Kenntnis derselben wird sich erst aus einer Durchforschung des Mgur ₒbum und Rnam thar gewinnen lassen. Als ziemlich gesichert mag feststehen, daß er ein Schüler des M a r p a war, mit dem er den Lehren der Sekte der Bka rgyud pa anhing. Wie der Name besagt, stützt sich diese Sekte auf den Glauben an die mündliche Ueberlieferung des Buddhawortes, die neben der schriftlichen Tradition der heiligen Bücher durch eine ununterbrochene Reihe von Lehrern und Schülern fortgepflanzt sein soll. Als ihre Begründer werden die drei G r u b c h e n N a r o p a , T e l o p a und M a i t r i p a genannt. Marpa's Lehrer N a r o p a oder N a r o , ein Zeitgenosse des Padmasambhava und Atisha, erhielt seine Belehrung von T e l o p a , der seinerseits eine unmittelbare Inspiration vom Buddha Vajradhara empfing (vergl. Seite 37, Vers 26; Seite 38, Vers 6). M a r p a wird wiederholt als Bka rgyud bla ma oder kurz Bla ma, der Lama, bezeichnet; er wird am Anfang der meisten Lieder angerufen und stets mit dem Ausdrucke höchster Verehrung und innigster Liebe von Milaraspa genannt. Wenn sich auch Milaraspa's philosophische Anschauungen nicht eher werden systematisch darstellen lassen, als bis das ganze Buch bearbeitet ist, so läßt sich doch schon nach den hier mitgeteilten sieben Kapiteln einigermaßen sicher urteilen, daß er weit davon entfernt ist, ein eigenes System zu begründen; er gibt nur die Lehren seiner Vorgänger in einer seinem stark ausgesprochenen persönlichen Geiste entsprechenden Form wieder, mit einer übermäßig scharf hervorgekehrten Betonung der Meditation, wie sie in der von Asaṅga begründeten Yogacarya-Schule betrieben wurde. Viele Lieder beschäftigen sich eingehend mit den physiologischen und psychischen Grundlagen ekstatischer Beschauung und verlieren sich zuweilen in einen schwindelnden Abgrund bodenloser Mystik. Doch darf die große Wichtigkeit dieses Materials zur Erkenntnis und Beurteilung dieser merkwürdigen religiösen Richtung keineswegs unterschätzt werden. Aber gerade die interessantesten Züge an Milaraspa's Charakter sind die nichtbuddhistischen. So sehr er sich auch immer anstrengt, den starren asketischen Yogin hervorzukehren, gelangt doch der Tibeter und der Dichter in ihm wieder und wieder zum Durchbruch, und das bedingt eben einen der größten Reize dieses fesselnden Werkes. Die verschrobensten Meditationstheorien vermögen nicht den Menschen, das Ich in ihm zu unterdrücken; er ist Yogin weit mehr in der Theorie als in der Praxis, er vollzieht keine fabelhaften Bußübungen wie seine indischen Vettern, er ist nicht zum stumpfsinnigen Brüter herabgesunken, der den Zustand seligsten Glückes in stumpfer Gedankenlosigkeit erreicht zu haben vermeint. Nein, Milaraspa tut Dinge und entwickelt Eigenschaften, die sich mit den strenggläubigen Anforderungen an den heiligen Beruf des buddhistischen Mönches und Yogin nimmermehr vereinigen lassen; er dichtet, er singt, er lacht, er scherzt, ist froh mit den Fröhlichen, ja, er tanzt vor Freude über die Freude des Volkes, trinkt tibetischen Gerstensaft (chang) und genießt sogar Fleisch, er entwickelt Witz und Humor und zuweilen beißende Satire, d. h. kurz gesagt, er ist und bleibt Tibeter in seinem Herzen trotz allem Buddhismus. Hier bewahrheitet sich H e r d e r s schönes Wort über Tibet: „Sonderbar ist der Unzusammenhang, in welchem die Sachen der Menschen sich nicht nur binden, sondern auch lange erhalten. Befolgte jeder Tibetaner die Gesetze der Lama, indem er ihren höchsten Tugenden nachstrebte, so wäre kein Tibet mehr. Das Geschlecht der Menschen, die einander nicht berühren, die ihr kaltes Land nicht bauen, die weder Handel noch Geschäfte treiben, hörte auf; verhungert und erfroren lägen sie da, indem sie sich ihren Himmel träumen. Aber zum Glück ist die Natur der Menschen stärker als jeder angenommene Wahn. . . . Glücklicherweise hat die harte Mönchsreligion den Geist der Nation so wenig als ihr Bedürfnis und Klima ändern mögen. Der hohe Bergbewohner kauft seine Büßungen ab und ist gesund und munter; er zieht und schlachtet Tiere, ob er gleich die Seelenwanderung glaubt, und erlustigt sich fünfzehn Tage mit der Hochzeit, obgleich seine Priester der Vollkommenheit ehelos leben. So hat sich allenthalben der Wahn der Menschen mit dem Bedürfnisse abgefunden: er dung solange, bis ein leidlicher Vergleich ward. Sollte jede Torheit, die im angenommenen Glauben der Nationen herrscht, auch durchgängig geübt werden: welch ein Unglück! Nun aber werden die meisten geglaubt und nicht befolgt, und dies Mittelding toter Ueberzeugung heißt eben auf der Erde Glauben." In Milaraspa spricht sich ein starkes Nationalgefühl aus, sowohl in der Anhänglichkeit an den Boden seiner Heimat wie in der Liebe zu seinem Volke, dessen Lehrer, Prediger und Ermahner er ist, dessen Leiden und Freuden er gerne teilt. Sein stark ausgeprägtes Selbstgefühl gelangt besonders in der Erzählung von der Begegnung mit dem Inder D h a r m a b o d h i zum

Ausdruck, in der gerade die wiederholte Betonung des Nationalitätenunterschiedes einen eigenartigen Zug bildet.

Einen auffallenden Gegensatz bildet Milaraspa zu Padmasambhava, dessen Legendenbuch sich geradezu als die Ausgeburt einer wahnsinnigen Phantasie darstellt; es ist wie der tolle Tanz eines Schamanen, seine sogenannten Bekehrungen sind von rohestem Kannibalismus und widerlicher Unzucht begleitet. Nichts von solchen Zügen findet sich in Milaraspa's Buche. Seine Bekehrungen erfolgen nie unter Anwendung von Gewalt, sondern einzig und allein durch die Macht seines Wortes und seiner Predigt in der Form des Liedes. Alle Erzählungen tragen einen milden und ruhigen Charakter, und selbst seine Spukmären von bösen Geistern entbehren nie eines schalkhaften Humors und einer anmutigen Liebenswürdigkeit. In seinen beschaulichen Beobachtungen der Natur zeigt sich die Feinheit seiner Empfindungen, und zuweilen gerät in ihm, wie aus dem Schlusse des Liedes auf Seite 48 u. 49 hervorgeht, der Dichter mit dem Philosophen in Konflikt. Es ist schwer, eine Definition dieser Erzählungen aufzustellen. Mit den Jataka und Avadana haben sie nicht die mindeste Aehnlichkeit; auch liegt keine Uebertragung indischer Stoffe vor. Das Mgur ₒbum ist vielmehr von echt tibetischem Leben erfüllt und wird gerade durch seine lebendigen Schilderungen des Volkslebens zu einem wichtigen Beitrag zur Kulturgeschichte Tibets in der zweiten Hälfte des 11. Jahrhunderts. Einige Erzählungen könnte man als Heiligenlegenden im Sinne unseres Mittelalters bezeichnen, andere aber spiegeln wirkliche Erlebnisse Mila's wieder, wie z. B. unser erstes Kapitel, andere historische Ereignisse, wie die fünfte Erzählung. Zuweilen kann man sich kaum des Eindrucks erwehren, als wenn in einigen Gedanken abendländisch-christliche Einflüsse zutage träten.

I.

Milaraspa's Winteraufenthalt auf dem La phyi.

Verehrung dem Meister!

Auf die Kunde, daß der ehrwürdige Milaraspa bei seinem erstmaligen Besuche des Schneeberges La phyi die gefährlichen Geister und Kobolde bekehrt habe, strömten alle Einwohner von Gña nam herbei, um ihre Ehrerbietung zu bezeigen. Besonders Jo mo ∘Ur mo suchte religiöse Belehrung nach: „Gegenwärtig", sprach sie, „ist Ngam pa rgyags phu ba noch klein und jung; doch wenn dieser mein Sohn erwachsen sein wird, werde ich ihn dem Ehrwürdigen zum Diener geben." Infolge des in ihnen erwachten großen Glaubens luden ihn die Bewohner von Rtsar ma ein, und während Gsen rdor mo ihm ihre Dienste widmete, weilte der Ehrwürdige im Dorfe. Aus der Beobachtung der weltlichen Geschäfte gewann er eine außerordentlich frohe Stimmung. Als er nun sagte, daß er auf den Schneeberg La phyi gehen wolle, baten die Einwohner von Rtsar ma den Ehrwürdigen, indem Jeglicher sagte: „Es gibt nichts anderes als den Nutzen der Wesen; damit uns Nutzen entstehe, bringe doch den Winter hier zu und leite uns!" Dann fügten sie hinzu: „Auch wenn du zur Bekehrung der Dämonen später des Nachts ausziehst, genügt es; im Frühjahr, am Morgen, wollen wir dich als deine Diener begleiten." Besonders der Lehrer Çakyaguna und Gsen rdor mo wandten alle möglichen Bitten auf, um ihn durch den Hinweis auf die winterliche Jahreszeit und durch die Vorstellung der Lebensgefahr, der er sich auf dem Schneeberg aussetzte, zurückzuhalten; dennoch schenkte er ihnen kein Gehör und entgegnete: „Ich, der Sprößling des großen Pandita Naro, fürchte im allgemeinen die Luft der Schneeberge nicht; besonders muß ich nach dem Gebot des Marpa Weltgetümmel und Störung meiden und an einem menschenleeren, einsamen Orte verbleiben. Ueberdies bereitet mir der dauernde Aufenthalt in einem Dorfe tödlichen Ueberdruß." Als er so zur Abreise fest entschlossen war, beeilten sich die Leute von Rtsar ma, dem Ehrwürdigen Lebensmittel zu besorgen, und einige von ihnen baten ihn, wenigstens im Winter von da wegzugehen, und versprachen zu kommen. Der Lehrer Çakyaguna und Gsen rdor mo und vier andere Leute, die den Abschiedstrunk mitnahmen, geleiteten ihn bis über den Paß und gelangten, den Paß hinabsteigend, zum Teufelsteich. Darauf übergaben sie die für den Ehrwürdigen mitgebrachten Vorräte, zwei Maß Mehl, ein Maß Reis, ein Viertel Fleisch und einen Schnitt Butter. Dann trat er in die große Dämonen bekehrende Grotte, wo er zu bleiben gedachte. Jene traten den Rückweg an. Als sie die Paßhöhe überstiegen, umwölkte sich der Himmel und es brach ein wirbelndes Schneegestöber aus. Mit großer Schwierigkeit bahnten sie sich einen Weg, und bis über die Knie tief einsinkend, erreichten sie im Kampfe mit dem Schneesturm erst bei Einbruch der Nacht ihr Dorf. Von jenem Abend an fiel aber neun Tage und Nächte lang Schnee, so daß während sechs Monaten das handelstätige Gña nam von allem Verkehr abgeschnitten war. Die Anhänger des Lehrers waren daher der festen Ueberzeugung, daß der Ehrwürdige verschieden sei und brachten zu seinem Gedächtnis in vollkommener Weise Opfergaben dar. Als der folgende Saga-Monat kam, machten sich wiederum die früheren Schüler auf den Weg, indem sie das Eis mit der Axt spalteten, um des Ehrwürdigen Gebeine zu holen. Während sie gingen, um zu dem Orte zu gelangen, ließen sie sich, um sich zu erholen, auf einem Sitze nieder. Da stieg ein großer Schneeleopard einen Felsblock hinauf und gähnte. Er hielt lange Umschau und ließ seine Stimme ertönen. Da sprachen jene: „Der Schneeleopard dort drüben wird wohl den Leichnam des Ehrwürdigen verzehrt haben. Ob nun etwas da ist, was einem Stück seines Gewandes gleicht, etwas, was seinem Haupthaar gleicht? Etwas anderes werden wir wohl nicht finden." Sehr traurig und weinend schritten sie weiter. Auf dem leeren Platze, von dem der Schneeleopard herabstieg, zeigten sich weiter abwärts menschliche Fußspuren. Darauf verwandelte sich das Tier in einen Leopard und dann in einen Tiger; auf dem Querpfade, den sie nun einschlugen, erschien es, in der Längsrichtung gesehen, als Tiger, quer gesehen als Leopard. Da wurden jene nachdenklich und gingen in dem Gedanken, es möchte wohl ein Dämon sein, Zweifel nährend weiter. Als sie sich der großen Dämonen bekehrenden Grotte näherten, hörten sie den Ehrwürdigen Lieder singen. „Wie", dachten sie, „sollten ihm doch etwa Jäger Lebensmittel gebracht oder er selbst etwa Aasreste aus dem Vorrat eines Raubtieres gefunden haben, daß er nicht verschieden ist?" Da sprach der Lama: „Es ereignet sich zuweilen, daß die Toren alle draußen stehen. Was ist denn, daß ihr nicht eintretet? Brei und Gemüse werden euch ja kalt! So kommt doch schnell in die Grotte hinein!" Da hielten sie vor Freude kaum die Tränen zurück und stürzten sämtlich von allen Seiten herein auf den Ehrwürdigen, um ihm Hände und Füße zu drücken, und weinten. Der Ehrwürdige aber sagte: „Laßt das jetzt lieber sein und nehmt Speise zu euch!" Jene machten zuerst eine Verneigung und erkundigten sich nach seinem Befinden. Als sie sich umschauten, fanden sie, daß von den früheren Vorräten nicht mehr als ein Maß Mehl aufgebraucht und dazu das in den Reis gelegte Fleisch nicht einmal versucht war. Der Lehrer Çakyaguna sagte: „Mit unseren Nahrungsmitteln richtest du ja ein Totenmahl her. Hat der Ehrwürdige vermöge seines Seherblickes unsere Ankunft vorausgeschaut?" Der Ehrwürdige erwiderte: „Als ich oben von einem Felsblock her Umschau hielt, sah ich euch, wie ihr euch ausruhtet." Wiederum fragte der Lehrer Çakyaguna: „Wir haben oben auf dem Felsblock nur einen Schneeleoparden, den Ehrwürdigen aber nicht bemerkt, wo weilte denn der Ehrwürdige zu jener Zeit?" Der Ehrwürdige entgegnete: „Eben jener Schneeleopard war ich ja selbst: Da nämlich die Yogin, welche Adepten der höchsten Mystik sind, die auf den vier Elementen beruhenden Sinnesorgane durch ihren Glanz besiegen, so habe ich die Fähigkeit erlangt, mich in einen beliebigen Körper verwandeln und so Trugbilder zeigen zu können. Da ihr nun dessen würdig seid, habe ich euch diese Verwandlung meines Leibes sehen lassen. Doch sagt den Leuten nicht ein Wort davon, daß ich solches tue." Darauf fragte Gsen rdor mo: „Das Aussehen des Ehrwürdigen ist ja weit besser als im vorigen Jahre. Haben etwa zu der Zeit, als die beiden Zugänge des Weges durch den Schnee abgeschnitten waren und kein Mensch dich bedienen konnte, Geister dir Dienste geleistet? Oder hast du etwa dergleichen wie Aasreste von Wild gefunden? Wie ging das wohl zu?" „Die meiste Zeit", antwortete er, „war ich ganz in Beschauungen versunken, so daß ich keine Speise zu essen brauchte. Zur Zeit der Festtage aber brachten mir die Daka von ihren Anteilen an den Opferspenden. Ich nehme immer eine Löffelspitze Mehl auf einmal. Neulich überdies, in der zweiten Hälfte des Pferd-Monats (Caitra), wurdet ihr, meine Schüler, so lebendig in mir, daß ein solches Gefühl der Sättigung durch reichliche Speise und Trank über mich kam, daß ich viele Tage lang gar kein Bedürfnis fühlte, Nahrung zu mir zu nehmen. Was habt ihr denn eigentlich zu jener Zeit getan?" Jene rechneten die Zeit nach und fanden sehr genau heraus, daß es zu ebenderselben Zeit geschah, da sie zu seinem Gedächtnis in vollkommener Weise Opfergaben dargebracht hatten. Da sagte der Ehrwürdige: „Ja, wenn weltlich Gesinnte den Grund zur Tugend legen, so gereicht das für den Zwischenzustand zum Heil. Deshalb ist es jetzt auch von großer Kraft, um den Zwischenzustand zu verhindern." Darauf drangen jene mit Bitten in den Ehrwürdigen, einer Einladung nach Gña nam zu folgen. Der Ehrwürdige erwiderte aber: „Ich bin von Freude darüber erfüllt, doch da ich mit der Förderung meiner Meditation beschäftigt bin, kann ich nicht gehen. Ihr selbst aber mögt ziehen." Da sagten jene: „Wenn der Ehrwürdige jetzt nicht mitkommt, werden uns alle Einwohner von Gña nam beschuldigen, daß wir den Ehrwürdigen dem Tode preisgegeben haben; so werden uns üble Nachreden und große Schmähungen erwachsen." Jo mo ∘Ur mo sagte insbesondere: „Da böse Nachrichten mit den Worten: ‚Berichte doch von meinem Ehrwürdigen!' wieder und wieder ausgestreut werden würden, so ist es besser, daß auch wir, wenn der Lama nicht mitkommt, hier bleiben und den Tod erwarten." Da vermochte der Ehrwürdige ihrem Drängen nicht länger zu widerstehen und versprach, mit ihnen zu gehen. Im Winter hatten die Daka zu dem Ehrwürdigen gesagt: „Für dich, Milaraspa, ist es zwar nicht nötig, doch in Zukunft wird es für das Geschlecht deiner Schüler nötig sein: nimm daher folgende Methode an, um einen Weg über das Eis zu bahnen." So trugen sie denn die nach Anweisung der Daka verfertigten Schneeschuhe, und als es mittlerweile Morgen geworden war, machten sie sich nach dieser Seite hin auf den Weg. Von der Paßhöhe an eilte Gsen rdor mo voraus und verkündete den Leuten von Rtsar ma und den übrigen Zuhörern die frohe Botschaft: „Der Ehrwürdige ist heil und gesund und ist im Anzuge hierher!" Der Ehrwürdige,

Lehrer und Schüler nahmen auf einem flachen weißen Felsblock ihr Frühstück ein. Dahin eilten nun die Zuhörer, welche die frohe Botschaft vernommen, Männer und Frauen, Alt und Jung, alle ohne Unterschied, dem Ehrwürdigen entgegen. Alle Besucher hängten sich an den Ehrwürdigen, weinten und erkundigten sich unter Freudentränen nach seinem Befinden. Sie verneigten sich vor ihm und umwandelten ihn. Der Ehrwürdige hatte ein Bambusrohr, das ihm als Stab diente; auf dieses stützte er beide Hände und das Kinn. Jene einen Weg über das Eis bahnenden Schneeschuhe hatte er an die Füße gebunden und trug von dem flachen weißen Felsblock herab den dort versammelten Zuhörern zur Antwort auf ihre Frage nach seinem Ergehen folgendes Lied vor:

Daß wir heute unter dem Thronbaldachin des Glückes,
Ihr gabenspendenden Männer und Frauen, die ihr mich ehrerbietig besucht,
Und ich, der Yogin Milaraspa,
Uns noch einmal vor dem Sterben getroffen haben, freut mein Herz!
Ich, alter Mann, bin eine Schatzkammer von Liedern:
So will ich mit diesem Liede die Antwort auf eure Frage nach meinem Ergehen zurückzahlen.
Ich bitte, die Ohrwärmer abzunehmen und mir geneigtes Gehör zu schenken.
Am Jahresende des Tigerjahres,
Im Jahresanfang des Hasenjahres,
Am sechsten Tage des Wa-rgyal-Monates
War ich des Kreislaufes Weise überdrüssig
Und begab mich in die Wildnis, das Gehege des Schneeberges La phyi,
Wo ich die ersehnte Einsamkeit fand.
Da hielten Himmel und Erde zusammen Rat
Und entsandten als Eilboten den Wirbelwind.
Die Elemente des Wassers und Windes wurden entfesselt.
Die dunklen Südwolken wurden zu einer Beratung versammelt.
Sonne und Mond, das Paar, wurden zu Gefangenen gemacht.
Die 28 Mondhäuser wurden festgebunden und gefesselt.
Auf Befehl wurden die acht Planeten in Eisenketten gelegt.
Die Milchstraße wurde unsichtbar.
Die kleinen Sterne wurden ganz in Dunst gehüllt.
Als endlich alles vom Nebelglanz bedeckt war,
Fiel Schnee, neun Tage, neun Nächte lang,

Gleichmäßig verteilt auf achtzehn Tag-Nächte fiel er.
War großer Schneefall, fielen die Flocken wie Wollflausche,
Wie fliegende Vöglein herniederschwebend;
War kleiner Schneefall, kamen sie wie Spinnwirteln herunter,
Wie Bienen herumkreisend;
Dann wieder fielen sie wie kleine Erbsen und Senfkörner,
Gleich Spindeln, die sich rund drehen.
Aber großer und kleiner Schnee zusammen wächst zu einer unermeßlichen
Schicht:
Schon berührt des hohen Schneeberges weiße Kuppe den Himmel;
Die niedrigen Bäume und Wälder liegen darniedergehalten am Boden.
Die schwarzen Berge kleiden sich in Weiß.
Auf dem wogenden See bildet sich eine Eisdecke.
Der blaue Brahmaputra ist wie in einer Höhle eingesperrt.
Der Boden, ob er sich hoch oder niedriger hebt, gleicht sich zu einer Fläche aus.
Und wie es denn bei so starkem Schneefall geschieht,
Sind insgesamt die schwarzköpfigen Menschen zur Gefangenschaft gezwungen.
Hungersnot trifft das vierfüßige Vieh.
Einzeln findet das arme Wild kein Futter mehr.
Den beschwingten Vögeln oben ist die Nahrung ausgegangen.
Unten verbergen sich Ziesel und Mäuse bei ihren Schätzen.
Den Raubtieren sind die Rachen wie durch Fesseln gesperrt.
In solchem allgemeinen Schicksal
War dies mein, Milaraspa's, besonderes Los.
Zwischen dem von oben herabfegenden Schneesturm,
Dem kalten Windstoß des vollendeten Winterneujahrs
Und meinem, des Yogin Milaraspa, Baumwollkleid, zwischen diesen dreien
Entspann sich auf dem Gipfel des hohen weißen Schneeberges ein Kampf.
Der herabfallende Schnee schmolz im Barte zu Wasser;
Der Sturm legte sich trotz seines tosenden Brüllens von selbst;
Das Baumwollkleid zerfiel wie von Feuer verzehrt.
Am Ringer nahm ich mir ein Vorbild und kämpfte sterbend um dieses Leben.

Mit siegreichen Waffenspitzen kreuzten wir die Klingen:
Des Feindes Stärke verachtend, blieb ich Sieger in diesem Krieg.
Im allgemeinen ist in alle Geistlichen e i n Maß (von Kraft) gelegt,
Und im besonderen sind allen großen Mystikern zwei Maße bescheert.
Noch mehr im einzelnen erklärte mir die innere Glut der Meditation den Vorzug eines einfachen Baumwolltuches.
Die vier Ansammlungen der Krankheiten wurden mir auf einer Wage zugewogen.
Als außen und innen der Aufruhr beigelegt war, wurde ein Vertrag geschlossen.
Der kalte wie der warme Wind, beides war mir gut bekommen.
Dann versprach mir der Feind, allen meinen Worten zu gehorchen.
Den Dämon, der das Gesicht des Schnees zeigte, habe ich niedergeworfen.
Später, als Bewegung und Erschütterung aufhörte, ließ ich mich in meinem Baue nieder.
Das Dämonenheer hatte die Lust zu handeln verloren:
So war damals der Yogin siegreich im Kampfe.
Als des Großvaters Enkel besaß ich einen Tigerpelz;
Vorher, als ich in ein Fuchsfell gekleidet war, war ich nie gegangen.
Als der dem Vater geborene Sohn gehöre ich zum Geschlecht der Ringer:
Dem übelgesinnten Feind bin ich nie entflohen.
Ich gehöre zum Geschlecht des Löwen, des Königs der Raubtiere:
Ich habe nie anders als mitten im Schnee der Berge gehaust;
Vorbereitungen sind dadurch unnütz gemacht.
Wenn ihr mir, dem Greise, Gehör schenkt,
Wird sich auch bei den künftigen Geschlechtern die Lehre ausbreiten.
Einige Siddha sind erstanden;
Ich, der Yogin Milaraspa,
Bin in allen Ländern bekannt;
Ihr Schüler seid durch Nachdenken gläubig geworden.
Eine frohe Botschaft wird auch noch in Zukunft verkündet werden.
Ich, der Yogin, fühle mich wohl.
Ist auch euer Befinden gut, ihr Gabenspender?

Als er so gesprochen, führten die Zuhörer in brausendem Frohlocken einen stampfenden Tanz auf, und auch der Ehrwürdige stimmte in das Frohlocken ein und tanzte mit hüpfendem Fuße. Dadurch empfing der Felsblock Eindrücke, wie wenn man in weiche Erde tritt, so daß der Felsblock oben von allen Fußspuren und den Spuren des Stabes voll war. Aus den Eindrücken des mittleren Teiles entstand ein umgekehrtes Bang rim; früher wurde der Stein „flacher weißer Felsblock" genannt, später war er als „Spuren der Schneeschuhe" bekannt. Als ihn darauf die Hörer nach Rtsar ma in Gña nam einluden, sagte Legs se ₒbum, die sich mit großen Gaben in der Reihe der Verehrer befand: „Darüber, daß der Ehrwürdige diesmal heil und gesund angelangt ist, ist großer Jubel entstanden. Dein Aussehen ist sogar vortrefflicher als früher. Ist es dir als Folge eines guten Gelübdes zuteil geworden oder sind es die Dienste, welche dir die Daka geleistet haben?" Als Antwort auf ihre Frage trug er folgendes Lied vor:

Mit dem Scheitel verneige ich mich dem Herrn, dem Lama zu Füßen!
Die Siddhi des Segens haben mir die Daka verliehen.
Der Nektar des Gelübdes ist von großem Nutzen,
Durch gläubige Opferspenden wurden die Sinne wieder belebt;
Die von den Schülern aufgehäuften Tugendverdienste entsprangen einem guten Herzen.
Im Sinne, welcher der Beschauung zugewandt ist, ist die Leerheit entstanden.
Das Wesen der Ursache der Beschauung ist nicht mehr als ein Atom:
Mit dem, der das zu Beschauende beschaut, ist es sonst vorbei;
Doch die Art der Erkenntnis der Beschauung ist einem guten Herzen entsprungen.
Die Erleuchtung durch die Meditation ist wie die Strömung eines Flusses.
Man sollte nicht zum Zweck der Meditation Nachtwachen halten:
Sonst ist es mit dem, der über das zu Meditierende meditiert, vorbei;
Doch die Standhaftigkeit der Meditation ist einem guten Herzen entsprossen.
Das Vollziehen der Uebung beruht auf Erleuchtung;
Durch die Ueberzeugung von der Leerheit der Nidana
Ist es mit dem, der das Auszuübende ausübt, vorbei:
Doch die Art des Vollziehens der Uebung ist einem guten Herzen entsprungen.
Die Bedenklichkeit, die aus der Eingenommenheit für die Welt entsteht, schwindet gänzlich;

Die acht Truglehren der Welt sind weder Gegenstand der Hoffnung noch der Furcht:
Mit dem Bewahrer des zu Bewahrenden ist es sonst vorbei:
Doch die Art, wie das Gelübde bewahrt wurde, ist einem guten Herzen entsprungen.
Durch die Erkenntnis der Nichtigkeit der eigenen Seele
Ist es mit dem, der um des Strebens nach dem eigenen Heil und dem anderer
Das zu Erstrebende erstrebt, vorbei:
Doch die Art des Strebens nach der Frucht ist einem guten Herzen entsprungen.
Dies mein, des Greises, Freudenlied
Ist die Antwort auf die Frage der gläubigen Hörer.
Der Schnee hatte mich von der Welt abgeschnitten.
Lebensmittel setzten mir die Daka dienend vor;
Das herabfließende Wasser des Schieferbergs war das trefflichste Getränk;
Ohne daß mich jemand damit versorgte, ward es mir infolge meiner Vorzüge zuteil.
Ohne daß Arbeit erforderlich war, wurde die Wirtschaft geführt.
Ohne die täglichen Bedürfnisse gesammelt und niedergelegt zu haben, war mein Vorratshaus gefüllt.
In meine Seele schauend, sah ich alles.
Auf niedriger Erde sitzend, nahm ich einen Thron ein.
Der Gipfel befriedigender Vollendung ist des Lamas Gnade.
Schüler, samt dem Gefolge der Gabenspender,
Den Dank für eure mir mit gläubigem Dienst
Erwiesene Güte habe ich euch durch religiöse Belehrung abgestattet.
Freut euch in eurem Herzen, die ihr hierher gekommen seid!

Als er so gesprochen hatte, sagte der Lehrer Çakyaguna, indem er sich verneigte: „Der Ehrwürdige ist dieses Mal heil und unversehrt von einem solchen großen Schnee hierher zurückgekehrt; auch uns Schülern ist außerordentlich große Freude darüber entstanden, daß wir vor dem Tode den Ehrwürdigen noch einmal getroffen haben. Für die Worte, mit denen du uns am heutigen Tage als Gegengabe den Nutzen eines religiösen Geschenkes gewährt hast, sei vielmals bedankt! Gewähre uns nun bitte als Geschenk eine religiöse Erbauung, wie sie im Winter dieses Jahres in des Ehrwürdigen eigenem Herzen erwuchs.“ Da trug der Ehrwürdige als Antwort auf die Bitte des Lehrers Cakyaguna zum Willkommengeschenk für die Hörer die sechs Kernpunkte, wie sie in seinem Herzen entstanden waren, in folgendem Liede vor:

Dem Lama, der drei Gebieter hat, zu Füßen verneige ich mich!
Ich, der sonst in der Einsamkeit lebt, weile mit den Kräften meiner Meditation
Am heutigen Abend in dieser Versammlung des Heils.
Du, Lehrer Çakyaguna, das Haupt aller,
Und ihr Schüler, deren Haus reich an allen Gütern ist,
Habt durch reines Gebet euern Geist vorbereitet.
Durch den Tischgenossen der Gabenspenderin Rdor mo
Und die Wirksamkeit der Vorschriften seines Gelübdes
Habt ihr, meine Schüler, ein Bittgesuch gestellt.
„Wir bitten um Gewährung religiöser Belehrung als väterliches Gastgeschenk“, lauteten eure Worte.
Zu diesem Zweck sei euch die Antwort mit folgender Rede erteilt:
Des Kreislaufs Weise überdrüssig,
Begab ich mich mißmutig zum Schneeberg La phyi.
Am einsamen Orte, in jener Dämonen bekehrenden Grotte
Erwuchs mir, dem Yogin Milaraspa,
Sechs Monate lang die Kraft der Meditation.
Nun singe ich das Lied von den sechs Kernsätzen.
Die sechs Gebiete der Sinne als Gleichnis nehmend,
Bringe ich die sechs inneren Mängel in Ordnung.
Für die sechs bindenden Fesseln der Nicht-Befreiung
Sehe ich die sechs Wege der Befreiung durch die verschiedenen Mittel.
Aus den sechs zuversichtlichen Unermeßlichkeiten
Entsteht das sechsfache geistige Wohlbefinden.
Wenn man dies Lied nicht auch später wiederholt singt,
Dringt sein Sinn nicht ins Herz;
Nun will ich ihn mit Worten erklären wie folgt:
Wenn sich ein Hindernis entgegenstellt, gibt es keinen Himmel;

Wenn sie gezählt werden können, gibt es keine kleinen Sterne;
Wenn Bewegung und Erschütterung ist, gibt es keine Berge;
Wenn Abnahme und Zunahme ist, gibt es kein Meer;
Wenn man über Brücken geht, gibt es keinen Fluß;
Wenn er ergriffen wird, gibt es keinen Regenbogen:
Das sind die sechs Gleichnisse der Außenwelt.
Wenn man mit reichen Vorräten lebt, ist keine Beschauung;
Wo Zerstreuung ist, gibt es keine Meditation;
Wo ein Schwanken zwischen Für und Wider ist, gibt es keine Uebung;
Wo Skepsis ist, gibt es keinen Yoga;
Wo Aufgang und Untergang ist, gibt es keine Weisheit;
Wo Geburt und Tod ist, gibt es keinen Buddha:
Dies sind die sechs inneren Mängel.
Wo großer Haß herrscht, ist die Fessel der Hölle;
Wo großer Geiz herrscht, ist die Fessel der Preta;
Wo große Unwissenheit herrscht, ist die Fessel der Tiere;
Wo große Leidenschaft herrscht, ist die Fessel der Menschen;
Wo großer Neid herrscht, ist die Fessel der Asura;
Wo großer Stolz herrscht, ist die Fessel der Götter:
Dies sind die sechs bindenden Fesseln der Nicht-Befreiung.
Großen Glauben hegen, ist ein Weg zur Befreiung;
Auf gelehrte Geistliche vertrauen, ist ein Weg zur Befreiung;
Ein reines Gelübde haben, ist ein Weg zur Befreiung;
In der Bergwildnis umherwandern, ist ein Weg zur Befreiung;
Allein leben, ist ein Weg zur Befreiung;
Bannungen vollziehen, ist ein Weg zur Befreiung:
Dies sind die sechs Wege zur Befreiung durch die verschiedenen Mittel.
Das Mitgeborenwerden ist die natürliche Unermeßlichkeit;
Die Uebereinstimmung des Aeußeren und Inneren ist die Unermeßlichkeit des Wissens;
Die Uebereinstimmung von Licht und Schatten ist die Unermeßlichkeit der Weisheit;
Das große Allumfassen ist die Unermeßlichkeit der Religion;

Das Unwandelbare ist die Unermeßlichkeit der Beschauung;
Das Ununterbrochene ist die Unermeßlichkeit der Seele:
Dies sind die sechs zuversichtlichen Unermeßlichkeiten.
Wenn im Leibe die innere Glut entfacht ist, fühlt sich der Yogin wohl;
Wohl, wenn die Luft aus der rechten und linken Ader des Herzens in die mittlere eintritt;
Wohl im Oberleib durch das Herabfließen der Bodhi;
Wohl im Unterleib durch die Verbreitung des Chylussamens;
Wohl in der Mitte durch die Liebe des Erbarmens beim Zusammentreffen des weißen Samens der rechten Ader und des roten Blutes der linken Ader;
Der ganze Leib fühlt sich wohl durch die Befriedigung des glücklichen Gefühls der Sündlosigkeit:
Dies ist das sechsfache geistige Wohlbefinden des Yogin.
Dies ist mein Lied vom Sinn der sechs Kernsätze;
Dies ist der Sang der Seele, die sechs Monate meditiert hat.
Ihr, meine Schüler, veranstaltet den Versammelten ein Fest!
Jetzt, wo die Männer in froher Stimmung versammelt sind,
Schlürft alle freudig den Nektar des Bieres!
Mit diesem alten Lied, das ich alter Mann gesungen,
Habe ich nun euren Wunsch, meine Schüler, erfüllt.
Es ist denen ein Schatz, welchen jetzt erst die Lehre aufgegangen ist.
Werft euch freudigen Sinnes mit Macht auf die Religion des Segens!
Möge das Gebet, das reiner Tat entspringt, in Erfüllung gehen!

Als er so gesprochen, sagte Gsen rdor mo: „Es gibt Leute, die, obwohl sie mit dem ehrwürdigen Rin po che, vergleichbar dem Buddha der drei Zeiten, zusammentreffen, törichter als Tiere ihm nicht die geringste Achtung bezeigen, geschweige denn Dienste leisten und in seinem Gefolge als Diener einen religiösen Wandel führen.“ Der Ehrwürdige erwiderte: „Mögen sie mir immerhin keine Hochachtung bezeigen! Wer zur Zeit der Ausbreitung von Buddha's Lehre in dem kostbaren menschlichen Leibe wiedergeboren wurde und keinen religiösen Wandel führt, ist sehr töricht.“ Darauf trug er folgendes Lied vor.

Dem sprachgewandten Marpa zu Füßen verneige ich mich.
Du, merke auf und auch ihr, gläubige Gabenspender!
Jetzt, da die heilige Religion wie in einer Ebene ausgebreitet vor euch liegt,
Ist es sehr töricht, durch irriges Reden Sünden zu begehen.
Das Leben dieses Leibes, dessen innerer Friede schwer zu erlangen ist,
Nutzlos zu vergeuden, ist sehr töricht.
Auf dem Begräbnisplatz einer Stadt mit verfallener Erdmauer
Beständig seine Zeit hinzubringen, ist sehr töricht.
Unter Ehegatten, Marktleuten, Fremdlingen, unter solchen
Durch schlechte Scherze Streit zu erregen, ist sehr töricht.
Mit hochtönenden, trügerischen Worten der eigenen Stimme
Die Bosheit hochschätzen, ist sehr töricht.
Wenn der übelgesinnte Feind wie eine Blume verschwindend abzieht,
Ist es sehr töricht, durch einen Kampf sein Leben aufs Spiel zu setzen.
Wenn die Verwandten in dem trügerischen Hause der Tücke und Lüge
Verschieden sind, ist es sehr töricht, zu klagen.
Da der Reichtum nur ein geliehener Gegenstand, einem Tautropfen gleich, ist,
Ist es sehr töricht, vom Knoten des Geizes umschlungen zu sein.
Da der Leib nur ein mit unreinen Stoffen gefüllter Sack ist,
Ist es sehr töricht, sich mit Eitelkeit zu schmücken.
Der Nektar guter Lehren ist die köstlichste Speise:
Solche Speise gegen Geld zu verkaufen, ist sehr töricht.
Ja, in großer Zahl sind die Toren angesammelt;
Wenn sie unschädlich gemacht sind, wird die Götterreligion ihr Haupt erheben.
Wenn ein Mann klug ist, möge er handeln wie ich, der Yogin!

Als er so gesprochen, wurden die dort versammelten Hörer von Mitleidsgefühl ergriffen. „Wir sind zu der Ansicht gelangt", sagten sie, „daß wir alle Gefährten der Toren sind, ausgenommen du allein. Doch da ja ein so gescheiter und kluger Mann wie der Ehrwürdige sich nicht von selbst entwickelt, so bitten wir dich, weil du dich ja doch wohl in dieser Gegend anzusiedeln gedenkst, beständig hier zu wohnen, um uns Lebenden durch Verkündung der Religion Geistlicher und Opferpriester zu sein und die Toten mit dir himmelwärts emporzuziehen." Der Ehrwürdige erwiderte: „Ich muß auf dem Schneeberg La phyi der Beschauung obliegen, so will es das Gebot meines Lama; da oben will ich eine Zeit lang wohnen. Ich kenne nicht die Sitte der Leute, die gegen euch Gabenspender zu viel Rücksicht walten lassen, indem sie ihren Wohnsitz unter ihnen aufschlagen. Weil ihr mich am Ende mit verächtlichen Augen anschauen könntet, komme ich nicht." Darauf trug er folgendes Lied vor:

Dem Gebieter Marpa aus Lho brag zu Füßen verneige ich mich.
Ihr gabenspendenden Männer und Frauen, die hier versammelt sind,
An mich, den Yogin Milaraspa,
Hegt von ganzem Herzen unwandelbaren Glauben!
Mögt ihr ungeheuchelte Bitten erlangen!
Wenn man beständig an einem und demselben Wohnsitz lebt,
Wird man bald durch das Aufgehen in der Alltäglichkeit der Erscheinungen müde.
Wenn sich Menschen sehr aneinander gewöhnen, entsteht durch die Natur der Sache schließlich Verachtung.
Wenn die Gemeinschaft lange dauert, ergeben sich viele Reibungen.
Streit, der aus ungünstigen Verhältnissen entsteht, stört das Gelübde.
Schlechte Freunde verscheuchen die gute Andachtsübung.
Durch unüberlegte Worte des Stolzes werden böse Werke angehäuft.
Durch das Schwanken zwischen Recht und Unrecht schafft man sich Feinde.
Betrug zu eigenem Vorteil ist eine große Sünde.
Wenn schon bei der Danksagung für die eigene Nahrung böse Gedanken in Aufruhr geraten,
Wie ist es dann möglich, die Sporteln des Leichenschmauses zu empfangen!
Es wäre sehr leichtsinnig, wenn ich Geistlicher und Opferpriester von euch Laien sein wollte.
Wenn aus dem Zusammenleben Verachtung entstände, würde mich Verzweiflung ergreifen.
Noch mehr, als der Besitzer vieler Häuser sich in der Todesstunde unglücklich fühlt,
Empfindet der Yogin der Bergwildnis
Bei der Tätigkeit im Dorfe große Reue.
Ich gehe, um gleichgültig gegen die Welt in der Bergwildnis herumzuwandern.

Ihr Gläubigen, seid für eure Gaben und Dienste bedankt!
Ihr gabenspendenden Männer und Frauen, die hier versammelt sind,
Eure Liebe, die sich in den Worten bekundet: „Sei unser Geistlicher und Opferpriester!" ist schön.
Dank der Wirkung eures Gebets werde ich euch wiedersehen!

Als er so gesprochen, sagten die Zuhörer: „Unsere Ohren sind noch gar nicht müde zu hören, doch der Ehrwürdige wird wohl müde geworden sein. Obwohl wir nun mit Bitten in dich gedrungen sind, besorgen wir doch, daß dir nicht der Gedanke kommen wird, denselben Gehör zu schenken. Daher bitten wir dich, wieder bald auf den La phyi zurückzukehren." Allein die zahlreichen Geschenke, die sie ihm geben wollten, nahm er nicht an. Da gerieten alle dort versammelten Schüler in das größte Staunen und erlangten, von höchster Freude ergriffen, unerschütterlichen Glauben an den Ehrwürdigen.

Dieser Abschnitt enthält den Sang vom Schneeberg.

II.

Die Felsen-Rakshasi von Ling ba.

Verehrung dem Meister!

Obwohl an den ehrwürdigen Milaraspa die Zuhörer von Rtsar ma in Gña nam und die anderen Leute von Gña nam Bitten richteten, noch länger zu verweilen, hörte er nicht auf sie, sondern machte sich, um das Gebot seines Lama zu erfüllen, zum Zweck der Meditation auf den Weg nach dem Berge Dpal obar des Dorfes Skyid grong. Da fand er Gefallen an der Felsgrotte von Ling ba und verweilte dort einige Zeit lang, der Beschauung obliegend. Eines Abends, als die Dämmerung schon hereingebrochen war, ließ sich zur Linken des Platzes, wo sich der Ehrwürdige befand, in der Ritze eines Felsspaltes zu wiederholten Malen ein Pfeifen vernehmen. Er erhob sich und schaute hin, doch es war nichts da. In dem Gedanken: „Ich großer Meister in der Meditation muß mich wohl in einer Täuschung befinden", legte er sich wieder auf sein Lager hin. Da erschien in der Felsspalte auf dieser Seite ein heller Lichtstrahl, an der Spitze ein roter Mann, welcher auf einem schwarzen Moschustiere ritt, das von einer schönen Frau geführt wurde. Jener Mann versetzte dem Ehrwürdigen einen Stoß mit dem Ellenbogen und verschwand dann in einem Wirbelwinde. Jene Frau verwandelte sich in eine rote Hündin und packte den großen Zeh seines linken Fußes, so daß er ihn nicht los machen konnte. Da er wußte, daß dies das Zauberspiel einer Felsen-Rakshasi sei, trug er folgendes Lied vor:

Dem huldreichen Marpa zu Füßen verneige ich mich!
Die du auf eine Gelegenheit lauertest, mir Hohn zuzufügen,
Und dich in häßlich verwandelter Gestalt zeigst,
Du Felsen-Rakshasi vom Ling ba-Felsen,
Bist du nicht eine übeltuende Dämonin?
Des Liedes Wohllaut zu gewinnen, verstehe ich nicht,
Doch höre du auf die Worte wahrer Rede!
In der Mitte des blauen Himmels dort oben
Erwächst das Glück des Sonne-Mondpaares;
Jenes wundervollen Götterpalastes
Strahlen scheinen zur Wohlfahrt der Geschöpfe;
Wenn sie in ihrem Amte die vier Kontinente umkreisen,
Möge Rahu nicht als Feind wider sie aufstehen!

Unter dem Laubdach im Dickicht des Südwaldes
Erwächst das Glück der buntgestreiften Tigerin;
Unter allen Raubtieren ist sie die Ringerin;
Zum Zeichen ihrer Stärke schont sie des eigenen Lebens nicht;
Wenn sie auf den schmalen Felspfaden umherwandelt,
Möge die Falle nicht als Feind wider sie aufstehen!
In dem länglichen Türkissee Manasarovara im Westen
Erwächst das Glück der weißbauchigen Fische;
Sie sind die Tänzer im Element des Wassers;
Wunderbar rollen sie die Goldaugen;
Wenn sie der Lust und der Nahrung wegen einander folgen,
Möge der Angelhaken nicht als Feind wider sie aufstehen!
Auf dem roten Felsen Bsam yas im Norden
Erwächst das Glück des Geiers, des Königs der Vögel;
Er ist der Einsiedler (Rishi) unter den beschwingten Vögeln;
Wunderbarerweise nimmt er andern nicht das Leben;
Wenn er auf den Gipfeln der drei Berge seine Nahrung sucht,
Möge die aus Stricken gewundene Schlinge nicht als Feind wider ihn aufstehen!
Auf dem den Horst des Felsengeiers bergenden Ling ba
Erwächst Milaraspa's Glück;
Er trachtet nach seinem eigenen Heil und dem anderer;
Zum Zeichen der Wahrheit hat er auf dies weltliche Leben verzichtet
Und zur Grundlage die höchste Bodhi gemacht;
Zu derselben großen Zeit, da dieser eine Leib
Unablässig nach der Buddhawürde strebt,
Mögest du, Felsen-Rakshasi, nicht als Feind wider ihn aufstehen!
So enthält dies Lied fünf wunderbare Gleichnisse und sechstens deren Sinn.
Ich, der goldene Fesseln trägt im Augenblick, wo das Gedicht entsteht,
Verstehe ich wohl seinen Sinn? Die Felsen-Rakshasi
Ist durch Ansammlung ihrer Werke mit schweren Sünden behaftet.
Da dein Eifer nicht groß ist,
Ist das wilde Gift deiner Bosheit wieder zu bändigen.

Doch wenn alle ihre Seele nicht kennen,
Gibt es eine unendliche Zahl von Dämonen der Wahrnehmung.
Wenn man nicht erkennt, daß die Seele selbst leer ist,
Wird sich dann wohl der Frevel der Dämonen abwenden?
Du Schadenstifterin, schadende böse Dämonin,
Ohne mir zu schaden, kehre zurück!

Als er so gesprochen, ließ sie ihn, den sie am Fuße gepackt hatte, dennoch nicht los und sprach mit geisterhafter Stimme folgende Worte zur Antwort:

O du Sprößling eines würdigen Geschlechts,
Der den Mut hat, allein zu wandeln,
In der Bergwildnis umherziehender Yogin,
Wunderbar in den Bußübungen!
Das von dir gesungene Lied ist wohl ein Gebot des Königs;
Des Königs Gebot ist schwerer als Gold.
Doch gegen Gold Messing auszutauschen, ist ein großes Vergehen;
Wenn man nicht versteht, sich von solchem Vergehen abzuwenden,
Erschöpfen sich meine vorhergehenden Worte in einer Lüge.
Im ersten Teil will ich das Lied vom Königsrecht behandeln;
Nun bitte ich, eine aus jenen Gleichnissen ausgeschnittene Rede halten zu dürfen.
Geruhe einen Augenblick mit Aufmerksamkeit zuzuhören!
In der Mitte des blauen Himmels dort oben
Erwächst das Glück der leuchtenden Sonne und des Mondes;
Jener wundervolle Götterpalast
Vertreibt die Finsternis der vier Kontinente, sagt man;
Wenn sie in ihrem Amte die vier Kontinente umkreisen,
Nehmen sie sich nicht in Acht mit ihrer Helligkeit:
Denn wenn Rahu nicht durch das Licht der Scheibe verführt würde,
Weshalb sollte er als Feind wider sie aufstehen?
Im Haaraufsatz des Ost-Schneebergs „Weißer Kristall"
Erwächst das Glück des starken weißen Löwen;
Er ist der die Tiere beherrschende König;

Die untertänigen Tiere hält er durch Gesetze nieder, sagt man;
Wenn er am Rande des blauen Schieferfelsens hinabsteigt,
Ist sein Zornwüten und Stolz groß:
Denn wenn er nicht mit seiner blauen Türkismähne den Schneesturm siegreich bestände,
Weshalb sollte er als Feind wider ihn aufstehen?
Unter dem Laubdach im Dickicht des Südwaldes
Erwächst das Glück des gestreiften jungen Tigers;
Unter allen Raubtieren ist er der Ringer;
Die Krallentiere besiegt er durch seinen Glanz, sagt man;
Wenn er auf den schmalen Felspfaden umherwandelt,
Ist er von Stolz auf seine hohe Geschicklichkeit aufgeblasen:
Denn wenn die Falle nicht durch sein lächelndes Bildnis verführt würde,
Weshalb sollte sie als Feind wider ihn aufstehen?
In dem länglichen Türkissee Manasarovara im Westen
Erwächst das Glück der weißbauchigen Fische;
Sie sind die Tänzer im Element des Wassers;
Göttern und Einsiedlern sind sie ein großes Schauspiel, sagt man;
Wenn sie der Lust und der Nahrung wegen einander folgen,
Trachten die Menschen nach dem Erwerb von Nahrung:
Denn wenn der Angelhaken nicht durch ihren Täuschungsleib verführt würde,
Weshalb sollte er als Feind wider sie aufstehen?
Auf dem roten Felsen Bsam yas im Norden
Erwächst das Glück des Geiers, des Königs der Vögel;
Er ist der Einsiedler unter den beschwingten Vögeln;
Alle Geflügelten besiegt er durch seinen Glanz, sagt man;
Wenn er auf den Gipfeln der drei Berge seine Nahrung sucht,
Trachtet er nach Fleisch und Blut zur Nahrung:
Denn wenn die aus Stricken gewundene Schlinge durch sein Flügelschlagen nicht verführt würde,
Weshalb sollte sie als Feind wider ihn aufstehen?
Auf dem den Horst des Felsengeiers bergenden Ling ba

Erwächst dein Glück, Milaraspa;
Nach deinem eigenen und dem Heil anderer trachtend,
Hast du die höchste Bodhi zur Grundlage gemacht;
Während zu derselben großen Zeit dieser eine Leib
Unablässig nach der Buddhawürde strebt,
Bist du der Pfadführer der sechs Klassen der Wesen, sagt man;
Während du im Dhyana unablässig meditierst,
Wenn nicht infolge der starken Leidenschaft
Durch die ursprüngliche Ursache der Täuschung der eigenen Seele
Die Skepsis als Feind in dir aufstände,
Weshalb sollte ich Felsen-Rakshasi als Feind wider dich aufstehen?
Dieser Dämon der Leidenschaft aber entsteht aus der Seele.
Auch wenn man nicht das Wesen der Seele kennt,
Weiche ich nicht, nachdem du gesagt hast: „Gehe fort!"
Auch wenn man nicht die Leere der eigenen Seele erkennt,
Gibt es eine unendliche Zahl von anderen Dämonen außer mir.
Auch wenn man selbst die eigene Seele kennt,
Erscheinen ungünstige Umstände als Freunde.
Auch ich Felsen-Rakshasi bin andren untertan.
In deinem Gemüt jedoch herrschen Bosheit und Begierde;
Bring daher deine Seele noch besser in Ordnung!

Als die Felsen-Rakshasi diese Rede vorgetragen hatte, entstand darob im Herzen des Ehrwürdigen große Befriedigung, und er trug zur Antwort folgendes Lied von den acht Gleichnissen des Bewußtseins vor:

Du hast Recht, hast Recht, böse Dämonin!
Wahrere Worte als diese gibt es überhaupt nicht.
Auf meinen einstigen Kreuz- und Querzügen durch die Welt
Habe ich ein wohltönenderes Lied als dieses nie gehört.
Wenn auch hundert Gelehrte zum Vergleich herangezogen würden,
Etwas Vorzüglicheres als den Sinn desselben gibt es nicht.
Aus deinem Munde, Dämonin, ist eine treffliche Erklärung gekommen;
Die goldene Stange deiner trefflichen Erklärung

Schlägt mich mitten in meine menschliche Seele.
Der Herzenskummer, der aus dem Glauben an die Wirklichkeit der Existenz entspringt, ist entfernt;
Die schwarze Finsternis der auf Unwissenheit beruhenden Täuschung ist verscheucht;
Die weiße Lotosblume des Verstandes öffnet ihren Kelch;
Die Fackel des klaren Selbstwissens ist angezündet;
Die Weisheit des Bewußtseins erwacht deutlich.
Ist das Bewußtsein wirklich erwacht oder nicht?
Wenn ich hinauf zur Mitte des blauen Himmels schaue,
Kommt die Leere der Existenz deutlich zum Bewußtsein;
Ich fürchte nicht die Lehre von der Wirklichkeit.
Wenn ich den Blick auf Sonne und Mond richte,
Kommt die geistige Erleuchtung deutlich zum Bewußtsein;
Ich fürchte nicht Ermattung und Erschlaffung.
Wenn ich den Blick auf den Gipfel der Berge richte,
Kommt die unwandelbare Beschauung deutlich zum Bewußtsein;
Ich fürchte nicht immer wechselnde Grübelei.
Wenn ich nach unten in die Mitte des Flusses schaue,
Kommt das Ununterbrochene deutlich zum Bewußtsein;
Ich fürchte nicht plötzliche Geschehnisse.
Wenn ich das Bild des Regenbogens sehe,
Kommt die Leere der Erscheinungen im Zung ◦jug deutlich zum Bewußtsein;
Ich fürchte nicht das Dauernde und das Vergängliche.
Wenn ich das Spiegelbild des Mondes im Wasser sehe,
Kommt die von Interessen freie Selbsterleuchtung deutlich zum Bewußtsein;
Ich fürchte nicht die an Interessen geknüpften Erwägungen.
Wenn ich in meine eigene Seele schaue,
Kommt die Lampe im Innern des Gefäßes deutlich zum Bewußtsein;
Ich fürchte nicht Torheit und Dummheit.
Da ich jene Worte aus deinem Munde, Dämonin, vernommen,
Kommt meine Natur und meine Seele deutlich zum Bewußtsein;
Ich fürchte nicht den Widersacher.

Da du so viele treffliche Erklärungen weißt
Und das Wesen der Seele so gut verstehst,
Bist du jetzt als Dämonin in einem schlechten Leibe wiedergeboren.
Durch böse Taten stiftest du Schaden und Harm.
Das ist der Lohn für die Verachtung der Wiedervergeltung.
Nun halte dir die Strafe vor, die im Durchwandeln des Kreislaufs besteht!
Meide die zehn Todsünden ganz und gar!
Ich bin der einem Löwen gleiche Yogin.
Ohne Angst und Furcht bin ich.
Ich hatte nur einen Scherz gemacht:
Ich glaube nicht, daß du Recht hast, böse Dämonin!
Da du, Dämonin, mich heute Abend verhöhnst,
Habe ich zuvor gemäß der Stärke der fünf Yaksha- und Piçaca-Geschwister
Und des liebreichen Königs
Durch die Kraft des Gebets einen Entschluß gefaßt;
Wenn du den Sinn der Heiligkeit erlangt hast,
Möge später mein Bekehrungswerk erfolgen!

Als er so gesprochen, wurde die Felsen-Rakshasi sehr gläubig und ließ ihn, den sie am Fuße gepackt hielt, los. Wiederum mit geisterhafter, aber wohlklingender Stimme richtete sie folgende Bitte an ihn, der sich auf dieser Seite in der Luft befand:

O würdiger Yogin,
Tugendverdienste ansammelnd, übst du die heilige Religion.
Es ist wunderbar, daß du allein auf den Bergen wohnst.
Soweit das Auge des Erbarmens reicht, sorgst du für die Wesen.
Ich besitze den Lehrer mit dem Lotoskranz als Kopfschmuck;
Ich höre den Kranz der Worte der heiligen Religion.
Obwohl ich solche Worte gehört, ist meine Sehnsucht noch groß.
Ich will von einem Versammlungshaus der Yogin zum andren ziehen.
Wer durch frühere Werke glücklich ist, verweilt in der Tugend;
Die Würdigen pflegen auch die Wahrheit.
Obwohl die Begriffe meines guten Sinnes rein sind,
Empfindet mein schlechter Leib, wenn er nicht genährt wird, großen Hunger;

In meiner Tätigkeit durchwandre ich die Städte von Jambudvipa
Und nehme gern Fleisch und Blut zur Nahrung.
Ich werde in die Seele des ersten besten Mannes fahren, der mir begegnet;
Eine schöne Frau erregt Allen Kummer;
Ein schöner Mann ist Allen wie ein Bild in roter Farbe gemalt;
Mit den Augen bietet er Allen ein Schauspiel dar;
Im Reich der Seele unterdrückt er die Neigungen;
Mit dem Leibe stützt er Allen die Unbeständigkeit.
Auf dem Felsen von Ling ba hause ich.
Das ist mein Wandel.
An dich habe ich eine Bitte:
Noch einmal sei ein Lied gesungen.
Möge aus unsrer Begegnung Freude erstehen!
Jetzt habe ich eine schöne Bitte getan.
Zu deinen Diensten bin ich, Yogin.
Ich selbst bin des Glaubens klares Wort.
Gib ein aufrichtiges Lied zum Besten, an dem wir uns freuen mögen!

Als sie so gesprochen, richtete der Ehrwürdige eifrige Fragen an sie über diese Klasse von Geistern, und indem er dachte, daß er sie bannen müsse, trug er zur Antwort auf die Bitte des Weibes folgendes Lied vor:

Wohlan, merke auf, o Frau!
Der Lehrer ist gut, doch der Schüler ist schlecht.
Die, welche nachgedacht haben über das, was sie von der Götterlehre gelernt,
Haben den Kranz der Worte verstanden, deren Sinn unbegreiflich ist.
Obwohl ihr Mund treffliche Fragen stellt, fehlt es ihnen an Uebung.
Indem ihr lügnerischer Mund das Leere erklärt,
Können die Flecken ihrer eigenen Seele nicht gereinigt werden.
Du hast wegen deiner früheren bösen Leidenschaft
Und gegenwärtig angesammelten bösen Werke
Verpflichtung, Gelübde und Versprechen verletzt
Und bist kraft dessen als niederes Wesen wiedergeboren,
In unseliger Körperform in der Sphäre der Piçaca wandelnd.

Unwahre Rede, Trug, viele Lügen,
Sündiger Sinn schaden dem Leben der Wesen.
Diese Wiedergeburt in der Gestalt eines schlechten Leibes
Ist der Lohn für die Verachtung der Lehre von der Vergeltung der Werke.
Wenn du dir nun die Strafe vor Augen führst, die im Durchwandeln des
Kreislaufs besteht,
Sühne deine Sünden nach dem Maße, wie du sie begangen hast,
Und versprich, nach der Tugend zu streben.
Ich bin wie ein Löwe, aber nicht furchtbar;
Ich bin wie ein Elefant, doch habe ich keine Furcht;
Ich bin wie rasend, doch nicht der Wirkung der Planeten anheimgefallen.
Ich spreche wahre Worte zu dir.
Laß du noch einmal aufrichtige Rede hören!
Denn mir Hindernisse zu bereiten und mich zu verhöhnen,
War dein beständiges Geschäft, du Dämonin.
Erfasse nun die Kraft der Wirkung der Religion und des Gebets
Und bewahre sie in Zukunft im Gedächtnis!
Erwäge und prüfe das, du Verwandlung eines Weibes!

Als er so gesprochen, zeigte sich die Felsen-Rakshasi wie zuvor und richtete mit aufrichtiger Rede ihre Bitte an ihn in folgendem Liede:

Herr, Gebieter aller Buddha der drei Zeiten,
Der den Leib des Einsiedlers des Mahavajrapani hat,
In den Besitz der wunderbaren Lehre bist du gelangt.
Daß du in mir den Sinn erstaunlicher Bodhi geweckt hast, ist vortrefflich.
Für mich, das Weib, und meine Geschwister ist es ein großes Glück.
Verständnis der Reden, die ich von dir vernommen, ist mir erwachsen.
Zum ersten Male habe ich durch des Lehrers strenges Wort
Ueber das, was ich von der heiligen Religion, der Religion der Götter gehört,
nachgedacht,
Doch inzwischen bin ich ins Unheil der Folgen meiner Werke geraten;
Das wilde Gift der Sünden ist unerträglich.
Deshalb bin ich als Weib in schlechtem Leibe wiedergeboren.

Allen Geschöpfen
Erwies ich bald Nutzen, fügte bald Schaden zu.
Im verflossenen Jahre
Bist du, großer Meister der Beschauung auf dem Felsen von Ling ba,
Hierher gekommen und vollzogst Bannungen ganz allein.
Bald freute ich mich, bald freute ich mich nicht.
Da ich mich freute, besuchte ich dich heute Abend;
Da ich zürnte, ergriff ich dich am Fuße.
Da ich dir so schadete, habe ich darauf meine Sünden bekannt.
Von jetzt ab will ich das wilde Gift des Weibes meiden,
Will von Herzen die Religion üben und ihr Freund sein.
In Zukunft werden wir
Im kühlen Schatten des Baums der Seligkeit ruhen.
Nun Tag und Nacht von den fünf Giften gequält,
Bitten wir übeltuende Dämoninnen um Hilfe.
Wenn wir auf dein Wort vertrauen
Und damit beginnend bis zur Bodhi fortschreiten,
Ist der Hang zur Wildheit gänzlich beruhigt.
Der Yogin ist unser helfender Schützer.
Wir sind Freundinnen des Bannenden;
Wir sind untertänige Dienerinnen aller großen Meister der Meditation;
Wir sind der Beistand aller Pfleger der Religion;
Wir sind die Helferinnen der durch ihr Gelübde Geweihten.
Die Lehre bewachend, leisten wir Dienste.

Als sie so ein treffliches Gebet gesprochen, wurde sie sehr gläubig durch das Versprechen und Gelübde, Schützerin aller Bannenden zu sein und ihnen keinen Schaden mehr zuzufügen. Da trug der Ehrwürdige folgendes Lied vor, damit es die Felsen-Rakshasi im Gedächtnis bewahre:

Ich bin der den Kreislaut meidende Bandhe.
Ich bin des heiligen Lama Sohn.
Ich bin die kostbare Schatzkammer der guten Lehren.
Ich bin der die heilige Religion mit ganzem Herzen Uebende.

Ich bin der das Sein erkennende Yogin.
Ich bin der Wesen alte Mutter.
Ich bin der mutvolle heilige Mann.
Ich bin der Nachfolger der Gewohnheit Çakyamuni's.
Ich bin die verkörperte Bodhi.
Ich bin der von Anfang an über die Liebe Meditierende.
Mit Erbarmen vernichte ich das Gift.
Auf dem Ling ba-Felsen hause ich.
Ohne Ablenkung liege ich der Meditation ob.
Bist du von freudigen Gedanken erfüllt, du Verwandlung eines Weibes?
Wenn du keine Freude empfindest, bist du schlecht.
Größer als ihr Dämonen ist die Eigensucht.
Zahlreicher als ihr Dämonen sind die Wahrnehmungen.
Schlechter als ihr Dämonen sind schlechte Gedanken.
Wilder als ihr Dämonen ist die Skepsis.
Hartnäckiger als ihr Dämonen ist die Leidenschaft.
Wenn man die Dämonen für Dämonen hält, ist es ein Schaden;
Wenn man weiß, daß die Dämonen leer sind, ist es erfreulich;
Wenn man das Wesen der Dämonen begreift, ist es Befreiung.
Wenn man die Dämonen gleichsam als Eltern auffaßt, ist es Besessenheit.
Wenn man den Geist der Dämonen kennt, erweist es sich als Segen.
Wenn man ihre Umstände kennt, sind alle befreit.
Du, niedere Dämonin, hast ein Bittgesuch dargelegt;
Ich habe dich durch die Bekehrung gebannt.
Du erfülle nun dein Versprechen!
Unseres Vajradhara
Wort des strengen Gelübdes brich nicht!
Mache mein großes Erbarmen nicht zu Schanden!
Begehe keine Frevel des Leibes, des Worts und des Gedankens!
Wenn du dein Gelübde brichst,
Wirst du sicher in die Vajra-Hölle fahren,
Da dies von großer Wichtigkeit ist, wiederhole es dreimal.
Um das zu begreifen, wende dich der Uebung zu.

Wir haben uns mit trefflichem Gebete zusammengefunden.
In Zukunft wird dir Seligkeit zuteil:
Unendlich und unfaßbar
Von dem Sinn der Heiligkeit erfüllt,
Wirst du als Erste aus dem Gefolge der Bekehrung wiedergeboren
Und zum Weib des Vajrasattva gemacht werden.

Als er sie durch solche Worte gebannt hatte, bezeigte die Felsen-Rakshasi dem Ehrwürdigen Verehrung und umkreiste ihn oftmals. Mit dem Versprechen, alle seine Worte zu erfüllen, entfernte sie sich endlich wie ein verschwindender Regenbogen. Als bei Tagesanbruch die Sonne aufging, erschien die Felsen-Rakshasi mit dem Gefolge ihrer Geschwister, schön geschmückter Männer und Frauen mit lieblichem und anmutigem Wesen, und brachten eine Menge von Dingen und viele Opfergaben. Sie überreichten dem Ehrwürdigen die Opfergaben, und die Felsen-Rakshasi sprach: „Ich habe durch meine schlechten Werke den Leib einer Dämonin empfangen; böse Leidenschaft beherrscht mich. Geruhe, mir zu verzeihen, was ich dir mit böser Absicht geschadet habe. Von jetzt ab will ich auf das von dir, Yogin, verkündete Wort hören und dir untertänig dienen. Nun, Ehrwürdiger, teile uns bitte religiöse Belehrung mit über die mystische Erkenntnis, die in deinem eigenen Geiste entsteht." So richtete sie in folgendem Liede ein Bittgesuch an ihn:

O Sohn aus vornehmem Adelsgeschlecht,
Würdiger, der verdienstliche Handlungen angesammelt hat,
Der den Segen eines trefflichen Lehrers genießt,
Der Standhaftigkeit besitzt in der Vollziehung von Bannungen,
Der den Mut hat, allein zu wohnen!
Nach der Erreichung der tiefen Wahrheit strebend,
Können dir die Hindernisse der Dämonen nichts anhaben.
Durch Ursache und Wirkung der Anzeichen der im Innern befindlichen Aderluft
Lehrst du auf Grund des Tanzspiels der Illusion die Wahrheit.
Wir vereinigen uns mit deiner Seele:
Obwohl wir durch die uranfängliche Wirkung guten Gebets
Schon früher mit vielen Siddha zusammengetroffen sind,
Haben wir erst durch dich den Segen der Gnade erlangt.
Nun richtet ein Weib eine Bitte an dich.

Durch die hinsichtlich der Wahrheitserkenntnis aus den heiligen Schriften
veranlaßte Täuschung des Hinayana
Ist es schwer, die sündigen Werke zu zähmen.

— — — — — — — — — — — — — — — —

— — — — — — — — — — — — — — — —

Ich, ein Lehrer, der sich vor der Religion fürchtet,
Bin erzürnt, daß mein eigener Nutzen noch nicht erreicht ist.
Gebieter, Nirmanakaya der Buddha der drei Zeiten,
Der du das Sein und die mystische Erkenntnis begreifst,
Ein Sang der Unterweisung, der die in deinem Geist entstehende tiefe Wahrheit
In den Kernpunkten zusammenfaßt,
Sei unser Geleite bis ans Endziel der mystischen Erkenntnis!
Mir samt dem Gefolge meiner Geschwister
Geruhe den heiligen Sinn des Vajra-Geheimwortes,
Die große erleuchtende Weisheit,
Die höchste Erleuchtung zu gewähren.
Da wir das Zeichen des tiefen Geheimnisses der mystischen Erkenntnis
Vernommen, können wir nicht mehr zur Hölle fahren.
Da wir nachgedacht haben, können wir nicht mehr im Kreislauf herumwandern.
Teile bitte deine Gaben verschwenderisch aus!

Als sie diese Bitte getan hatte, sagte der Ehrwürdige: „Du bist nicht imstande, in der mystischen Erkenntnis der Lehre zu meditieren; wenn du aber dazu imstande bist, so mußt du mir dein Leben zum Pfande einsetzen und ein starkes Versprechen geben. Nachdem sie unabänderlich ihr Leben zum Pfande eingesetzt und das Versprechen gegeben, in Zukunft alle Gebote des Ehrwürdigen zu erfüllen und allen Pflegern der Religion in Freundschaft beizustehen, trug er zur Antwort auf die Bitte des Weibes die Lehre von der mystischen Erkenntnis in folgendem Liede von den 27 verschwindenden Dingen vor:

Dem Gebieter, der Buddha's verborgene Menschengestalt hat,
Dem Uebersetzer schwer auszusprechender Namen,
Dem huldreichen Vater zu Füßen verneige ich mich.
Ich bin kein berufsmäßiger Sänger;
Ihr Dämonen sagt: Singe, singe ein Lied!

Deshalb singe ich jetzt ein Lied vom Wesen der Dinge.
Donner, Blitz und Südwolke,
Wenn sie entstehen, entstehen aus dem Himmel selbst,
Wenn sie verschwinden, verschwinden am Himmel selbst.
Regenbogen, Nebel und Dunst,
Wenn sie entstehen, entstehen aus der Luft selbst,
Wenn sie verschwinden, verschwinden in der Luft selbst.
Fruchtsaft, Ernte und Frucht,
Wenn sie entstehen, entstehen aus dem Boden selbst,
Wenn sie verschwinden, verschwinden im Boden selbst,
Wald, Blumen und Laubwerk,
Wenn sie entstehen, entstehen aus den Bergen selbst,
Wenn sie verschwinden, verschwinden in den Bergen selbst.
Flüsse, Wasserschaum, Wogen,
Wenn sie entstehen, entstehen aus dem Meere selbst,
Wenn sie verschwinden, verschwinden im Meere selbst.
Leidenschaft, Begierde und Habsucht,
Wenn sie entstehen, entstehen aus der Seele selbst,
Wenn sie verschwinden, verschwinden in der Seele selbst.
Selbstwissen, Selbsterleuchtung, Selbstbefreiung,
Wenn sie entstehen, entstehen aus dem Geiste selbst,
Wenn sie verschwinden, verschwinden im Geiste selbst.
Nicht-Wiedergeboren-Werden, das Unbehinderte, das Unaussprechliche,
Wenn sie entstehen, entstehen aus dem Sein selbst,
Wenn sie verschwinden, verschwinden im Sein selbst.
Was als Dämon erscheint, was als Dämon gilt, was als Dämon erkannt wird,
Wenn es entsteht, entsteht aus dem Yogin selbst,
Wenn es verschwindet, verschwindet im Yogin selbst.
Da nun die Dämonen ein Täuschungsspiel der Seele sind,
So ist ein Yogin, wenn er, ohne die Leere der eigenen Gedanken zu erkennen,
Die Dämonen in seiner eigenen Seele begreift, im Irrtum befangen.
Die Wurzel des Irrtums aber ist aus der Seele entstanden;
Aus der Erkenntnis der Natur der Seele

Ersieht man, daß die Erleuchtung weder geht noch kommt.
Wenn die in den Erscheinungen der Außenwelt sich täuschende Seele
Die Theorie der Erscheinungen erkannt hat,
So erkennt sie, daß zwischen den Erscheinungen und der Leere kein Unterschied besteht.
Wenn man in der Erkenntnis der Beschauung
Die Nicht-Beschauung erkennt,
Ist zwischen Beschauung und Nicht-Beschauung kein Unterschied.
Zwei Dinge als verschieden betrachten, ist die Ursache des Irrtums.
Dann sind die Gedanken nicht auf den Nutzen der Erreichung des Endzieles gerichtet.
Wenn man die Natur der Seele
Mit der Natur des Aethers vergleicht,
Ist das Wesen der Wahrheit in Ordnung gebracht.
Sorge du nun als Dienerin der Beschauenden für ihren Nutzen!
Schließe dich der Sache der unablenkbaren Meditation an!
Schütze die unbehinderte Kraftentfaltnug ihrer Ausübung!
Meide das Gerede von Hoffnung und Furcht der Vergeltung!
Gib den Dämonen Anteil an deiner Religion!
Ich kann keine Gelehrsamkeit in Lobliedern entfalten.
Ohne viele Fragen und Erwägungen sitzt ihr schweigend da.
Nur weil die Dämonen sagten: „Singe ein Lied!" habe ich gesungen.
Jetzt habe ich ein törichtes Wort gesprochen.
Wenn ihr, Dämonen, es zu beherzigen vermögt,
Verzehrt als Nahrung Seligkeit
Und trinkt als Trank kummerfreien Nektar!
Seid in euerm Tun Freunde des Yogin!

Als er so gesprochen, wurde die Felsen-Rakshasi samt ihrem Gefolge sehr gläubig, bezeigte ihm Verehrung und umkreiste ihn oftmals. „Vielen Dank!" sagte sie und entfernte sich wie ein verschwindender Regenbogen. Von da an tat sie nach dem Gebot des Ehrwürdigen, so lange er an jenem Orte in großer Beschauung verweilte, keinen Schaden mehr und war eine Helferin im Einklang mit der Religion.

Dies ist der Abschnitt von der Felsen-Rakshasi von Ling ba.

III.

Milaraspa in Rag ma.

Verehrung dem Meister!

Als der ehrwürdige Milaraspa von Ling ba brag auf den Berg Dpal ₒbar zum Zweck der Meditation zu gehen gedachte, entstand, während er sich auf den Weg machte, unter den Gabenspendern von Rag ma über seine Wanderung nach dem Berge Dpal ₒbar ein Gerede. Sie sprachen: „Was den Berg Dpal ₒbar betrifft, so bietet Dgon rdzong skyid po selbst den einzigen Durchgang zum Berge Dpal ₒbar. Dort verweilt er auch gerne, aber einen Führer auf den Berg Dpal ₒbar haben wir nicht. Wenn er dort auf Dgon rdzong verweilen will, wollen wir ihm einen Führer senden." In dieser Erwägung blieb er anfangs an diesem Orte. Darauf kam ihm der Gedanke, daß er auf den Berg Dpal ₒbar gehen müsse, aber eines Führers von jenen nicht bedürfe, und sagte: „Ich brauche euern Führer nicht, ich werde selbst den Weg finden." „Wir forschen nicht nach, da es keinen Führer gibt. So hast du denn wirklich einen Führer?" fragten sie. „In der Tat", sagte er. „Wer ist es, und wie ist sein Name?" Darauf gab er Antwort in folgendem Liede:

Jener ausgezeichnete heilige Lama
Ist ein Führer, um die Finsternis zu vertreiben.
Dies einfache Baumwolltuch, das nicht kalt noch warm ist,
Ist ein Führer, um auf Wünsche zu verzichten.
Dies dreimalige Umwandeln der Berggipfel, mit guten Lehren verbunden,
Ist ein Führer, um den Zwischenzustand zu vernichten.
Diese vollkommene Errungenschaft höchster Ekstase
Ist ein Führer, um die Länder zu durchwandern,
Dieser geringschätzig behandelte Leib
Ist ein Führer, um die Selbstzucht zu bezwingen.
Dies Meditieren in der Einsamkeit
Ist ein Führer, um die Bodhi zu erreichen.
Da ich von sechs Führern geleitet werde,
Begehre ich, auf Byang chub rdzong zu verweilen.

Als er so gesprochen, begab er sich an das obere Ende von Rag ma. Der Ort unterhalb davon heißt Byang chub rdzong. Darauf verweilte dort der Ehrwürdige, in die Betrachtung des fließenden Wassers vertieft. Einstmals geschah es, als die Mitternacht heranrückte, daß Töne ven Kriegstrompeten und Losungsworte in großer Zahl hörbar wurden. Da dachte er, ob vielleicht ein Feind über die Einwohner hereingebrochen sei, und als er sich darüber einer sein tiefstes Mitleid regenden Betrachtung hingab, kam es immer näher und näher. Ein großes rotes Licht leuchtete vorbeifahrend auf, und während er sich noch besann, was es sein möchte, schaute er hin: siehe, da war die ganze Fläche von Feuer bedeckt, unter dem Erde und Luft ganz verschwanden. Schrecken erregende Heerscharen waren da, die das Feuer anzündeten, Wasser in Bewegung setzten, Berge hinabschleuderten, Erdbeben, stechende Waffen und andern mannigfachen Spuk erscheinen ließen. Insbesondere zerstörten sie seine Grotte und stießen mißtönende Worte aus. Da stieg dem Ehrwürdigen der Gedanke auf: „Das sind die Nachstellungen und Verhöhnungen der Geister." Und weiter dachte er: „Ach! Seit undenklicher Zeit haben sie schlechte Werke aufgehäuft und wandeln am Orte der sechs Klassen der Wesen. Unter ihnen gibt es welche, die unter den in der Luft herumziehenden Preta wiedergeboren werden. Böse Gedanken sinnend, voll roher Schadenfreude, stellen sie dem Leben vieler Wesen nach und gehen nur darauf aus, Harm zu stiften. Ach, ich bedaure sie, die ohne Befreiung in der Hölle wiedergeboren werden und unerträgliche Qualen erdulden müssen." In diesem Gedanken trug er folgendes Lied vor:

Der an dem weiten Himmel seiner Liebe
Die Wolken des Erbarmens versammelnd,
Den Regenguß der Taten herabströmen
Und die Ernte der Bekehrung reifen ließ,
Dem sprachgewandten Marpa zu Füßen verneige ich mich!
Auf daß ich dessen, welcher die dem Himmel gleich zahllosen Geschöpfe
Kennt, des Buddha Würde erlange, möge er mich segnen!
Ihr hier versammelten Yaksha und Gespenster,
Ihr in der Luft wandelnden Scharen der Daka,
Ihr Preta, deren Sinn nur auf Speise gerichtet ist,
Durch die gereifte Vergeltung euerer sündigen Werke
Seid ihr gegenwärtig im Leibe der Preta wiedergeboren,
Und weil ihr zu dieser Zeit anderen Schaden zufügt,
Werdet ihr später im Ort der Hölle wiedergeboren werden.

Dieses die Vergeltung nur in kurzem Umriß behandelnde Lied
Habe ich, soweit für das Verständnis erforderlich, verfaßt.
Ich, der Sohn des Lama, welcher der Ueberlieferung anhängt,
Bin in den geistlichen Stand getreten, als die Grundlage, der Glaube, erzeugt war;
Habe Bußübungen vollzogen, als ich die Werke und ihre Frucht erkannte.
Nun, da ich den rechten Pfad durch meinen Eifer gewonnen, sehe ich kraft der Meditation
Das innere Wesen des Sinnes der Vergeltung.
Ich weiß, daß alle Erscheinungen eine Täuschung sind.
Von der Krankheit der Eigensucht bin ich befreit,
Die Fesseln des das Ergriffene festhaltenden Samsara sind durchschnitten.
Den Thron des unveränderlichen Dharmakaya habe ich in Besitz genommen.
An dem vom Denken losgelösten Yogin
Scheitert eure Schadenfreude.
Sein Leib ist zwar ermattet und erschöpft,
Doch das bewirkt, daß in seinem Sinn aufs neue der Zorn erwacht.
Da mein Wesen die treibende Ursache ist,
Wenn auch zahlreicher als ihr Dämonen, von der Welt des Brahma
Bis zu den achtzehn Höllenreichen
Die sechs Klassen der Wesen als Feinde wider mich aufstünden,
Könnte mir der bloße Gedanke an Furcht nimmermehr kommen.
Die ihr nun hier versammelt seid, Yaksha und Gespenster,
Ihr habt durch Zauberkraft und Gaukelei ein Heer hervorgebracht;
Wenn ihr, ohne mir Schaden zuzufügen,
Von hier anderswohin abziehen werdet,
So sind all eure früheren Bemühungen unnütz,
Immer wieder müßt ihr Scham empfinden.
Auf, frisch ans Werk, frisch ans Werk, ihr Koboldscharen!

Als er so gesprochen, brachte er sie durch den Glauben an die Lehre zur Ruhe. Die Koboldscharen wurden gläubig, bezeigten ihm Verehrung, umwandelten ihn oftmals und setzten seinen Fuß auf ihren Scheitel. „Daß du ein Yogin bist, der die Festigkeit erlangt hat, wußten wir nicht. Wir bitten daher, uns die früheren Verhöhnungen zu verzeihen. In Zukunft wollen wir alle deine Befehle vollziehen; gewähre uns bitte doch ein wenig religiöse Belehrung", sagten sie. Der Ehrwürdige sprach: „Wohlan, begeht keine Sünden, übt herrliche Tugenden!" Als er so gesagt hatte, erzählten sie ihm ihre Umstände. Dann weihten sie ihm Leben und Herz und versprachen ihm, Diener seines Wortes zu sein, worauf sie an ihren Ort zurückkehrten. Es waren die nährende Göttin von Mang yul und die Ortsgottheit des Berges Dpal ₒbar. Der Ehrwürdige dachte: „Nachdem die Ortsgottheit des Berges Dpal ₒbar hier erschienen ist, brauche ich nicht absichtlich zur Meditation auf den Berg Dpal ₒbar zu gehen." So verweilte er denn einige Tage an jenem Orte, und als seine Meditation sehr stark wuchs, stimmte er folgendes Lied an:

In der Einsamkeit der „Feste der Bodhi"
Weile ich, Mila, der die Bodhi erreicht hat.
Den auf die Bodhi gerichteten Sinn besitzend,
Bin ich der Meditation des auf die Bodhi gerichteten Sinnes ergeben.
Da ich die große Bodhi schnell erlangt habe,
Mögen alle diese wie eine Mutter geliebten Wesen
Mit der höchsten Bodhi vereinigt werden!

So sprach er und beeiferte sich der Meditation. Darauf kam nach einigen Tagen ein Gabenspender, der ihm eine Ladung Holz und einen halben Scheffel Mehl brachte. Er sagte: „Da dein Gewand so dünn ist, wirst du frieren. Zwischen den Südbergen ist dieses Rag ma gelegen, und unter diesen ist gerade dieses Felsdach außerordentlich kalt. Willst du ihn nehmen, so gebe ich dir einen Pelzmantel. Lama, was sagst du dazu?" Der Ehrwürdige fragte: „Gabenspender, wie ist dein Name?" „Lha ₒbar heiße ich", erwiderte er. Der Ehrwürdige sagte: „Der Name ist gut. Obwohl du nicht für Mehl, einen Mantel und anderes zu sorgen brauchst, so sage ich dir doch für die Mehlspende meinen Dank, den Mantel aber brauche ich nicht, Ich bin nun einmal ein solcher." Dann trug er dem Lha ₒbar folgendes Lied vor:

Ich wanderte umher in der Sphäre der Täuschungen der sechs Klassen der Wesen
Als kleiner Knabe, der sich in den Wahrnehmungen täuscht.
Manche Illusionen des Tuns habe ich erfahren:
Bald entstand in mir die Illusion des Hungers,
Milde Gaben und Almosen bildeten meine Nahrung.
Bald habe ich zur Bußübung Steinchen gemacht;
Bald habe ich die Leerheit zur Speise genommen,

Bald erlangte ich Notbehelfe und Abhärtung.
Bald entstand in mir die Illusion des Durstes.
Das blaue Wasser des kühlen Schiefers trank ich,
Bald genoß ich selbst-entstandenes duftiges Wasser,
Bald trank ich von dem Strome des Erbarmens,
Bald trank ich von den heiligen Spenden der Daka.
Bald entstand in mir die Illusion des Frierens.
In ein einfaches Baumwollgewand kleidete ich mich.
Bald entfachte ich die glückliche Wärme der inneren Glut.
Bald erlangte ich Notbehelfe und Abhärtung.
Bald erhub sich die Illusion der Freundschaft in mir.
Auf Wissen und Weisheit vertraute ich als meinen Freunden
Und übte die zehn reinen Tugenden.
Vollkommene Beschauung habe ich mir angeeignet,
Den Sinn des Selbstwissens habe ich genau geprüft.
Ich, der Yogin, bin der Löwe der Menschen,
Dessen Türkisglanzmähne trefflicher Beschauung ausgebreitet ist,
Der die Fangzähne und Krallen trefflicher Meditation hat;
Merkverse mache ich auf dem Gipfel der Berge
Und hoffe, die Frucht der Tugend zu erlangen.
Ich, der Yogin, bin der Königstiger der Menschen,
Der die drei vollendeten Fertigkeiten des Bodhi-Sinnes besitzt,
Der in der unzertrennlichen Verbindung von Stoff und Geist (gra ?) lächelnd,
In dunklen, an Arzneien der Erleuchtung reichen Tälern gelebt hat.
Ich hoffe, daß die Frucht der Uneigennützigkeit (parartha) erzielt wird.
Ich, der Yogin, bin der Geier der Menschen,
Der die Flügel der lichten Meditationsstufe des Utsakrama ausgebreitet hält,
Der mit den Schwingen der starken Meditationsstufe des Sampannakrama schlägt,
Der am Himmel der Lehre vom zung ₒjug schwebt,
Der auf dem Felsen der mystischen Erkenntnis ruht,
Der hofft, daß die Frucht in den beiden Arten der Erkenntnis erlangt wird.
Ich, der Yogin, bin der Heilige der Menschen,

Ich bin Milaraspa.
Ohne Rücksicht auf andere verfolge ich meinen Weg.
Ich bin der Ratschaffer in allen Fällen.
Ich bin der heimatlose Yogin.
Ich lasse nicht fahren, was mir zu teil geworden.
Ich bin der Bettler ohne Nahrung,
Ich bin der Nackte ohne Kleidung,
Ich bin der Besitzlose, der von Almosen lebt.
Ich bin der nicht an Berechnung Denkende.
Ich bin nicht hier wohnend, hier weilend.
Ich bin der König derer, denen die Uebung der Meditation obliegt.
Ich bin der erfreuliches Glück Besitzende.
Ich bin, der nichts hat und nichts braucht.
Wenn ich die notwendigen Lebensbedürfnisse selbst erwerben müßte,
Wäre das eine schwere Sünde von dir.
Die Mühen der Gabenspender werden sich abwenden und dahingehen,
Denn der Yogin weiß für alle Fälle Rat zu schaffen.
Durch die guten und tugendhaften Gedanken deines Herzens
Geschenke darbringend, hast du dich meiner liebevoll angenommen.
Mögest du in dieser Existenz langes Leben, Freisein von Krankheit,
Das Glück inneren Friedens genießen!
Möchten wir uns in Zukunft in dem reinen Gefilde
Wieder begegnen! Möge religiöser Wandel sich ausbreiten
Und darauf das Heil anderer erzielt werden!

Als er so gesprochen, wurde jener sehr gläubig und sagte: „Da du Mila bist, der die Heiligkeit erlangt hat, und das eben mich befriedigt, so bitte ich, der nur ein einfacher Laie ist, um die Ansammlung der Verdienste zu vollenden, dich während deines Aufenthaltes hier mit wertvollen Lebensmitteln versehen zu dürfen und alles anzunehmen." So gab ihm, während er in Byang chub rdzong verweilte, Lha ₒbar wertvolle reichliche Lebensmittel. Als darauf dem Ehrwürdigen große Förderung der Meditation zu teil wurde, wodurch er in frohe Stimmung geriet, erschienen einige seiner Anhänger von Rag ma und sagten: „Findest du Wohlgefallen an diesem Orte, und ist die Beschauung gewinnreich gewesen?" Der Ehrwürdige erwiderte: „An dem Orte habe ich meine Freude, auch in der Kunst der Meditation habe ich gewonnen." „Vortrefflich!" sagten sie, „geruhe uns ein Preislied auf diesen lieblichen Ort und die Art, wie du deine Beschauung vollziehst, zum besten zu geben." In Erwiderung ihrer Bitte trug er folgendes Lied vor:

Das ist die Einsiedelei von Byang chub rdzong.
Oben ragt der weiße hohe Gletscherberg mächtiger Geister.
Unten stehen viele gläubige Gabenspender.
Der Berg hinter mir ist mit einem weißen Seidenvorhang bedeckt.
Vor mir dehnen sich wunschstillende Wälder aus.
Da sind große und weite Rasengründe und Matten.
Auf den duftenden lieblichen Blumen
Schweben summend die sechsfüßigen Insekten.
Am Strand der Teiche und Weiher
Späht der Wasservogel, den Hals drehend.
Im weitverästelten Gezweig der Bäume
Singt lieblich die schöne Vogelschar.
Vom Düfte tragenden Winde bewegt,
Wiegen sich die Zweige der Bäume tanzend hin und her.
Im Wipfel der hohen, weit sichtbaren Bäume
Zeigen Affen und Aefflein ihre mannigfachen Geschicklichkeiten.
Auf dem grünen, weichen, weiten Wiesenteppich
Breitet sich weidend vierfüßiges Vieh hin.
Die das Vieh hütenden Hirten
Singen und entlocken der Flöte liebliche Töne.
Die Knechte weltlicher Habsucht
Stapeln auf dem Boden ihre Waren auf.
Wenn ich, der Yogin, darauf hinabschaue,
Auf meinem weithin sichtbaren herrlichen Felsen,
Betrachte ich die vergänglichen Erscheinungen als ein Gleichnis.
Die sinnlichen Genüsse sehe ich als ein Spiegelbild im Wasser an.
Dieses Leben halte ich für die Täuschung eines Traumes.
Gegen die Unverständigen hege ich Mitleid.
Den leeren Raum nehme ich zur Speise,

Ungestörter Contemplation weihe ich mich.
Wie alle Bilder, die in unserem Geiste aufsteigen,
Ach, nach dem Gesetz des Kreislaufs der drei Welten
Nicht vorhanden sind, so auch nicht die herrlichen Erscheinungen der Welt.

Als er so gesprochen, kehrten jene gläubig zurück.

Dies ist der erste Abschnitt von Rag ma.

IV.

Milaraspa auf dem Rkyang phan namkha rdzong.

Verehrung dem Meister!

Als der ehrwürdige Milaraspa von Rag ma nach dem Rkyang phan namkha rdzong gekommen war, geschah es eines Nachts, als er dort verweilte, daß ein auf einem Hasen reitender Affe vor ihm erschien, der einen Pilz als Schild trug und Bogen und Pfeil aus Stroh hielt, als wenn er auf eine Gelegenheit lauerte. Der Ehrwürdige stieß ein Gelächter aus. Jener sprach: „In der Hoffnung, dich zu vernichten, bin ich hierher gekommen; doch wenn du nicht zu vernichten bist, gehe ich weg." Der Ehrwürdige erwiderte: „Da ich die sichtbare Welt für imaginär halte und meine eigene Wesenheit in der Nichtexistenz wahrnehme, sind deine, eines Kobolds Trugspiele, welche auch immer du zeigen mögtest, ein Gegenstand des Gelächters für mich, den Yogin!" Da verpflichtete sich jener, ihm Dienste zu leisten und verschwand wie ein verblassender Regenbogen. Es war der Herrscher von Gro thang. Darauf kamen die Gabenspender von Gro thang, um den Ehrwürdigen zu besuchen, und sagten: „Welche Vorzüge haften an diesem Orte?" Als Antwort auf ihre Frage trug er folgendes Lied vor:

Zu meinem Lama, dem Gebieter, sende ich mein Flehen!
Kennt ihr, kennt ihr nicht die Vorzüge dieses Ortes?
Wenn ihr die Vorzüge dieses Ortes nicht kennt,
So wißt, die Einsiedelei ist Rkyang phan namkha rdzong.
In dem Palast dieser Himmelsfestung
Sammeln sich oben purpurfarbene Südwolken,
Unten strömt der blaue Brahmaputra dahin,
Hinter mir rote Felsen gleich dem Himmelsraum,
Vor mir Wiesen mit buntfarbigen Blumen;
Am Rande stößt das Raubtier sein Gebrüll aus,
An den Seiten schwebt der Geier, der König der Vögel;
In der Luft fallen Insekten wie ein feiner Staubregen herab,
Und ohne Unterlaß summen die Bienchen ihr Liedchen.
Hirsche und Wildesel, Mutter und Junges, spielen und springen wie im Tanz;
Affen und Aefflein üben ihre Geschicklichkeit.
Lerchen, Mutter und Junges, trillern im Wechselsang.

Der Göttervogel, das Schneehuhn, singt sein Lied.
Ueber seinem Tonschieferbett murmelt der Bach mit melodischem Plätschern.
Das sind die jeweiligen Stimmen (der Natur), die Freunde des Herzens!
Die Vorzüge dieses Ortes sind unermeßlich:
Deshalb singe ich ein Lied der Herzensfreude auf ihn.
Belehrung fließt aus meinem Munde:
Ihr Gabenspender, Männer und Frauen, die ihr hier versammelt seid,
Folgt mir um meines Namens willen und tut wie ich!
Meidet sündige Werke und strebt nach der Tugend!

So sang er. Unter den Anwesenden befand sich ein Kenner der Mantra, welcher sagte: „Ehrwürdiger! Als ein Fest oder Gastgeschenk für unsere Pilgerfahrt hierher gewähre uns bitte einige zu Herzen gehende, leicht verständliche und gut zu bewahrende Belehrung über die Ausübung der Meditation!" Als Antwort auf diese Bitte trug er folgendes Lied vor:

Möge der Segen meines Lama bei mir einziehen!
Möge er mich segnen, daß ich die Leere erkenne!
Zum Dank für die Verehrung der gläubigen Gabenspender
Will ich ein Lied singen, an dem sich Götter und Schutzgottheiten erfreuen.
Die Erscheinungen, die Leere, das Unzertrennliche:
Diese drei bilden den Kern der Beschauung.
Das Verständliche, das Unbegreifliche, die Nichterregbarkeit des Geistes,
Diese drei bilden den Kern der Meditation.
Leidenschaftslosigkeit, Wunschlosigkeit und Standhaftigkeit:
Diese drei bilden den Kern der Lebensführung.
Hoffnungslosigkeit, Furchtlosigkeit, Irrtumslosigkeit:
Diese drei bilden den Kern der Vergeltung.
Reinheit im öffentlichen wie im privaten Leben, Freisein von Betrug:
Diese drei bilden die Grundlage des Gelübdes.

Als er so gesprochen, kehrten jene gläubigen Sinnes zurück. Nach einigen Tagen erschienen wiederum zahlreiche Zuhörer, um ihm die Verehrung zu bezeigen. Jene, die vorher dagewesen waren, fragten: „Ist das Befinden und der Zustand des Ehrwürdigen gut?" Auf ihre Frage nach seiner Gesundheit erwiderte er in folgendem Liede.

Zu den Füßen des heiligen Lama verneige ich mich.
Im unbewohnten einsamen Walde
Ist Milaraspa's Beschauungsart segensreich.
Glücklich ist, wer da wandelt frei von der Leidenschaft zu besitzen,
Glücklich, wessen Leib frei ist von brennendem Schmerz.
Glücklich, wessen Wesen der Trägheit bar ist.
Glücklich, wer einfachen Herzens Beschauung übt.
Glücklich, wer, ohne kalt zu sein, die innere Glut besitzt.
Glücklich, wer ohne Furcht Bußübungen vollzieht.
Glücklich der Landwirt, der nach nichts trachtet.
Glücklich, wer die ungestörte Einsamkeit wählt:
Das alles sind Vorzüge des Leibes.
Glücklich, wer das Fahrzeug der Materie und des Geistes hat,
Wer die Meditationsstufen des Skyed rdzogs und Zung ◦jug erreicht hat,
Glücklich, wer sich bewußt ist, daß zwischen Ausströmen und Einatmen der Luft kein Unterschied besteht.
Glücklich, wer, ohne Freunde, zu reden nicht gebunden ist:
Das alles sind Vorzüge des Wortes.
Glücklich, wessen Anschauungen frei von Selbstsucht sind,
Glücklich, wer sich beständig ununterbrochener Betrachtung weiht.
Glücklich, wessen Wandel frei von Furcht ist.
Glücklich, wer die Belohnung der Hoffnungs- und Furchtlosigkeit gewonnen:
Das alles sind Vorzüge des Sinnes.
Glücklich, wer unwandelbar, einfachen Herzens und erleuchtet ist.
Glücklich, wer in der reinen Sphäre des höchsten Segens weilt.
Glücklich, wer in der Tiefe schrankenloser Gedanken weilt.
Diesen sehr segensreichen kleinen Sang,
Dies Lied der Wonne habe ich gesungen.
Beschauung und Uebung sind eng miteinander verbunden.
Die ihr durch eure Bitten nach der Bodhi strebt,
Nehmt es euch zu Herzen und handelt so!

Als er so gesprochen, sagten die Zuhörer: „Wir sind von großem Staunen befangen über den glücklichen Zustand, den der Lama in Leib, Wort und Sinn genießt." Als sie darauf fragten, woher derselbe entstanden sei, antwortete er, daß das alles von der Erkenntnis seiner selbst herrühre. Da sagten sie: „Wenn uns nun auch eine solche Glückseligkeit nicht selbst zuteil wird, so könnten wir doch wenigstens hoffen, daß uns ein Teil davon zufallen wird. Geruhe daher, uns eine leicht verständliche und gut zu bewahrende Lehre vorzutragen, in der du die Art und Weise der Meditation vermittelst der Selbsterkenntnis auseinandersetzest." Zur Erwiderung auf ihre Bitte trug der Ehrwürdige ein Lied von den zwölf geistigen Gütern vor:

Dem heiligen Lama zu Füßen verneige ich mich.
Ihr Gabenspender, die ihr der Seele Erkenntnis begehrt,
Wenn ihr sie euch zu Herzen nehmt, so handelt darnach!
Glaube, Klugheit und Güte:
Diese drei sind der Lebensbaum der Seele.
Wenn ihr diesen pflanzt und sorgfältig behütet, seid ihr glücklich;
Macht den Lebensbaum und handelt darnach!
Leidenschaftslosigkeit, Wunschlosigkeit, Freisein von Torheit:
Diese drei sind der Panzer der Seele.
Wenn ihr ihn anlegt, seid ihr stichfest.
Macht diesen Panzer und handelt darnach!
Beschauung, Eifer und Abhärtung:
Diese drei sind der Hengst der Seele.
Wenn er dahineilt und ihr geschwind auf ihm flieht, seid ihr befreit.
Schafft euch diesen Hengst und handelt darnach!
Selbstwissen, Selbsterleuchtung, Selbstglückseligkeit:
Diese drei sind die Frucht der Seele.
Wenn ihr sie angepflanzt habt und sie reif verzehrt, ist sie Nahrung.
Erlangt diese Frucht und handelt darnach!
Diese zwölf geistigen Güter
Entstehen im Geiste des Yogin, der sie empfängt.
Ihr Gabenspender, verschafft sie euch zum Dank für eure gläubige Verehrung.

Als er so gesprochen, wurden jene gläubig und erwiesen ihm später ausgezeichnete Dienste. Dann entschloß sich der Ehrwürdige, auf den Yol mo gangs ra zu gehen. Dies ist der Abschnitt von Rkyang phan namkha rdzong.

V.

Dharmabodhi.

Verehrung dem Meister!

Als der ehrwürdige Milaraspa im Verein mit Ras chung pa und den übrigen Schülern in der Grotte des Dorfes von Gña nam verweilte und die Gebetsmühle, deren Nutzen den Kernpunkt (der Religion) bildet, drehte, gab es gleichzeitig fünf Siddha, nämlich den Guru Tshems chen oben auf dem Passe, den heiligen Sangs rgyas auf dem Ding ri, Shilabharo in Nepal, Dharmabodhi in Indien und Milaraspa in Gña nam. Shilabharo lud Dharmabodhi ein, und als sie die Gebetsmühle drehend in Bal po rdzong verweilten, kamen viele Männer aus Nepal und Tibet dorthin, um Dharmabodhi zu besuchen. Auch die Schüler des Ehrwürdigen hatten den Wunsch, dem Dharmabodhi ihre Aufwartung zu machen. Ras chung pa setzte daher dem Ehrwürdigen mit vielen Gründen auseinander, weshalb es sich empfehlen würde, Dharmabodhi zu treffen. Der Ehrwürdige trug zur Antwort folgendes Lied vor:

Dank dem Segen des Herrn, des Lama, gibt es viele Siddha;
Buddha's wundervolle Lehre breitet sich aus,
Seligkeit erscheint im Wohlbefinden der Wesen,
Durch die Begegnung mit den Siddha sind viele Menschen glücklich,
Und einige Würdige tragen die Vorzeichen an sich, daß sie es noch werden.
Der heilige Sangs rgyas vom Ding ri,
Der Guru Tshems chen auf der Paßhöhe,
Shilabharo aus Nepal,
Der Inder Dharmabodhi,
Milaraspa aus Gung thang,
Eines Jeglichen Geist ist so groß wie ein Berg.
Doch da ich durch meine Beschauung allein auf mich selbst gestellt bin,
wem gehöre ich an?
Wer kennt die Geistesklarheit meines eigenen Geistes?
Zu Verwandlungen und Zaubereien sind alle fähig,
Durch die Leere werden alle zum Mitleid hingerissen,
Jene sind schnell bereit, wunderbare Schauspiele zu zeigen,

Doch ich selbst bin geschickt, siegreiche Lieder zu singen, die meinem Herzen entsprungen sind;
Ich selbst bin groß in Standhaftigkeit und Resignation.
Die andern sind nicht so vortrefflich!
Ich gehe nicht, ihn zu besuchen.
Ihr, meine Söhne, mögt ihn auf jeden Fall besuchen.
Doch es ist nicht eines Andern Schuld,
Nur meines Alters wegen ziehe ich nicht von hier.
In das herrliche Land von Udyana möchte ich pilgern.
Zweifel hege ich nicht, Zuversicht ist da!

Als er so gesprochen, sagten die Männer, welche die Berufung auf das Alter nicht gelten ließen: „Milaraspa will aus Eifersucht nicht gehen." So schmähten sie ihn. Als sie nun meinten, daß es passend sei, jedenfalls hinzugehen, trug er zur Antwort folgendes Lied vor.

An die gebietenden Siddha richte ich mein Gebet!
Mögen durch ihren Segen die Sünden rein werden!
Wenn man unseliges Gerede nach dem Geschwätz der Leute führt
Und Zweifel hegt, täuscht man allein sich selbst.
Wenn zu der Zeit, wo man mit ganzer Seele der Beschauung obliegt,
Viele Wesen zugegen sind, so ist das ein Hindernis.
Wenn zur Zeit des Besuchs bei dem Lama, dem Gebieter,
Viel künstliches Zeremoniell ist, werden die Götterfreunde gestört.
Wenn man an dem Wege der rechten Mittel zu den tiefen Geheimzaubersprüchen
Zweifelt, ist man zur Vollendung nicht befähigt.
Obwohl der Segen der Siddha groß ist,
Herrscht Unwille, wenn die Schar des Gefolges zahlreich ist.
Du, mein Sohn Ras chung, brüderlicher Freund, geh jedenfalls hin!

Als er so gesprochen, sagte Ras chung pa: „Viele Männer, die doch Sünden ansammeln, gehen sicher zu ihm hin, so wird es auch segensreich für uns sein; deshalb wollen wir zur Uebung der Religion zu Dharmabodhi gehen." Ras chung pa und die übrigen Schüler sagten freudig: „Wenn der Ehrwürdige hingeht, mag er auf indisches

Gold rechnen; wenn er Gold gewonnen hat, wird ihm die Reise Ruhm bringen." Als Antwort darauf trug er folgendes Lied vor:

An die gebietenden Siddha richte ich mein Gebet!
Mögen durch ihren Segen die Wünsche des Bettlers erschöpft sein!
Möchte das rechte Maß des Tuns zum Gesetz werden!
Wenn das rechte Maß des Tuns nicht zum Gesetz wird,
Wäre das Verständnis vom Wesen der Beschauung des reinen Sinnes gering.
Die Mitwirkung von Freunden zur Erzeugung der Beschauung und Betrachtung begehre ich nicht;
Wenn ich Freunde begehrte zur Mitwirkung bei der Entstehung der Selbstbefreiung durch die Beschauung,
Wäre mein Verständnis vom Wesen der ununterbrochenen Beschauung wahrlich gering.
Milaraspa will keinen Reichtum gewinnen;
Wenn Milaraspa Reichtum gewänne,
Wäre sein Verständnis vom Wesen des Weltverzichts wahrlich gering.
Dharmabodhi's Gold begehre ich nicht;
Wenn ich Dharmabodhi's Gold begehrte,
Wäre mein Verständnis vom Wesen der Heiligkeit wahrlich gering.
Ras chung, nach dem Gewinn köstlichen Ruhmes verlange ich nicht;
Wenn ich nach dem Gewinn köstlichen Ruhmes verlangte,
Wäre mein Verständnis vom Wesen des Vertrauens auf den Lama wahrlich gering.

Als er so gesprochen, fügte er hinzu: „Geht ihr selbst zuvor, ich werde später kommen." So entsandte er seine Schüler zuerst. Indem sie unrichtige Zweifel hegten, ob der Ehrwürdige wohl kommen würde oder nicht, gelangten sie endlich nach Bal po rdzong. Der Ehrwürdige verwandelte seinen Leib in ein kristallenes Caitya und ging wie eine Sternschnuppe unbehindert am Himmel hin. Als Dharmabodhi, der unter seine Schüler getreten war, ihn erblickte, wurde er von Staunen ergriffen, seine Schüler nährten Zweifel. Doch als der Ehrwürdige vom Himmel herabstieg, freuten sie sich außerordentlich. Darauf traten der Lehrer und die Schüler vereint vor Dharmabodhi hin. Der Inder Dharmabodhi stieg von seinem Sessel herab und verneigte sich vor dem Tibeter Milaraspa. Alle glaubten, daß Milaraspa besser sei als Dharmabodhi. Gegenüber diesen beiden konnten sich alle Versammelten von der Vorstellung Buddha's nicht losreißen. Darauf ließen sich beide Siddha auf einen und denselben Sessel nieder und führten einander erfreuende Gespräche. Dharmabodhi sagte zu dem Ehrwürdigen: „Du, der allein dahinlebt und sich selbst Befriedigung gewährt, sei für dein Kommen bedankt!" Der Ehrwürdige trug zur Antwort folgendes Lied vor:

Zum Nirmanakaya des Lama sende ich mein Flehen!
Mögen die der Ueberlieferung anhängenden Siddha gesegnet sein!
Der Inder Dharmabodhi ist der erste hier
Unter der hier versammelten Menge würdiger Nepaler und Tibeter;
Ich, der tibetische Yogin Milaraspa,
Singe ein Lied der Wonne und Weisheit;
Ohne ein Lied der Wonne und Weisheit
Ließe ich es an der schuldigen Rücksicht gegen den Nirmanakaya der Siddha fehlen.
Die fünf krummen Adern werden durch den Wind gerade gemacht,
Die fünf widrigen Winde werden an ihrer Stelle vernichtet,
Die fünf unreinen Niederschläge des Körpers werden im Feuer verbrannt.
Der Stamm der fünf Sündengifte der Seele wird zu Boden gerissen:
Der Zweifel, der aus diesem Feinde kommende Wind, wird in der mittleren Ader vernichtet.
Der den schädigenden Feind bezwingende Held
Sollte gegen schlechte Freunde nicht ergeben sein.

Als er so gesprochen, sagte Dharmabodhi für das Lied von der Uebung der Meditation des Yogin seinen Dank. Der Ehrwürdige sprach zu Dharmabodhi: „Geruhe, mir die tiefen Hauptpunkte deiner Merkverse vorzutragen." Da trug Dharmabodhi folgendes Lied vor:

An den, der die Ausführung der trefflichen Beschauung lehrt,
Richte ich mein Gebet in dieser würdigen Versammlung.
Möge er durch die gute Wirkung der Werke
Uns segnen, daß wir uns bald wieder treffen!
Wenn der Hang zur Skepsis nicht zerstört wird,
Wozu nützt dann die Beschauung des Gemüts?

Wenn das Verlangen nach dem Glück der Eigensucht nicht gemieden wird,
Was nützt dann langdauernde Meditation?
Wenn man sich nicht um das Heil der Wesen bemüht,
Was nützt dann stolzer Wandel?
Wenn man auf des Lama Wort nicht hört,
Was nützt dann das festlich versammelte Gefolge?
Das ist die Ursache von Trug, Täuschung und Scham.
Wenn man als Frucht nicht die Uneigennützigkeit erwirbt,
Wird die unvergleichliche Bodhi nicht erreicht.
Doch durch die Darlegung deiner eigenen Umstände trifft dich Schaden;
Wenn man Krieg beginnt, ist große Ursache zum Verfall da.
Schweigend dasitzen ist eine tiefe Lehre.
Wohlklingend ist dein Gesang, tibetischer Yogin!
Mein Gesang aber ist nicht wohlklingend;
Im Drang der Freude nur habe ich gesungen.
Der du in den an Pracht und Glück reichen Gefilden
Lieder zu singen froh bist, besuche mich bald wieder!

Als er so gesprochen, führten sie noch weiter viele erfreuende Gespräche. Dann ging der Inder Dharmabodhi weiter, der Tibeter Milaraspa, der Lehrer samt seinen Schülern, zog aufwärts. Die Zuhörer von Gña nam boten dem ehrwürdigen Meister den Willkommenstrunk und fragten ihn nach den Umständen seines Besuches. Zur Antwort trug der Ehrwürdige folgendes Lied vor:

Durch die Begegnung von Sonne und Mond am Himmel werden die vier Kontinente erleuchtet,
Durch das Zusammentreffen von Mutter und Sohn wird die Familie beruhigt,
Durch die Begegnung von Wärme und Feuchtigkeit reift das saftige Obst,
Durch die Begegnung der Siddha wird das Weltreich beglückt.
In dem Wald von Bal po rdzong
War Dharmabodhi angelangt;
Milaraspa ging hin, ihn zu besuchen;
Er, Dharmabodhi, erhob sich
Und verneigte sich vor mir, Mila.

Die versammelten Männer hegten Zweifel.
Er mit der Mudra des Täuschungsleibes
Faltete in der Meditationsstufe des Zung ·jug die Hände und kniete nieder
Und wirkte im Dharmadhatu, der frei von Leiden und Flecken ist.
Noch einmal vollzog er die Mahamudra,
Dann führten wir aufrichtig in der reinen Einsamkeit
Klare unvergeßliche Gespräche.
Das ist die Wirkung trefflichen Gebets!
Sicherlich hat er schon früher zur Zeit des Jina gelebt.
Jene schöne Begegnung der brüderlichen Freunde
Gereicht heute dem ganzen Weltreich zum Ruhme.

Als er so gesprochen, wurden die Zuhörer von hoher Freude und Verwunderung ergriffen. Infolge der ihm von Dharmabodhi bezeigten Verehrung mehrte sich der Ruf und das Wohlbefinden des Ehrwürdigen.

Der Abschnitt vom Besuch bei Dharmabodhi.

VI.

Milaraspa, der Holzsammler.

Verehrung dem Meister!

Als der Gebieter der Yogin, der ehrwürdige Milaraspa, in der Art und Weise der Mahamudra prophetischer Erleuchtung in "Chong lung Khyung gi rdzong verweilte, geschah es einst, daß, als er sich erhob, um Vorbereitungen für seine Mahlzeit zu treffen, gar nichts vorhanden war, weder Mehl noch Salz noch Wasser noch Gewürz, ja nicht einmal Holz in der Vorratskammer noch Wasser und Feuer auf dem Herde. Da dachte er: „Mein Tun bedeutet doch wohl eine zu große Vernachlässigung meiner Person; ich muß doch Holz sammeln gehen." Er machte sich auf, und nachdem er so viel Holz gesammelt, als er gefunden hatte, erhob sich plötzlich ein großer Sturm, der, wenn er sein Baumwollentuch faßte, das Holz, und wenn er das Holz faßte, sein Baumwollentuch zu zerreißen drohte. Endlich kam ihm der Gedanke: „Obwohl ich früher in dieser Bergwildnis schon so lange verweilt habe, habe ich dennoch die Selbstsucht noch nicht aufgegeben. Was soll ich tun, um die Gewohnheit des Haftens an der Selbstsucht zu überwinden?" „Wünschst du das Tuch", so rief er, „nimm das Tuch weg; wünschst du das Holz, so nimm das Holz!" So gab er beides dahin und setzte sich nieder.

Unter dem Einfluß der schlechten Ernährung und infolge des kalten Windes sank er eine Zeit lang in Bewußtlosigkeit. Als er sich wieder erhob, hatte sich der Sturm gelegt, und sein Tuch hing flatternd an der Spitze eines Baumzweiges. Da regten sich in seiner Seele Gedanken des Ueberdrusses; er ließ sich eine Zeit lang auf einen Felsblock von der Größe eines Schafleibes nieder, bis er das Gleichgewicht des Gemütes wiedergewonnen.

Im Osten zog in der Richtung von Grobolung her am Himmel schwebend eine weiße Wolke herauf. „Unter dieser Wolke", dachte er, „liegt das Kloster Grobolung: dort hat mein Lama, der sprachgewandte Marpa, der Gelehrte, gelebt." Er erinnerte sich an die im Kreise des Lama, der ihm Vater und Mutter war, seiner geistlichen Brüder und Vajra-geschwister empfangenen kraftvollen Erklärungen der Tantra und an die Lehren in der Fingerstellung der Meditation. Er dachte: „Was mir auch begegnen möge, wenn er jetzt dort weilte, ich will ihm meine Hochachtung bezeigen." Da überwand er trotz seiner früheren Mißstimmung durch die lebhafte Erinnerung an den Lama das Uebermaß von Verdruß: ein Tränenstrom stürzte ihm aus den Augen, und in der Erinnerung an den Lama stimmte er folgendes sechsteilige Lied an, einen Sang wehmutsvoller Sehnsucht.

In der Erinnerung an den Vater, zu den Füßen des den Sehnsuchtschmerz stillenden Marpa, erhob er einen flehenden Sehnsuchtsang.

Marpa, Gebieter!
Im Osten des Rotenfels «Chong lung
Wogt eine Wasser bergende weiße Wolke.
Unter der schwebenden weißen Wolke,
Vor dem Berge hinter mir, der sich wie ein Elefant gähnend streckt,
Auf dem Berge vor mir, der sich wie ein großer Löwe gähnend streckt,
Liegt der heilige Ort, das Kloster von Grobolung.
In diesem steht ein Thron aus dem großen Steine Amolika;
Auf dem Polstersitz liegt ein Einhornfell;
Wer saß darauf, wer saß nicht darauf?
Darauf saß der sprachenbewanderte Marpa!
Wenn er jetzt dort weilte, wie freute ich mich!
Wie schwach auch meine Andachtsglut, mein Herz begehrte zu ihm zu eilen!
Wie klein auch meine Sehnsucht, mein Herz begehrte zu ihm zu eilen!
In meinen Gedanken erinnere ich mich des herrlichen Lama,
In meinen Betrachtungen erinnere ich mich an Marpa, den Gelehrten.
Er ist der die Mütter übertreffenden unverehelichten Frau gleich.
Wenn er jetzt dort weilte, wie freute ich mich!
Wenn der Umweg noch so weit, mein Herz begehrte zu ihm zu eilen!
Wenn der Weg auch mühsam zu wandeln wäre, mein Herz begehrte doch
zu ihm zu eilen!
In meinen Gedanken erinnere ich mich des herrlichen Lama,
In meinen Betrachtungen erinnere ich mich an Marpa, den Gelehrten.
Er ist der das Herz tief erfreuende Edelstein.
Wenn er jetzt dort weilte, wie freute ich mich!
Wenn auch mein Wissen arm, mein Herz begehrte es zu fassen!
Wenn auch mein Verstand gering, mein Herz begehrte des Lehrers Worte
zu wiederholen!
In meinen Gedanken erinnere ich mich des herrlichen Lama,
In meinen Betrachtungen erinnere ich mich an Marpa, den Gelehrten.
Wenn er die vier Kräfte der Worte mündlicher Unterweisung
Jetzt überlieferte, wie freute ich mich!

Obwohl meine Gaben nur klein sind, begehrte mein Herz zu bitten;
Obwohl ich nicht über Opferspenden verfüge, begehrte mein Herz zu bitten.
Wenn er die tiefe Belehrung, Naro's sechs Vorschriften
Jetzt verkündete, wie freute ich mich!
Wiewohl meine Ausdauer nur schwach ist, begehrte mein Herz zu bitten;
Wiewohl der Faden meiner Beschauung nur klein ist, begehrte mein Herz Beschauung.
In meinen Gedanken erinnere ich mich des herrlichen Lama,
In meinen Betrachtungen erinnere ich mich an Marpa, den Gelehrten.
Wenn er, sich in den reinen Kreis der geistlichen Brüder gesellend,
Jetzt dort weilte, wie freute ich mich!
Wie schlecht auch meine Erkenntnis, mein Herz begehrte doch sie zu sammeln;
Wie schwach auch mein Verständnis, mein Herz begehrte doch sie zu sammeln.
In meinen Gedanken erinnere ich mich des herrlichen Lama,
In meinen Betrachtungen erinnere ich mich an Marpa, den Gelehrten.
Obwohl mein Wesen von inbrünstiger Andacht unzertrennlich,
Ist durch die Erinnerung an des Lama Herz,
Durch das Verlangen und die Sehnsuchtqual meine Geduld erschöpft.
Der Atem stockt in meiner Brust, daß mir die Stimme versagt.
Stille deines Sohnes Sehnsuchtqual, du Gnadenreicher!

Als er so gesprochen, erschien auf der Spitze jener Wolke, welche die Gestalt eines fünffarbigen, ausgebreiteten Seidengewandes angenommen hatte, der erhabene Marpa selbst, in herrlichem Glanze wie nie zuvor, auf einer reich geschmückten weißen Löwin reitend, und sprach: „Du mein älterer Bruder, sprich, welcher heftige Schmerz ist dir widerfahren, daß du mich rufst? Verzweifelst du an dem Lama, den Schutzgottheiten und dem höchsten Gut? Spürst du den Dingen mit unheilvollem Skrupel nach? Oder haben sich in deiner Betklause Hindernisse durch die Lehren der Welt eingestellt? Bist du vom Dämon der Furcht und Hoffnung gequält und mißmutig gemacht? So sei denn wieder oben der Lama und des höchsten Gutes Diener! Bring unter den Wesen der sechs Klassen Gaben dar! Dann werden dir selbst im Zwischenraum die Sündenflecken rein und gute Gelegenheiten zur Erzeugung von Vorzügen zu teil werden. Oder was es auch sein mag, ich werde mich nicht von dir trennen. Indem du durch beschauliches Bannen lehrst, stifte den Wesen Nutzen!“ Diese Worte flößten ihm Mut ein, und in großer Freude sang er zum Dank für die Ermutigung folgendes Lied:

Da ich des väterlichen Lama Antlitz geschaut, seine Worte gehört habe,
Hat sich des Bettlers Herzenskummer aufgerichtet.
In der Erinnerung an des Lama Lebensgeschichte
Ist beschauliche Andacht in der Seele Tiefen erwacht.
Des Erbarmers Segen hat sich verkörpert,
Unfromme Bilder sind alle entrückt.
Dieser Sehnsuchtsang der Erinnerung an den Lama
Erklingt zwar in des Ehrwürdigen Ohren,
Doch dem Bettler bleibt nichts als die bloße Vorstellung davon.
Noch einmal will ich versuchen: Erbarmer, schütze mich!
Der ausharrende, gegen Ungemach abgehärtete Asket hier
Ist des väterlichen Lama freudiger Diener,
Ist der allein in der Bergwildnis umherwandelnden
Daka freudiger Diener,
Ist der dünkellosen heiligen Religion,
Der Buddha-Lehre, Diener,
Gleichgültigkeit gegen Leben und Streben
Bedeutet Gabenspenden für die schutzlosen Wesen;
Freudiges Ausharren im Wechsel von Krankheit, Alter und Tod
Erzeugt Reinigung von der Sündenbefleckung durch die Werke;
Sündige Speisen meidende Askese
Fördert die Entstehung guter Erkenntnis.
Des väterlichen Lama Wohltaten vergelte ich durch beschauliche
Bannungen;
Deinen Sohn schütze erbarmungsvoll, Lehrer und Herr!
Segne des Bettlers Wahl der Bergeinsamkeit!

So sprach er. Von großer Freude erfüllt, ergriff er sein Baumwollgewand und kehrte mit einer Hand voll Holz in seine Betklause zurück. Da waren in der Hütte fünf eiserne Kobolde (Atsara) mit Augen wie Gefäße hervorquillend. Einer saß auf des Ehrwürdigen

Lagerstätte und trug aus einem geistlichen Buche vor, zwei lauschten seinem Wort; der vierte bereitete Essen; der fünfte war damit beschäftigt, die übrigen Bücher zu durchstöbern. Anfangs überlief ihn doch ein kleiner Schauder. Dann kam ihm der Gedanke: „Das ist das Zauberspiel, an dem die Ortsgöttin ihre Freude hat; an welchem Orte ich auch weilen mag, ein Streuopfer brauche ich nicht darzubringen. Doch da ich die edle Herrin noch nicht gepriesen habe, muß ich wohl ein Loblied auf diesen Ort verfassen.“ Damit stimmte er folgenden Lobgesang auf den Ort an:

O du Einsiedelei in der Bergeinsamkeit,
Stätte, wo die herrlichen Jina die Bodhi erlangen,
Gefilde, wo die heiligen Männer weilen,
Ort, wo ich der einzige Mensch jetzt bin!
Rotfels ₒChong lung, Adlerhorst,
Ueber dir ballen sich des Südens Wolken,
Unten schlängeln Flüsse sich in schnellem Lauf,
In der Luft schwebt der Geier kreisend;
Die artreichen Waldbäume säuseln,
Prachtbäume wiegen sich nach Tänzerart;
Bienen summen ihr Liedchen khorroro,
Blumen strömen Duft aus chillili;
Vögel zwitschern wohllautend kyurruru;
Auf diesem Rotfels ₒChong lung
Ueben Vögel und Vöglein des Fittichs Behendigkeit,
Ueben Affen und Aefflein sich im Wettsprung,
Ueben Hirsch und Reh sich im Wettlauf;
Ich, Milaraspa, übe geistige Geschicklichkeit,
Geistige Geschicklichkeit und innere Heiligkeit übe ich:
Ich bin mit der Ortsgottheit der Einsiedelei in friedlicher Eintracht.
Gespenstige Unholde, die ihr hier versammelt seid,
Trinkt den Saft der Liebe und des Erbarmens
Und weicht, jeder an seinen Ort, von hinnen!

So sang er. Die Kobolde aber, aus Aerger über den Ehrwürdigen, rollten zornfunkelnd die Augen hin und her. Die Kobolde vermehrten sich um zwei, und als sie so sieben geworden, fletschten einige wutschnaubend die Zähne, andere knirschten und wetzten die Hauer, andere lachten und brüllten in grauenvollen Tönen. Dann stürzten alle, zu einem Knäuel geballt, in Kampfbereitschaft mit drohenden Geberden auf ihn zu. Da sann er, den Unholden Unheil zu bereiten. Mit magischem Zornesblick rezitierte er schreckliche Beschwörungen, aber sie gingen dennoch nicht. Voll von großem in seinem Herzen erwachenden Erbarmen trug er ihnen geistliche Lehren vor, dennoch bezeigten sie keine Lust zu weichen. Da dachte er: „Mir hat zwar Marpa aus Lho brag alle Lehren aus seinem Innersten erklärt; doch da mein Innerstes der Erleuchtung bar ist, werde ich von Verzweiflung ergriffen. An die Wirklichkeit der Kobolde und Dämonen muß ich wohl glauben; sie könnten mich daher töten, und ihnen diese Freude zu bereiten, wäre nutzlos." Indem sich seine unerschütterliche Zuversicht so offenbarte, stimmte er folgendes beschauliche und zuversichtliche Lied an:

Vor dem Vater, dem Sieger über das Heer der vier Mara,
Dem sprachgewandten Marpa zu Füßen verneige ich mich!
Meinen Namen nenne ich nicht.
Ich bin der starken weißen Löwin Sohn:
Im Mutterleibe besaß ich die drei Fertigkeiten vollendet;
In den Kinderjahren lag ich im Neste,
In den Jahren der Jugendblüte hütete ich des Nestes Tür,
In den Jahren der Mannheit wanderte ich über die Gletscher.
Wenn auch Schneesturm die Gletscher umwirbelt, bebe ich nicht;
Ist der Felsenabgrund auch groß, ich fürchte nicht.
Meinen Namen nenne ich nicht.
Ich bin des Königs der Vögel, des Adlers Sohn:
Im Inneren des Eies entwickelten sich die Fittiche;
In den Kinderjahren lag ich im Neste,
In den Jahren der Jugendblüte hütete ich des Nestes Tür,
In den Mannheitsjahren des großen Aars kreuzte ich des Himmels Höhe.
War der Himmel auch groß, ich bebte nicht;
Wenn die Täler der Erde auch dunkelten, ich fürchtete nicht.
Meinen Namen nenne ich nicht.
Ich bin der Sohn der Woge, des großen Fisches:
Im Mutterleibe rollte ich meine goldenen Augen;
In den Kindheitsjahren lag ich im Neste,
In den Jahren der Jugendblüte zog ich an der Spitze der wimmelnden Schar,

In den Mannheitsjahren des großen Fisches umkreiste ich die Grenzen des Meeres.
Waren auch schrecklich die Meereswogen, ich bebte nicht;
Waren der Fischer Eisenhaken auch zahlreich, ich fürchtete nicht.
Meinen Namen nenne ich nicht.
Ich bin der Sohn des Lama der mündlichen Ueberlieferung:
Im Mutterschoße erwuchs mir der Glaube,
In den Kindheitsjahren trat ich in das Tor der Religion ein,
In den Jahren der Jugendblüte widmete ich mich dem Studium,
In den Mannheitsjahren des großen Mystikers durchwanderte ich die Bergwildnis.
Ist auch der Geister Bosheit groß, ich bebe nicht;
Wie oft auch der Dämonen Gaukelspiel wiederkehrt, ich fürchte nicht.
Der auf dem Gletscher sich reckende Löwe friert nicht an den Pfoten:
Wenn der Löwe auf dem Gletscher an den Pfoten fröre,
Wäre der Nutzen der drei vollendeten Fertigkeiten wahrlich gering!
Der am Himmel fliegende Aar kann nicht herabstürzen:
Wenn der große Aar vom Himmel herabstürzen könnte,
Wäre der Nutzen der entwickelten Flugkraft wahrlich gering!
Der im Wasser schwimmende Fisch kann nicht ertrinken:
Wenn der große Fisch im Wasser ertrinken könnte,
Wäre der Nutzen der Geburt im Wasser wahrlich gering!
Ein eiserner Felsblock wird durch Steine nicht zerstört:
Wenn ein eiserner Felsblock durch Steine zerstört würde,
Wäre der Nutzen des Schmelzens wahrlich gering!
Ich, Milaraspa, fürchte keine Geister:
Wenn Milaraspa Geister fürchtete,
Wäre der Nutzen von der Kenntnis des Wesens der Dinge wahrlich gering!
Ihr, die ihr hier erschienen seid, Geister und Kobolde, Scharen der Widersacher,
Vor allem meinen Dank für eueren Besuch!
Wenn ihr keine Eile habt, weilet nur immer hier!
Ergeht euch nur in ausführlichen Reden und Diskursen!

Wenn ihr's auch eilig habt, bleibt unter allen Umständen heute Abend hier!
Im Wettstreit mit der Gewandtheit von Leib, Wort und Gedanke
Wollen wir einmal den Wert der weißen und schwarzen Lehre messen!
Ihr wollt nicht zurückkehren, ohne mir einen Streich zu spielen!
Doch wenn ihr, ohne mir einen Streich zu spielen, zurückkehren werdet,
Müßt ihr euch schämen, noch einmal wiederzukommen!

Mit diesen Worten erhob er sich im Stolz seiner Beschauung und ging ohne Zaudern hinein. Da wurden die Kobolde von Furcht und Schrecken ergriffen und blickten mit scheuen Augen um sich. Sie zitterten so am ganzen Leibe, daß die ganze Höhle im Innern zitterte und bebte.

Schleunigst schwanden die Sieben zu Einem zusammen, und auch dieser Eine fuhr schließlich in einem kreisenden Wirbelwind von dannen. Der Ehrwürdige dachte: „Der Dämonenkönig Vinayaka trachtet nach einer Gelegenheit, mir zu schaden; der Sturm vorher wird wohl auch sein Zauberwerk gewesen sein. Aber dank dem Erbarmen meines Lama wird er weiter keine Gelegenheit mehr finden." Darauf wurde ihm eine unermeßliche Förderung in der Meditation zu teil.

Dieser Abschnitt heißt: Des Dämonenkönigs Vinayaka Spuk und die glühende Sehnsucht nach dem Lama, oder der Rotenfels oChong lung, oder Mila, der Holzsammler, auch kurz der Abschnitt mit drei Namen.

VII.

Milaraspa am Flusse Chu bzang.

Nachdem der Gebieter der Yogin, der ehrwürdige Milaraspa, auf dem Felsen ₒChong lung verweilt hatte, überschritt er, um das Gebot seines Lama zu erfüllen, den Paß von Thong und machte sich auf den Weg, um sich auf dem gleichsam eingezäunten Schneeberg La phyi der Meditation hinzugeben. Als er im Dorfe Rtsar ma in Gña nang am Aufgang zum La phyi anlangte, da waren gerade die Männer von Rtsar ma zu einem großen Gelage in ihrer Bierschenke versammelt und ergingen sich in Gesprächen: „Da ist doch jetzt ein gewisser Milaraspa, der das Leben eines Büßers führt; ganz allein haust er dort nach den Vorschriften just in der menschenleeren Bergwildnis." So unterhielten sie sich in Lobessprüchen über den Ehrwürdigen. In diesem Augenblick trat der Ehrwürdige an ihre Tür. Eine junge, wohlgebildete und schön geputzte Frau, mit Namen Legs se ₒbum, trat gerade heraus und sagte: „Heiliger, woher kommst du?" Der Ehrwürdige erwiderte: „Ich bin der hier irgendwo im Gebirge hausende große Mystiker Milaraspa. Mildtätige, du solltest mir ein wenig Speise und Trank reichen, da ich dich darum bitte." Die Frau sprach: „Du magst Nahrung vorgesetzt erhalten. Bist du denn wirklich jener Milaraspa oder sagst du es bloß?" Der Ehrwürdige antwortete: „Welchen Vorteil hätte es, zu lügen?" In freudig aufgeregter Hast eilte die Frau hinein und sagte den Biergästen: „Jener heilige Mönch, von dem ihr vorhin gesagt habt, daß er weit entfernt wohne, ist hier angekommen und steht jetzt vor unserer Tür." Alle stürzten hinaus. Einige brachten ihm sogleich Verehrung dar, andere forschten ihn zuerst umständlich über seine Geschichte aus, und als sie die Ueberzeugung gewonnen, daß es wirklich der Ehrwürdige sei, führten sie ihn hinein und erwiesen ihm ausgezeichnete Ehrenbezeugungen. An der Spitze des Tisches befand sich die Wirtin, eine wohlhabende junge Frau mit Namen Gsen rdor mo, die den Ehrwürdigen mit aller Sorgfalt bediente und ihn schließlich fragte: „Wohin begibt sich jetzt der Lama?" „Ich gehe zu beschaulicher Uebung auf den La phyi", sagte er. Sie sprach: „Du solltest doch erst ein wenig in deinem ₒDre lung skyog mo bleiben und den Erdboden dort segnen; wir werden unermüdlich zu deinen Diensten sein." Darauf sagte ein gerade dort anwesender Lehrer: „Gabenspender und Gabenempfänger mögen sich einigen! Der La phyi ist die Bezeichnung von ₒDre lung skyog mo. Wenn nun der Lama dort verweilen wird, werde ich ihn nach Kräften mit allem Nötigen versehen und dafür religiöse Belehrung erbitten."

Der Sprecher war der Lehrer Çakyaguna. Darauf sagte die Wirtin: „Das ist wahrhaftig vortrefflich! Dir erscheint die Einsamkeit sogar als Genuß, während wir andere uns vor den Streichen der Kobolde schrecklich fürchten und gar keine Lust haben, uns in der Bergeinsamkeit anzusiedeln, vielmehr lieber schnell hinabsteigen." Die ganze Versammlung bezeigte ihm Verehrung. Da sprach der Ehrwürdige: „Ich muß bald gehen, aber der Einsamkeit wegen, wie du sagst, gehe ich nicht; ich habe nämlich ein Gebot von meinem Lama erhalten, und um dieses zu erfüllen, muß ich gehen." Sie erwiderten: „An Geschenken soll es dir nicht fehlen. Wir werden dir mit guten Nahrungsmitteln zu Diensten sein." Der Ehrwürdige sprach: „Ich habe in der Bergwildnis keinen Freund, keine gute Nahrung noch einen zuverlässigen Menschen, doch fürs erste will ich selbst allein gehen; für euere Dienste danke ich euch, später werden wir das Weitere sehen." Da machte sich der Ehrwürdige selbst zum Schneeberge La phyi auf.

Als er die Paßhöhe überstiegen, entstand ein schrecklicher Gespensterspuk. In dem Augenblick, als er den Berggipfel erreichte, umwölkte sich der Himmel. Furchtbarer Donner krachte, Blitze zuckten rings umher, Berge zu beiden Seiten der Täler zerbröckelten, und Gießbäche stürzten wirbelnd zusammen und bildeten einen großen See mit furchtbar tosendem Strudel. Der Ehrwürdige entsandte seinen magischen Blick und schlug mit seinem Stabe auf. Da floß der See auf dem Boden ab und verschwand. Er ist bekannt als der „Teufelsteich" (rmu rdzing).

Als er darauf ein wenig weiter nach unten kam, schleuderten die Geister zu beiden Seiten des Berges in einer Staubwolke viele Felsblöcke wie tosende Wogen hinab. Da erschienen zu beiden Seiten des Tales die Daka und machten gleich herabschießenden Bergschlangen den Weg frei. So hörte denn das Trümmergeröll auf. Dieser Weg ist bekannt als der Pfad „des Felsvorsprungs der Daka".

Darauf ließen die Geister von schwacher Kraft von selbst ab; die stärkeren indessen suchten, obwohl sie keine Gelegenheit fanden, noch einmal nach einer Gelegenheit, ihm beizukommen. Da wurden sie gerade an der Stelle, wo der Pfad des Felsvorsprungs der Daka endete, durch den die Anfechtungen besiegenden Zauberblick des Ehrwürdigen zurückgeschlagen. Nun war der ganze Spuk zu Ende. An jener Stelle blieb in einem Steine seine Fußspur zurück.

Darauf schritt er eine Strecke weiter. Der Himmel heiterte sich auf, und in froher Stimmung ließ er sich auf einem Felsvorsprung nieder. Dort versenkte er sich in eine Betrachtung der Liebe zu allen Geschöpfen, wodurch ihm reicher Gewinn an innerer Förderung zu teil ward. Davon heißt dieser Platz der „Felsvorsprung der Liebe".

Darauf kam er zu dem Flusse Chu bzang (gutes Wasser). Als er sich in die Betrachtung des strömenden Wassers versenkte, ging der 10. Tag des ersten Herbstmonats des Feuer-Tiger-Jahres (10. September 1086) zur Rüste.

Da erschien, von dem großen nepalesischen Dämonenfürsten Bharo angeführt, ein ganzes Heer von Geistern und erfüllte Fluß, Tal, Himmel und Erde. Sie schleuderten auf den Ehrwürdigen einen schrecklichen Regen von Felstrümmern, Donnerkeilen und anderen Waffen herab, riefen ihn beim Namen und stießen ein Geschrei wie: fange! töte! und andere gräßliche Worte aus. Auch ließen sie viele häßliche Gestalten erscheinen. Im Gedanken an die Nachstellungen der Geister trug er ihnen die wahrhaftige Lehre von der Frucht der Taten in folgendem Gesange vor:

Verehrung sei dem Lama!
Dir, huldreicher Herr, befehle ich mich!
Indem das Auge die Illusionen gewahrt,
Entstehen die Gaukelspiele der männlichen und weiblichen Dämonen und Yaksha.
Ach, ihr armseliges Hungervolk der Hölle,
Mir könnt ihr keinen Schaden zufügen!
Die angesammelte Frucht euerer früheren schlechten Taten
Ist jetzt durch die gereifte Vergeltung in euerem Körper wirksam.
In der Form der in der Luft wandelnden Seelen
Tun sie durch die Erregung der Sünden schlechter Gedanken
Und durch das scharfe Gift von Leib und Wort
Nichts anderes als töten, schneiden, schlagen und stechen, sagt man.
Einfachen Herzens wandle der Einsiedler Ras pa,
Einfältigen Sinnes, voll Zuversicht auf seine Beschauung,
Dem Wandel des tapferen Löwen gleich!
Der Leib sehnt sich nach dem Hort des Götterbildes,
Die Rede sehnt sich nach dem Hort des Zaubers,
Das Herz sehnt sich nach dem Hort der Erleuchtung.
Die sechs Arten der Sinneswahrnehmung preise ich.
Mich, der ein solcher Yogin ist,
Kann euer, der Preta, Höhnen nicht verletzen.
Da die Belohnung der Tugenden und Sünden wahr ist,
Sammeln die Wesen Vergeltung durch Mißgeschick
Und verzweifeln in der Hölle.
Ach, ihr Sündenvolk der Hölle,
Armselige, die ihr Sinn und Wesen der Dinge nicht wißt!
Ich, der abgemagerte Milaraspa hier,
Habe die Religion mit einem Liede belehrender Rede erklärt,

Das den durch das Lebensprinzip geeinigten Wesen
Gleichsam ein elterlicher Beistand geworden
Und sie kraft seiner Wohltätigkeit vor Unglück bewahrt.
Wendet euch nun ab von der Bosheit!
Wenn ihr über die Vergeltung der Werke nachdächtet, wäre vortrefflich;
Wenn ihr die zehn Tugenden der Religion übtet, wäre vortrefflich.
Versteht mein Wort gut und prüft es mit Ueberlegung!
Höret noch einmal, erfaßt seinen Sinn!

So sang er. Doch die Dämonenheere riefen: „Mit deiner Zungenfertigkeit kannst du uns nicht den Kopf verdrehen; wir sind nicht gekommen, um unsere Magie einzustellen und dich hübsch in Ruhe zu lassen." Ihre Heere wurden zahlreich und der Spuk mehrte sich. Der Ehrwürdige überdachte die Sache und rief den Dämonenheeren zu: „Der dank der Gnade seines Lama das Wesen der Dinge durchschauende Yogin empfindet den Teufelsspuk der Widersacher als einen Segen für seine Seele. Meinetwegen treibt es noch schlimmer! Ich verharre hier mit der höchsten Bodhi vereinigt." Darauf trug er den Sang von den sieben großen Zierden vor:

Zu des Herrn, des sprachgewandten Marpa, Füßen verneige ich mich.
Ich, der das Wesen der Dinge durchschauende Yogin,
Singe ein Lied von den großen Zierden.
Ihr hier versammelten männlichen und weiblichen Schadenstifter (Yaksha),
Leiht mir aufmerksam euer Ohr.
In einem Caitya des in der Mitte der Welt gelegenen herrlichsten Berges
Strahlt im Süden der Lichtglanz des Lasurs:
Dem Himmel von Jambudvipa ist er eine große Zierde.
Auf der Oberfläche des Berges Yugamdhara
Strahlt der Lichtglanz des Sonne-Mond-Paares:
Allen vier Kontinenten ist es eine große Zierde.
Durch den Zauber der Nagabodhi
Fällt vom Himmelsgewölbe der Regen:
Der engen Erde ist er eine große Zierde.
Aus dem Wasserdampf, der aus dem Ozean aufsteigt,
Bilden sich im ganzen Luftraum die Südwolken:
Die Südwolken sind eine große Zierde des Luftraums.

Aus der Verbindung der Wärme und Feuchtigkeit der Elemente
Entsteht im Sommer gegenüber den Almen der buntfarbige Regenbogen:
Der Regenbogen ist eine große Zierde der Almen.
Durch den im Westen aus dem Manasarovara-See hervorströmenden Fluß
Gedeihen die Obstbäume von Jambudvipa im Süden:
Allen Wesen sind sie eine große Zierde.
Mir, dem die Bergeinsamkeit wählenden Klausner,
Sind durch die Kraft der Meditation über die Leere
Die Gaukelspiele männlicher und weiblicher Kobolde und Yaksha entstanden:
Die Gaukelspiele sind eine große Zierde des Klausners.
So hört mir richtig zu, ihr Geister!
Ihr wißt doch wohl, wer ich bin?
Wenn ihr mich noch nicht kennt,
So vernehmt: ich bin der Klausner Milaraspa,
Geboren aus dem Blumenschoß der Liebe!
Mit des Liedes Wohllaut weiß ich zu deuten,
Mit den Worten wahrhaftiger Rede die Lehre zu erklären
Und hilfreichen Sinnes Rat zu erteilen.
Richtet eueren Sinn auf die höchste Erkenntnis,
Und wenn auch ein Nutzen für andere Wesen nicht daraus erwächst,
Wie solltet ihr nicht, wenn ihr die zehn Sünden meidet,
Seelenfrieden und Befreiung erlangen?
Wenn ihr auf meine Worte hört, werdet ihr großen Segen erzielen;
Wenn ihr sogleich die heilige Lehre übt, werdet ihr in Zukunft glücklich sein!

So sang er. Da waren die Geister der Mehrzahl nach schon so weit gebracht, daß sie dem Ehrwürdigen gläubige Verehrung bezeigten und ihr Gaukelspiel einstellten. „Einsiedler, sei bedankt", sprachen sie, „doch wenn wir ohne Erklärung der Natur der Dinge deine Argumente nicht erfassen, müssen wir in unserer Unwissenheit verharren. Da wir dir ja nun gar keinen bösen Streich gespielt haben, so bitten wir, wenn auch der Segen der verkündeten Lehre von der Frucht der früheren Taten groß war, um einige neue kurzgefaßte, nützliche, leichtverständliche und freundlich anziehende Lehren; denn unser Hang zur Schlechtigkeit ist stark, und unser Verstand ist klein und von schwacher Fassungskraft." Auf ihre Bitte hin trug der Ehrwürdige folgendes Lied von den sieben rechten Dingen vor.

Dem sprachgewandten Marpa zu Füßen verneige ich mich!
Gewähre deinen Segen, daß die Bodhi vollendet werde!
Wenn nicht mit Worten wahrhaftigen Sinnes verbunden,
Sind Lied und Stimme, wenn auch noch so wohltönend, nur einer Zither gleich.
Wenn nicht durch Gleichnisse, im Einklang mit der Religion, erläutert,
Sind Worte, wenn auch mit Geschick gesetzt, nur ein leeres Echo.
Wenn die praktische Ausübung der Religion nicht von Herzen kommt,
Ist es, auch wenn man sagt, daß man sie kennt, eine bloße Täuschung.
Wenn man über die Vorschriften der Unterweisung in der Wahrheit nicht nachdenkt,
Ist es, auch wenn man die Bergeinsamkeit wählt, eigenes Mißgeschick.
Wenn man die heilige Religion des wahren Heils nicht übt,
Ist die Magie, auch wenn man darin sehr stark ist, nur ein Unglück.
Wenn man die Vergeltung der Werke nicht genau berechnet,
Sind Worte und Ratschläge, wenn auch noch so groß, nur ein frommer Wunsch.
Wenn man den Sinn der Worte nicht betätigt,
Ist man, auch wenn man im Augenblick viele Worte macht, nur ein Lügner.
Reinigt euch von den Lastern, und sie werden sich allmählich vermindern;
Erstrebt die Tugend, und ihr habt innere Kraft gewonnen.
Richtet euere Gedanken auf einen Punkt und nehmt es euch zu Herzen!
Aus wortreichen Erklärungen entsteht nur geringer Nutzen;
In diesem Sinne nehmt es euch, bitte, zu Herzen!

So sang er. Da zollten sie, die Belehrung nachgesucht, dem Ehrwürdigen gläubige Verehrung, verneigten sich und umkreisten ihn viele Male, dann entfernten sich alle. Nun erschien aber erst jener Fürst Bharo selbst mit einigen Begleitern und ließ noch einmal wie vorher einen Spuk losgehen. Da trug der Ehrwürdige wiederum die Lehre von der Frucht der Taten in einem Gesange vor:

Dem huldreichen Marpa zu Füßen verneige ich mich!
Ihr Geister höret mich noch einmal!
Euer in der Luft wandelnder Leib ist ungehindert;
Böse Gedanken und schlimme Leidenschaften haben in euerem Herzen Wurzel geschlagen;

Mit den Fangzähnen der Sünde stürzt ihr euch auf andere;
Unter der Sünde, anderen Uebles zugefügt zu haben, müßt ihr selbst leiden:
Die Wahrheit von der Vergeltung der Werke geht nicht verloren,
Die Macht der vergeltenden Gerechtigkeit kann euch nicht fahren lassen,
Ihr selbst bereitet euch die Qualen der Verdammnis!
Ach, Aermste! Wenn ihr über den Wahnsinn der Höllengeschöpfe,
Ueber die Verzweiflung, die durch schlechte Werke bewirkt wird,
Wenn ihr über dergleichen selber nachdenkt, erhebt sich der Wind solcher Gedanken.
Ihr habt schon früher Schlechtigkeit auf Schlechtigkeit aufgehäuft,
Und noch immer begehrt euer Herz, Böses zu sammeln:
Gepackt von der Sünde des Schlachtens,
Liebt ihr zur Nahrung Fleich und Blut;
Ihr raubt, als übtet ihr ein Gewerbe, das Leben der Wesen:
Darum werdet ihr unter den sechs Klassen der Wesen in der Gestalt der Preta,
In Sünden wandelnd, zur Hölle fahren.
O laßt ab! Wendet eueren Sinn der Lehre zu!
Wenn ihr ohne Hoffnung und Furcht seid, ist die Glückseligkeit schnell erreicht.

Darauf sprachen jene: „Durch deine uns gegebenen Erklärungen der Lehre sind wir nun gar kenntnisreich und gutbegreifend. Solchem Verständnis entsprechend, sage uns jetzt, wie du dir selbst diese Zuversicht erworben hast.“ Da trug der Ehrwürdige folgenden zuversichtsvollen Sang vor:

Zu den Füßen des vortrefflichen Marpa verneige ich mich!
Da ich, der das Wesen der Dinge durchschauende Klausner,
Darauf vertraue, daß es keinen Ursprung und kein Entstehen gibt,
Bin ich durch die Stufen der den Weg freimachenden Fertigkeiten vollendet.
Die rechten Mittel mit großem Erbarmen erläuternd,
Singe ich, von den Lehren der Wahrheit selbst angespornt, ein Lied.
Da ihr, mit Sünden böser Taten dicht bedeckt,
Der mystischen Erkenntnis Wesenheit nicht begreift,

Soll euch die Wahrheit der Schrift noch einmal erklärt werden.
Einstens hat der allwissende Buddha
In den Sutra und Tantra des fleckenlosen Wortes
Die Lehre von der Vergeltung der Werke eifrig verkündet:
Diese ist allen Wesen ein großer Helfer,
Ein Wort von untrüglichem Wahrheitssinn.
So lauscht auch ihr der Verkündigung des Erbarmers!
Wenn ich, der Einsiedler, mit gereinigtem Geiste,
Mich nach den Dämonen der äußeren Illusionen umschaue,
So begreift der Geist, daß die Gaukelspiele keine Entstehung haben.
Wenn ich in das Innere meiner Seele schaue,
So ist der ursprungslose Geist vollständig leer.
Da ich den Segen genoß, jenen Lama zum Lehrer zu haben,
Da ich den Vorzug habe, ganz allein der Beschauung obzuliegen,
Da ich den großen Gebieter Naro zum Lehrer hatte, begreife ich
Und übe in der Wahrheit untrügliche, siegreiche Beschauungen.
Da die Meditationen der Tantra über die tiefen Mittel
Der Lama, der Gebieter, in der Hauptsache erklärt hat,
So kenne ich kraft der durch die Erlangung des rechten Pfades vollendeten starken Meditation
Die Verbindung des inneren Fasernetzes
Und fürchte daher nicht die Dämonen der äußeren Illusionen.
Heute sind die Heiligen, die große Brahmanen zu Lehrern haben,
Zahlreich wie der prächtige Himmel.
Wenn man beständig den Sinn eines einfachen Gemütes übt,
Schwinden die Illusionen aus dem Bewußtsein:
Weder Harm noch Harmstifter sehe ich!
Die heiligen Schriften des Pitaka haben sich euch geöffnet:
Daß es keinen anderen als diesen Segen gibt, steht fest.

So sang er. Bharo und sein Gefolge legten ihre Turbane ab, verneigten sich und umkreisten ihn oftmals.

Mit dem Versprechen, ihm einen Monat lang Nahrungsmittel zu liefern, entfernten sie sich, wie ein Regenbogen verschwindet. Am dritten Tage, bei Sonnenaufgang, erschien Bharo, von zahlreichem, mit Bharimaschmuck schön geschmücktem Gefolge umgeben, und brachte dem Ehrwürdigen Wein, Bier und viele undere Arten gegorener Getränke, in kostbare Gefäße gefüllt, Reisbrei, Fleisch und viele andere mannigfaltige Speisen in schönen Bronzeschüsseln. Mit dem Versprechen, von jetzt ab seinem Gebote zu gehorchen und seinen Worten Folge zu leisten, verneigten sie sich, umkreisten ihn viele Male und verschwanden. Der König derselben ist der sogenannte Thang ₒgrem, d. i. der große Gott Ganapati (Tshogs kyi bdag po).

Dadurch wurde die innere Förderung des Ehrwürdigen so groß und seine Gesundheit so vortrefflich, daß ihm einen Monat lang nicht einmal der Gedanke an Hunger kam.

Als nun der Ehrwürdige die Gegend des La phyi und des Chu bzang genugsam kannte, machte er sich auf den Weg, um Gnas ₒthil am La phyi zu besuchen.

Als er sich eine Weile auf einem großen Felsblock niederließ, der sich unter einem überhängenden Felsen inmitten der großen Ebene von Ram bu befand, brachten viele Daka unter Verneigungen Spenden mannigfacher begehrenswerter Gaben dar und umwandelten ihn erfurchtsvoll, so daß in einem Steine zwei Fußspuren der Daka zurückblieben, worauf sie wie ein verblassender Regenbogen verschwanden.

Deinde cum aliquantum itineris progressus esset, in illa, qua larvae lemuresque transformationes magicas ostenderant, via, muliebria multa et magna ubique videbantur. Milaraspa ille venerabilis vi magica sua videndo peni sedato iter perrexit. Eodem loco quo vestigia novem evanescebant, cum membrum, ut sucus eius conflueret, in saxo perfricasset, vultu magico oculos in locum convertit, quoad praestigiis finem afferret, unde locus ille appellatur La dgu lung dgu, id est „Furcae novem, Valles novem".

Als er seinen Weg fortsetzte, um nach Gnas ₒthil zu gelangen, ging ihm wieder der vorher genannte Bharo entgegen, brachte ihm Ehrengaben dar und errichtete in dem Verlangen nach einem religiösen Vortrag eine Kanzel. Wiederum predigte der Ehrwürdige über viele Dogmen der Vergeltungslehre, bis er endlich in einem vor der Kanzel befindlichen großen Felsblock verschwand.

Darauf begab sich der Ehrwürdige nach Gnas ₒthil, wo er in froher Stimmung einen ganzen Monat lang zubrachte. Dann kehrte er nach Rtsar ma in Gña nam zurück, und als er in ₒDre lung skyog mo selbst war, sagte er zu den Gabenspendern: „Ich habe die Dämonen bekehrt und will jetzt in meine Betklause zurückkehren; zuerst will ich gehen und mich der Beschauung hingeben." Da wurden sie alle des Glaubens voll.

Dies ist das Kapitel von der Reise nach dem La phyi oder von dem Flusse Chu bzang.

Anmerkungen.

Zur vorliegenden Neuausgabe bemerken wir im Einzelnen noch Folgendes:

Die Einleitung ist der Vorrede des Übersetzers in den Denkschriften der Wiener Akademie, phil.-histor. Kl., Bd. 48 (1902), Abhandlung II entnommen.

Die Legenden I-V des Neudrucks entsprechen den 1902 in den „Denkschriften" erschienenen Erzählungen, denen die beiden im Archiv für Religionswissenschaft, Bd. 4 (1901) veröffentlichten Legenden als VI. und VII. angeschlossen sind.

Um Wiederholungen in den Überschriften zu vermeiden, ist der Titel der VII. Legende: „Milaraspa auf dem La phyi" in: „Milaraspa am Flusse Chu bzang" geändert worden.

Wo der Text der Neuausgabe sachlich von dem des Erstdrucks abweicht, beruhen diese Abweichungen auf brieflich mitgeteilten Verbesserungen des Übersetzers, die hier verwertet wurden.

Die Numerierung der einzelnen Gesänge wurde weggelassen.

Endlich wurden der Neuausgabe unter Mitbenutzung der Noten des Erstdruckes die folgenden erklärenden Anmerkungen hinzugefügt:

Zu Seite 7: H. A. Jäschke: Handwörterbuch der tibetischen Sprache (Gnadau 1871-75), Seite 419a.

8: Jäschke: Zeitschrift der deutschen morgenländischen Gesellschaft, Bd. 23, Seite 543.

10: Rnam thar: Eine ganz in Prosa geschriebene Biographie des Milaraspa.

11: Herder: Ideen zur Philosophie der Geschichte der Menschheit, Buch 11, Kap. 3.

13: Pandita: Gelehrter, Professor.

Naro oder Naropa: der Lehrer des Marpa, der wiederum Lehrer des Milaraspa war. (Vgl. Einleitung, Seite 10.)

14: Saga-Monat: Der auf den Pferd-Monat folgende 4. Monat des Jahres 1086. (Vgl. Einl. Seite 9.)

Lama: Oberer, Abt, Titel der tibetischen Mönche besonders höheren und höchsten Grades.

Yogin: Yoga-Anhänger, Heiliger, Asket.

Caitra: d. h. der Monat, in dem der Vollmond in einer bestimmten „Citra" genannten Konstellation steht.

15: Daka: In der Gebirgswelt hausende und in der Luft wandelnde mythische Wesen.

17: Brahmaputra: d. h. „Sohn des (höchsten indischen Gottes) Brahman", einer der größten Ströme Asiens von 2900 km Länge, der an der tibetischen Seite des Himalaya entspringt und sich schließlich mit dem Ganges vereinigt.

Schwarzköpfig: häufiges Attribut der Menschen (besonders auch in mongolischen und türkischen Heldensagen).

18: Die Übersetzung der beiden Verse: Dem übelgesinnten Feind bin ich nie entflohen und Vorbereitungen sind dadurch unnütz gemacht ist unsicher.

Siddha: Ein Vollendeter, ein großer mit übernatürlicher Macht begabter Heiliger. (Vgl. Seite 54.)

19: Bang rim: Der treppenförmige Teil eines Stupa oder buddhistischen Grabmals.

Die Übersetzung: mit großen Gaben (?) ist zweifelhaft.

Siddhi: Vollendung, Erfüllung.

Nidana: Die Grundlagen der Existenz.

20: Die Übersetzung: Der Gipfel befriedigender (?) ist unsicher.

Zu Seite 22: Wenn er ergriffen wird: wohl von dem Epilepsie verursachenden Planetendämon, der herabsteigt, um von dem Wasser zu trinken, in welches das eine Ende des Regenbogens getaucht ist.

Yoga: Die als praktische Seite der dualistischen Samkhya-Philosophie gelehrte „Yoga-Praxis", das Streben, durch Unterdrückung der Sinnlichkeit und durch Meditation die Vereinigung mit Gott und damit die Herrschaft über die Materie zu erlangen.

Buddha: ein Erleuchteter, Weiser, der durch vollendete Ergründung der Wahrheit jenseits von Gut und Böse steht.

Preta: Die Seelen der Abgestorbenen, Manen.

Asura: Halbgötter, Dämonen.

23: Bodhi: Die Erleuchtung, die dem Buddha zuteil gewordene Erkenntnis.

Chylus: Der Milchsaft der Lymphgefäße des Darmes.

Rin po che: „Edelstein", Titel jedes höheren Lama's.

27: Felsen-Rakshasi: eine auf Felsen hausende Teufelin, Dämonin.

Rahu: berüchtigter Dämon der indischen Mythologie, ein Feind der Sonne und des Mondes, die er nach indischem Glauben beizeiten verschlingt. (Vgl. Seite 29.)

28: Manasarovara: Ein vielgepriesener heiliger See am Götterberge Kailasa im Himalaya. Der Name bedeutet: Manasa („der aus dem Geiste [Brahman's] Erzeugte"), der schönste der Seen. Wie berechtigt diese Bezeichnung ist, beweisen auch neuerdings die begeisterten Schilderungen Sven Hedins (Transhimalaja, Leipzig 1909-12, Bd. 2, Seite 91, 95; Bd. 3, Seite 170 ff.). Man vergleiche auch das wundervolle Titelaquarell Hedins zum 3. Bande seines Werkes! Interessant ist, daß Milaraspa in bezug auf das Epitheton ornans „Türkissee" mit dem großen Forschungsreisenden übereinstimmt, welch letzterer aaO., Bd. 3, Seite 170 schreibt: „Auf Erden gibt es keinen schöneren Ring als den, welcher den Namen Manasarovar, Kailas und Gurla-mandatta trägt; er ist ein Türkis zwischen zwei Diamanten."

Rishi: Heiliger, Weiser, Seher.

31: Dhyana: Meditation, religiöse Betrachtung, mystisches Schauen.

32: Zung ojug: Kunstausdruck der praktischen Mystik, das Hineinzwängen des Geistes in die Hauptarterie, um bei der Meditation der Zerstreuung vorzubeugen. (Vgl. Seite 23, Vers 5-9.)

33: Yaksha: Halbgötter, Elfen.

Piçaca: Kobolde.

34: Jambudvipa: Indien.

35: Mahavajrapani: „Mit einem großen Donnerkeil in der Hand", ein bestimmter himmlischer Bodhisattva, d. h. ein auf dem Wege zur Buddhaschaft begriffener, im Himmel weilender und noch nicht inkarnierter Buddha.

36: Bandhe: Priester.

37: Çakyamuni: „Der Weise aus dem Çakya-Geschlechte, d. i. Buddha.

Vajradhara: „Donnerkeilträger", Name eines buddhistischen Heiligen. - Der Donnerkeil (vajra) ist ursprünglich ein Attribut des indischen Nationalgottes Indra; er ist dann später von Vajrapani (Vajradhara) her auch auf Buddha selber übertragen worden. Daher rührt dann weiter der Gebrauch kleiner Messing-Donnerkeile, die zu den unvermeidlichen Ausstattungsstücken eines jeden Lama gehören. Von hier aus sind auch Ausdrücke wie „Vajra-Geschwister", „Vajra-Geheimwort" usw. zu verstehen.

Vajra-Hölle: „Donnerkeil- oder Diamant-Hölle", eine der zahlreichen Höllen des Mahayana-Buddhismus.

38: Vajrasattva: Name eines Bodhisattva oder Dhyani-Buddha, eines „Buddha der Meditation".

39: Zwischen dem 2. und 3. Vers sind zwei Verse als teilweise unübersetzbar weggelassen.

Zu Seite 39: Hinayana: „kleines Fahrzeug", d.h. der südliche Buddhismus im Gegensatz zum Mahayana, „dem großen Fahrzeug", d. h. dem nördlichen Buddhismus, der in stark entstellter Form durch Padmasambhawa im 8. Jahrhundert nach Chr. in Tibet eingeführt wurde.

Nirmanakaya: Einer der drei Leiber des Buddha im Mahayana-System.

40: Daß die Erleuchtung weder geht noch kommt: d. h. beständig an ihrer Stelle bleibt.

42: Jener ausgezeichnete heilige Lama: gemeint ist Milaraspa's Lehrer Marpa. (Vgl. Einleitung, Seite 10.)

43: Samsara: Der Inbegriff alles Vergänglichen, der Kreislauf des Werdens, dem der Buddhist zu entgehen strebt.

Dharmakaya: Einer der drei Leiber des Buddha im Mahayana-System.

45: Mang yul: Eine an Nepal angrenzende Provinz Tibets.
Die Übersetzung des Verses: „Bald habe ich zur Bußübung Steinchen gemacht" ist nicht ganz sicher.

46: Utsakrama und Sampannakrama: Bestimmte Grade oder Stufen der Meditation.

51: Mantra: Heilige Hymnen, Zaubersprüche.

52: Skyed rdzogs: Terminus technicus für einen bestimmten Akt der praktischen Mystik. (Vgl. die Anmerkung zu Seite 32.)

55: Ras chung pa: d. h. „Der mit einem Baumwolltuch", ein Schüler Milaraspa's. Der Name Milaraspa bedeutet: „Mila mit dem Baumwolltuch". An anderer Stelle werden Milaraspa's Schüler „Die unvergleichlichen mit einem Baumwolltuch Bekleideten" genannt. (Vgl. Seite 17, 18, 42, 46.)

Udyana: Kafiristan, die Heimat des Padmasambhawa, des „Mannes aus Udyana". (Vgl. Einleitung Seite 12 und Anmerkung zu Seite 39).

56: Caitya: Ein Monument wie ein Stupa. (Vgl. Anmerkung zu Seite 19.)

57: Hinter dem Verse: Sollte gegen schlechte Freunde nicht ergeben sein hat der Erstdruck folgende Note, die im Neudruck hier Platz finden möge: „Darauf trägt Milaraspa auf Dharmabodhi's Bitte noch zwei kurze Lieder von 18 und 11 Versen vor, in demselben, Stil wie das vorhergehende und gleichfalls Gegenstände der Mystik behandelnd. Da das Verständnis derselben durch die Dunkelheit des Themas an sich wie durch unsere Unkenntnis mancher technischer und anderer Ausdrücke sehr erschwert ist, so verzichte ich hier auf deren Wiedergabe. Dann fährt der Text folgendermaßen fort:"

59: Mudra: Wörtlich „Siegel", wohl eine bestimmte Art der Fingerhaltung.

Dharmadhatu: Eines der buddhistischen Elemente der Existenz.

Mahamudra: Wörtlich „großes Siegel", eine bestimmte Haltung der Hände oder Füße in der Yoga-Praxis.

Jina: Der „Sieger", ein Name Buddha's.

60: Tantra: Wörtlich die „Netze", d. i. die Lehrbücher der Zauberei und Magie.

61: Amolika: Eine Art Achat.

65: Mara: Todesgott.

Nagabodhi: Wörtlich „Schlangen-Erleuchtung", eine bestimmte Stufe der mystischen Erkenntnis.

75: Sutra: Leitfäden, Lehrbücher.

Brahmanen: Angehörige der altindischen Priesterkaste.

Pitaka: „Korb", d. i. die Gesamtheit der hl. Schriften des Mahayana-Buddhismus.

Tibet. Malerei.
Besitzer Herr Moers, Amsterdam.

1

Tibet. Malerei.
Besitzer Herr Moers, Amsterdam.

2

Tibet. Malerei.
Besitzer Herr Moers, Amsterdam.

3

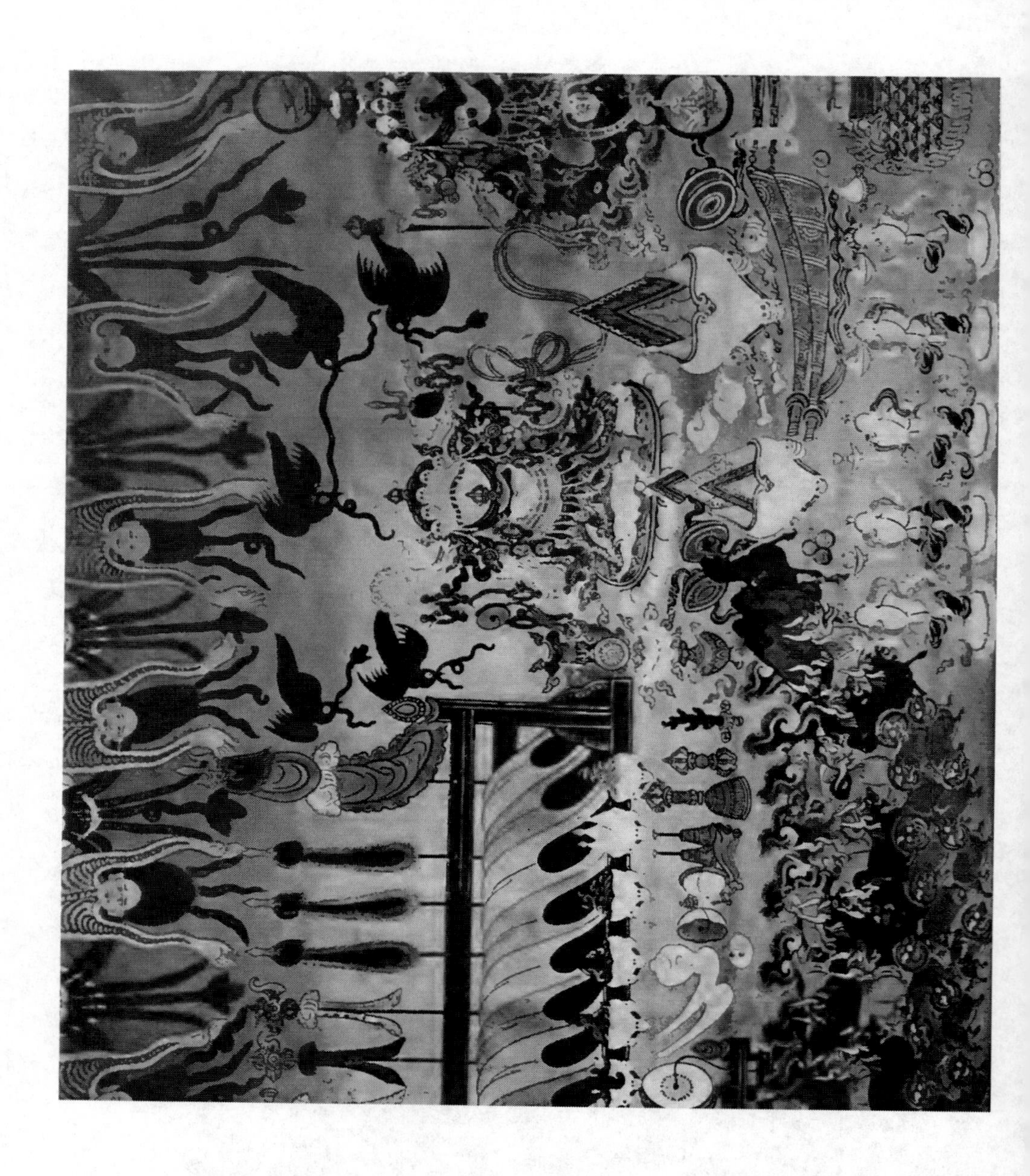

4

Tibet. Malerei. Ausschnitt, rechter Teil.
Phot. von E. E. Schlieper.

Tibet. Malerei. Ausschnitt, linke Hälfte, nochmals vergrößert.
Phot. von E. E. Schlieper.

6

Tibet. Malerei. Mittelstück, vergrößert.
Phot. von F. E. Schlieper.

7

Leh am oberen Indos, alte ehrwürdige Königstadt, die Hauptstadt des früheren westtibetischen Königreiches.

8

Ruinenstadt Klar itsac in einem linken Seitentale des Shäyok, das vom Wege Tirit-Khardang abbiegt. Das Charakteristikum dieses Platzes sind wiederum die m-chod r-ten, in denen sich potted lamas aus der sogenannten „vorbuddhistischen Periode" fanden.

9

Dorfstraße in Tag cha am oberen Nubrafluß mit zu einem Tor ausgebautem m-chod r-ten.

10

Wundervoller Typus eines m-chod r-ten mit vier dazugehörenden kleineren. Es führt ein Weg unten hindurch, wie links sichtbar ist. Dieser m-chod r-ten steht nordwestlich von Leh.

11

Tibet. Steine mit Schrift.

12

Tibetisches Dorf Tirit, am Shäyokfluß, etwas unterhalb der Mündung des Nubraflusses; rechts der typische, stets wiederkehrende m-chod r-ten (eigtl. „Opfer- [m-chod] Behälter [r-ten]"), jetzt der Aufbewahrungsort der potted lamas, jener kleinen aus Lehm und menschlicher Asche geformten Pasten. Die m-chod r-ten in Tibet sind den indischen Stupas ebenbürtig.

13

Tibetische Götterdarstellungen. Bronze. In der Nähe von Peking ausgegraben und seit der Aufnahme nicht wieder gemeldet.
Phot. von E. E. Schlieper.

161

中国艺术概观

THE JOURNAL OF THE AMERICAN INSTITUTE OF ARCHITECTS

JUNE
1922

A Bird's-Eye View of Chinese Art

Illustrated by Examples in the Collections of Field Museum of Natural History, Chicago

By BERTHOLD LAUFER

Curator, Department of Anthropology, Field Museum, Chicago

IN 1907, Mrs. T. B. Blackstone of Chicago provided the Field Museum of Natural History with an endowment for carrying on researches and securing collections in China and Tibet. The work was entrusted to the writer of this article, who spent over a year in Tibet and almost two years, from 1908 to 1910, in the interior of China. The results of this expedition are shown in the east and west galleries of the new museum building and aim at giving a general survey of the development of Chinese civilization from earliest times until the beginning of the nineteenth century, as well as a representative picture of the culture of Tibet. China is a world in itself, and its civilization is a complex, as vast as the ocean. Within the brief compass of an article, only a few phases in this rich development could be selected for discussion, and the objects serving as illustration are very few examples picked from thousands in the Museum's collections.

All emanations of ancient Chinese art must be interpreted from the religious conceptions and ideals of the nation. Worship of the great elementary forces of nature, deep reverence for the departed, unlimited devotion to ancestors and their ethical traditions, an insatiable yearning for salvation and immortality, combined with a sound and practical philosophy of life and moral standards, form the keynote of the mentality of Chinese society. Like that of Egypt, the art of ancient China is one of the dead, and the monuments discovered in the graves bear a distinct relation to the beliefs in a future life entertained by the people and simultaneously reflect the actual state which their civilization had reached.

The Han period covering the time around the Christian era (206 B.C.-A.D. 220) marks the transition from the impersonal art of the archaic epoch to the middle ages. It denotes the culminating point of idealistic art in that religious sentiments or ideas are expressed in a straightforward manner with an intimate personal and human touch. During this epoch, the graves were laid out in large sepulchral chambers composed either of stone slabs or enormous bricks, which formed a vault sheltering the coffin and the paraphernalia interred with the departed spirit. The slabs were usually adorned with pictures traced in flat relief and illustrating favorite incidents of ancient history or mythological lore in a narrative of almost epic style. Somewhat naïve and primitive in the representation of human figures and in the expression of emotions, they are nevertheless full of life and movement in their records of battles, hunting scenes, court processions, royal receptions, or domestic affairs. These engravings in stone come down from the second century A.D., and present an important source for the study of ancient civilization. The bricks were impressed with elaborate compositions of geometric designs (see Fig. 1). On the same plate may be seen roofing-tiles coming from the ancient palaces of the Han dynasty, of which no other remains are left. These tiles consist of a long half-cylinder to the front of which a disk is attached, and are made of a burnt steel-hard clay. The cylindrical portion rested on the lower end of the roof so that the disk projected over the eaves, and was visible to the passer-by from below. This part was therefore embellished with designs or ornamental characters in relief, forming a saying of good omen or giving the name of the palace to which they belonged.

The most conspicuous feature of the Han dynasty graves is formed by a magnificent display of plain or green and brown glazed pottery, of which a very comprehensive and representative collection is assembled in the Field Museum. This mortuary fictile art presents a microcosm of the life and culture of that age, and makes a substantial contribution toward a reconstruction of China's past. All the property dear to the living was then reproduced in clay objects of miniature size and confided to the grave, as houses, granaries, watch-towers, farm-sheds, barn-yards, mills, grain-crushers, sheepfolds, stoves, as well as the favorite domestic animals like dogs and swine. The likeness of an object suggested a living reality, and the inmate of the tomb was believed to enjoy the possession of the durable clay offerings as though they were the real thing. The meaning of death was, to the Chinese, a continuation of this life in almost the identical surroundings. The spirits of the deceased, though they had relinquished their bodily form, still were compelled to partake of food and drink. Hundreds of models of cooking-stoves have been discovered in the ancient graves, which goes to show that cooking, symbolically at least, was believed to be continued in the other world.

Pottery jars bearing out the conception of a draw-well (Fig. 2) were lowered into the grave to furnish the dead with a constant supply of fresh water. The square body of the jar represents the well-curb, the well-frame being erected over its edges on which leans a water-bucket, resembling in shape the one still used at present. The frame above makes space for the insertion of a pulley over which a rope passes, the buckets being suspended from the ends of the rope. In some specimens the pulley is actually moulded in clay, in the form of a small wheel. The pulley is protected from rain by a sloping, tiled roof.

This type of draw-well jar affords an interesting example of how realistic objects assumed idealized expression and artistic form under the hands of unnamed potters. It is still more interesting to observe how this realistic subject gradually became conventionalized to shrink into a mere ceramic type of basket shape with a handle. In Fig. 3 the draw-well idea is plainly manifest; the well-curb is a cylindrical vessel; the bucket is there, resting on the edge, but the structural framework is replaced with an elegantly curved handle, while the pulley and roof have developed into a merely ornamental function. There are other specimens like this one, devoid of the well-bucket; while in others, ultimately, even pulley and roof have disappeared, leaving solely a jar surmounted by a rounded handle. One of the chief attractions of Han pottery is the beauty and color variety of the glazes and the decomposition of the glaze subsequently brought about by oxidation underground of the metals mixed in the glaze and frequently resulting in golden and silvery iridescence.

The clay modelers of the Han period chiefly depended for their forms on the contemporaneous bronze-founders. Most of the large pottery vases interred with the dead were in fact derived from corresponding types of bronze, which, on their part, served the purposes of the living generation. Of all nations, the ancient Chinese remain unexcelled in the art of bronze-casting, both as to beauty of form and technical excellence. The process of casting was *à cire perdue,* of which Benvenuto Cellini has left so classical a description; and it is amazing that large vessels, many of them of great complexity, were in ancient times produced in a single cast, inclusive of bottom and handles. The bottle-shaped bronze vase (Fig. 4) is a good example of Han art, exquisite in shape, admirable in proportion, and striking by its simplicity. It is coated with a layer of fine, deep green patina.

Some fifteen years ago, graves of the middle ages were first opened during the construction of railroads, yielding an unexpected harvest of clay figurines of a bewildering variety of forms. Under the Tang dynasty (A.D. 618-906), from which the majority of these figurines come down, culture had made a considerable advance, and life was enriched by a noble refinement of social customs, as well as by a vast progress in poetry, painting, and sculpture. It was China's Augustan age. In distinction from the art of the Han, this epoch is characterized by naturalistic tendencies of art, as notably evidenced by the work of the great landscapists. In the Tang graves we encounter an overwhelming number of human figures, and this personal element that makes a direct appeal to us is the most remarkable feature cropping out of these new discoveries. The feminine ideal of that age is illustrated by numerous statuettes of graceful women, who were loyal companions of their masters in the beyond. They exhibit a large variety of costumes and hair-dressings (Fig. 5), as suggested by local usage; this trait renders them a live source for the study of former fashions.

In viewing the Chinese exclusively under the influence of the Confucian doctrines and the rigid ethical system based on them, we are prone to make them out as a rather serious and even pedantic people. It should be borne in mind, however, that there was at all times a merry old China fond of good shows and addicted to entertaining games. Dominoes and playing-cards were invented in China, football was played as early as the Han, and polo was introduced from Central Asia under the Tang. Figures of acrobats, jugglers, musicians, and dancers are carved on the walls of the Han grave-chambers; and skilfully modelled clay statuettes of quaint mimes and actors, providing entertainment for the souls of the deceased, have arisen from the graves of the middle ages (Fig. 6). Some are represented in the midst of reciting a monologue, others are shown in highly dramatic poses, gesticulating with lively motions as if acting on the stage. Others are portrayed with such impressive realism and individuality of expression that we feel almost tempted to name them after favorite casts familiar to us. The countenance of these actors displays decidedly Aryan features. It seems almost certain that they are intended to represent performers hailing from Kucha in Turkestan. We know from the contemporaneous Chinese annals that music and art were eagerly cultivated by the people of Kucha, and that their actors paid frequent visits to China, being favorite guests at the imperial court residing in Chang-an, Shen-si Province. From manuscripts discovered in the sandy deserts of Turkestan, and recently deciphered, it is now ascertained that the inhabitants of Kucha spoke an Indo-European language, designated as Tokharian B.

It is a far cry from the Tang (A.D. 618-906) to the Ming dynasty (1368-1643), which in the main was a retrospective period of art. The styles of ancient masters were then copied; and while novel ideas were no longer created, technical skill and perfection, as well as grace and refinement prevailed. A unique set of ten bronze figures, each representing a renowned beauty of the Han period, in the collections of Field Museum, is thoroughly characteristic of these tendencies (Fig. 7). Each is shown in a different posture of dancing, dressed in a long flowing robe with embroidered collar. The bases, of bronze also, are moulded in the shape of rocks surrounded by waves. This work is exceptional in that it is the individual

FIG. 1

FIG. 2

FIG. 3

FIG. 4

Fig. 5

Fig. 6

FIG. 7

Fig. 9

conception of an artist, widely differing from the conventional figure-productions of contemporaneous craftsmen.

From the third and fourth century onward a new current of thought and art swept over eastern Asia. This novel influence emanated from India, the land of speculative dreams, poetic imagery, good stories, philosophical doctrines, and many theories of religious salvation. The expansion of Indian influence throughout the far east, inclusive of the Malay Archipelago,

Fig. 8

is one of the outstanding and fundamental facts in the history of the world, and disseminated the elements of Indian civilization to the mountain tribes of the Himalayas and the poorest jungle tribes of Indo-China, in the same manner as it enriched and deepened the life of the civilized nations like the Chinese, Koreans, and Japanese. This intellectual conquest is connected with the name of Buddha, the first who founded a universal religion that made a world-wide appeal. Many Buddhist missionaries travelled from India to China, either over land by way of Kashmir and Turkestan,

FIG. 10

or over the maritime route leading to Canton, to preach the gospel of Buddha, to assist in the translation into Chinese of their sacred books written in Sanskrit, and to promote the foundation of monasteries and temples. Again, numerous Chinese monks who had taken the vows in the new order wended their way to India to learn Sanskrit, study Buddha's law, and return to their native country, loaded with palm-leaf manuscripts, pictures, and statues. The records which these undaunted pilgrims have left to us are of the highest value for reconstructing the history of Central Asia and India during the middle ages, and felicitously supplement what India lacks,—a sound chronology, as well as geographical and historical accounts.

The Buddhist art of India reached the climax of its development during the first and second centuries A.D. along the northwestern frontier of the country, in a stretch of territory comprising the modern district of Peshawar and the western portion of the Panjab, anciently known as the kingdom of Gandhāra. Here arose a school of artists, chiefly excelling in stone sculpture and working under the influence of late Hellenistic traditions, which had filtered into India under the successors to Alexander the Great. It was

FIG. 11

the Gandhāra school which for the first time created a statue of Buddha after the model of the Greek Apollo, with such modifications as were compatible with the Indian national spirit. A remarkable collection of early Gandhāra sculptures was recently acquired by the Field Museum. The great historical importance of this art centre rests on the fact that it became the mother and fountain-head of all Buddhist art in the east, that its models and lessons were propagated to Nepal, Tibet, China, Korea, Japan, as well as to Siam, Cambodja, and Java; the Indo-Hellenic style, though sometimes modified, overlaid or even obscured by native traditions, has persisted to this day, and is still plainly discernible in the most recent productions of all those countries.

In China, particularly, the advent of Buddhism tended to revolutionize many long-inherited conceptions of art. National Chinese art, as conditioned perhaps by its appliances—paper, silk, ink, and soft hair-brush—is one of line and color, unrivalled in surface decorations; while the sense of plasticity is by far less developed. Here remains a significant psychological problem for investigation. There are analogous phenomena in other lines of mental activity; glancing at mathematics, we observe that the Chinese have successfully cultivated and signally advanced plane geometry, arithmetic, and trigonometry; they had no understanding, however, of the laws governing stereometry. They embarked on scientific surveying and map-making as early as the Tang period, their early maps are fairly accurate and in many ways excellent; boundaries, routes, rivers, canals, and other water-courses, in short, the configuration of plain surface areas, are registered with a high degree of fidelity, but mountain-ranges always presented a task beyond their capacity, and were simply neglected. The mountain scenery was enjoyed esthetically, but its structure was not grasped, its characteristics not expressed, its elevations not measured; in fact, I dare say, as I often made the experiment, no ordinary Chinese has any conception of orography, or is capable of giving a half-way intelligent description of any mountain formation. To return to Buddhism, however,—its exponents opened the astonished eyes of Chinese artists, for the first time, to the beauty of the human body and its personification in free sculpture. The nude has always been alien to Chinese spirit, and is not merely timidly dodged, but stringently tabooed by the great painters. They never accentuate sex and passion, and the eternal love-theme does not furnish them with any inspiration. What attracts them is the richly decorated and easily flowing silk robe with its graceful movements, and face and hands are the sole organs finding expression. And now Buddhism came with an art showing undercurrents of Hellenic thought and, despite the pessimistic keynote of its teachings, making a free display of the nude and of bodily forms. While, in the beginning, the Chinese merely copied the models, as transmitted from India, they gradually learned how to assimilate them to their own national consciousness, and the masters of the chisel during the Tang period have handed down to us monuments which not only vie with their Indian counterparts, but even surpass them in spirit and fervor of faith.

The praying Bodhisatva (a saint on the way to the dignity of a Buddha, a future Buddha), represented in the marble relief of Fig. 8, reminds us, in the naïve, pious simplicity of the conception, of our own mediæval tomb-sculptures of saints and devout kings. The two recumbent lions by which he is flanked symbolize the saint's power over the king of the beasts. The nimbus, foreign to the East, was early derived in India from western Asia. He is equipped with a five-leaved diadem, a necklace falling crosswise over the chest, and a mantle covering the shoulders and reaching the ground; above all, the fine pose, the hands devotedly

Fig. 12

FIG. 13

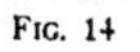

FIG. 14

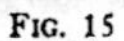

FIG. 15

FIG. 16

folded for prayer (a custom attested in the Buddhist community at least three centuries B.C.), and the tranquil, contemplative countenance make this marble a little masterpiece.

The marble statuette in Fig. 9 represents the Bodhisatva Maitreya, the Messiah of the Buddhists, who will appear at the end of this world-period as the future Buddha for the salvation of mankind. He stands on a base formed by lotus petals, the lotus being an emblem of purity and virtue. In his left he holds a holy-water bottle. He appears adorned with the regalia of a prince and decked with elaborate jewelry. The high-relief carving of a seated Bodhisatva (Fig. 10) is laid out on a square marble block, which served as a building-stone in a temple near Si-ngan fu, the capital of Shen-si Province and the ancient metropolis and imperial seat under the Tang. It was chiefly in this centre that most of the Buddhist scriptures were translated into Chinese. The unconventional freedom with which the shawl is treated, its ends fluttering in the air and in their motion strangely contrasting with the motionless repose of the meditating saint, is a noteworthy feature of this sculpture. The detached marble head of a Bodhisatva, shown in Fig. 11, allows one to view in detail the somewhat extravagant hair-dressing with which most of these statues are adorned; a very decorative fillet holds the hair in order, and the closed eyes of the Bodhisatva indicate the state of religious contemplation.

The black marble image, illustrated in Fig. 12, reveals a quite different aspect of Buddhist art. This is a guardian deity, a defender of the faith, the Buddha Acala; that is, the Immovable, who combines features of the Hindu gods Civa and Indra, and who reappears in Japan as Fudo. Immovable and stern, he is seated cross-legged on a lotus-base, clad only with a sash running over his left shoulder. A powerful, double-edged sword, ready to strike, is firmly clenched in his right fist to ward off demons and the enemies of the faith, and which demon could gather courage enough to approach him? His stanch countenance with the sturdy muscles of an athlete would assuredly deter any one.

FIG. 17

Fig. 18

His hair is combed from the forehead upward, bound up in a top-knot, where it is held by an ornament of floral shape, and then falls down in front in a single long tress, kept in shape by three rings. This is one of the most vigorous examples of modelling in Buddhist art that has ever come under my notice. This stone was excavated in the village of Yang-kia, three miles north of Si-ngan fu, and comes down from the period of the Wei dynasty (A.D. 386-532).

While Buddhism obtained a firm grip over the masses of China, modifying to a considerable extent their beliefs and hopes of the hereafter, it never succeeded in wiping out the old national religion, commonly known as Taoism. As a practical people, the Chinese were always intent on having several roads to salvation open to them. In its esoteric aspect, Taoism is a sort of pantheistic philosophy; in its popular garb, it is essentially worship of nature gods, spirits, and fairies. Alchemy, the quest for the philosophers' stone and the elixir of life, yearning for eternal youth and immortal life on the isles of the blest far away in the eastern ocean, are essential articles in the faith of its adepts. They worshipped among others a trinity of gods, termed the Three Pure Ones, the first of whom was Yüan Shi Tien-tsun ("the Venerable One of Heaven"), the supreme god, who was regarded as the personification of the beginning and creation of all things. His cult was highly developed under the Tang dynasty. A life-size head chiseled from a marble of beautifully yellowish tinge (Figs. 13-14) testifies to the fact that the Taoists of that period mustered artists of the same high calibre as the Buddhists, whether they may have learned the art of sculpture from the latter or not. The virile, majestic forehead strongly modelled and the spiritual expressiveness of the face divulge the presence of a god; and with all its conventional features, as, for instance, evidenced by the beard and the ears, this bust nevertheless makes a strong impression, and is distinguished by high artistic qualities. This impression would be still more favorable, were the nose not mutilated. The crown shaped lotus leaves or lotus-petals superposed in three layers is noteworthy.

The same god we encounter again in the votive image represented in Fig. 15. Here he is seated cross-legged on a railed throne skillfully draped. This conception is rather naturalistic; the right hand, full of

FIG. 19

life, rests leisurely on the top of the rail; the left hand, unfortunately broken off, was raised in the act of preaching. The face is profoundly spiritualized and decidedly noble and beautiful. The whole composition of this sculpture (note the pedestal built in four sections) is harmonious and monumental, and stamps it as a work of art of the first order. It is provided with an inscription, which yields a date corresponding to our year A.D. 709; that is, the early Tang period. The inscribed stones were all dedicated to temples by faithful laymen, the usual occasion for such an event being cases of sickness in the family, especially on the part of a man's parents; it was accordingly an act of filial piety, and the donor invoked, in the inscription, the deity figured on the stone, that his mother or father might soon recover. In large votive stones we occasionally have a lengthy pedigree or a long list of a whole village community carved on the reverse of the slab, and in some cases the founders had their portraits incised in the stone or represented in relief. Such a votive stone is illustrated in Fig. 16. The god Yüan Shi Tien-tsun occupies the centre on a throne flanked by two lions. The founder of the stone is portrayed to his right; the founder's wife, to his left, in the attitude of worshipping the god. An inscription of eleven lines is spread over the socle and contains a date, which answers to A.D. 665 (Tang dynasty).

In northern China it was customary to erect a mound or tumulus over the grave. In front of this mound was usually placed an altar of stone on which were arranged five vessels for sacrificing—an incense-burner in the centre, surrounded by two flower-vases and a pair of candlesticks, all carved from solid stone. In the ancestral and other temples this set of vessels was ordinarily cast from bronze, and the funerary sets were modelled in stone after the bronze vessels. Such a funerary stone flower-vase of the Ming period (1368-1643) is shown in Fig. 17. Its surface is elaborately decorated with dragons soaring in clouds and conceived of as the messengers of fertile rain. The handles are

Fig. 20

Fig. 21

shaped into elephant-heads holding dead rings carved in relief, while the rings are alive and movable in the corresponding bronze vases.

Characteristic of the Ming period also are small portable bronze stoves neatly decorated with relief pictures and very scarce at present. The Chinese are in the habit of heating their bodies rather than their rooms by laying onion-like suits of clothes one on top of the other, adding on, as the severity of the winter increases. Open braziers of copper or clay pans filled with burning charcoal, as a rule, are the only means of heating employed in the average man's home, while asbestos stoves are now largely used in Peking. Artistic bronze stoves, such as illustrated in Figs. 18 and 19, are at present things of the past and no longer made. The former is decorated on both sides with a picture of cranes wading through a lotus-pond; the latter, posed on four dragon-heads, is elaborately adorned with four-clawed, imperial dragons skillfully moulded in high relief, testifying to the fact that this stove was made for and actually utilized in the palaces of the Ming emperors. It is divided into three cylindrical compartments, the fire being built in the central larger one, the opening being formed by the gaping jaws of a powerful dragon-head.

The most prominent and conspicuous feature of a Chinese mansion, public building, or temple, is the roof on which an exuberant wealth of ornamentation is usually lavished, and which glitters in dazzling colors of green, blue, or yellow brought out in the glazes of the tiles. Under the Manchu dynasty yellow tiles were reserved for the imperial palaces and temples; and green tiles for the buildings of the ministries; color symbolism of this kind played a significant rôle in Chinese society from time immemorial. The roof is the index of the position, rank, and taste of the tenant; and, according to the simplicity, grandeur, massiveness, or elegance of construction, as the case may be, foreshadows the scope and importance of the building.

The tiles on the gables are frequently surmounted by figures of fantastic animals or guardian gods, set up in long rows and giving the roof a very picturesque appearance. In Fig. 20 a green-glazed lion is shown with curly mane, tail and lower part of the mane being glazed yellow. Fig. 21 represents a winged griffin or dragon-horse of a brilliant, pure yellow. The statuette of a mail-clad warrior, glazed a beautiful torquoise blue (Fig. 22), was placed on the roof to ward off from the house malignant spirits and any evil influences. The same function was shared by the devil-exorciser in Fig. 23, glazed green and yellow, who, with his magic sword and ferocious grimaces, made an efficient guard and tutelary saint of a fine temple structure in Shn-sai Province.

FIG. 22

FIG. 23

162

中国玄关

THE CHINESE GATEWAY

FIELD MUSEUM OF NATURAL HISTORY

CHICAGO

1922

FIELD MUSEUM OF NATURAL HISTORY
DEPARTMENT OF ANTHROPOLOGY
CHICAGO, 1922

LEAFLET NUMBER 1

The Chinese Gateway

(At South End of Stanley Field Hall)

Large gateways of high architectural order occupy a prominent place in the streets of Chinese cities, in the courtyards of temples, or on avenues leading to a tomb or mausoleum. As a type of architecture, they are based on the so-called Torana of India, plans of which were introduced into China and Japan as a sequel of Buddhism. In the Buddhist art of ancient India, ornamental stone rails were built as enclosures around the topes (mounds or structures containing sacred relics), four gateways of highly decorative style being placed in these rails. The Chinese, however, did not slavishly imitate these monuments, but merely took them as models and lavished on them the wealth of their own decorative motives.

While the Romans erected triumphal archways in commemoration of military successes, while the people of India built them in honor of their greatest man, Buddha, the minds of a philosophical nation like the Chinese drifted in a different direction. The exaltation of military victories had no room in their thoughts; they raised sanctuaries to glorify their philosophers and statesmen, their sages and scholars, who shaped and advanced the mental and ethical culture of the nation. The character of Chinese art is impersonal, nor does it glorify the individual. China has no statues or portraits of emperors and generals.

Honorary gateways were erected in memory of deserving servants of the state and virtuous women. Widows, who did not remarry after their husband's death, and who faithfully nursed their parents-in-law, were entitled to this honor; likewise children for unusual acts of filial piety, persons who had reached the age of one hundred, and statesmen for loyal devotion to the throne. If such a canonization was recommended, the emperor, on receiving the petition, issued a "holy edict," which was chiseled in stone on the top of the monument, and he contributed the sum of thirty ounces of silver. The balance of the cost was subscribed by the family of the honored person or by the grateful community. The ideal purpose of these gateways, accordingly, was to perpetuate to posterity the memory of excellent men and women and to act as an influence on the conduct of the following generations.

The number of openings or passages in gateways is usually three, as in the gateway of the Museum, while more rarely five arches occur. Such may be seen, for instance, on the tomb of Confucius in Shantung or on the avenue leading to the mausoleum of the Ming emperors, north of Peking. Marble, granite, sandstone, and wood are employed for gateways. Wooden structures, as a rule, are less elaborate than those of stone. The gateway shown in the Museum is a very exact reproduction in teakwood of one in stone, with all the rich details of ornamentation.

The harmonious construction, the ingenious composition, the pleasing proportions, and skill in workmanship, are notable features. The four pillars are each adorned with a powerful, scaly dragon, rising from the depths of the sea heavenward into clouds and making for a flamed pearl: the aspiration for an ideal that is set before man, but that can never be reached. The pedestals carry figures in the round of lions playing with their young; altogether forty-two lions are represented. It is remarkable that each and every lion is different in posture and action. Some are provided with collar-bands and bells; some have their manes rolled up in spirals, while others have two bearded tips. In some, the eyebrows are conventionalized into spiral designs; in others, they hang over the eyeballs. In the two corners is represented a drum on each side of which are shown three lion-cubs chasing a ball.

A great deal of symbolism is connected with the lion. The Chinese recognize him as the king of all beasts, and his roaring dispels phantoms. Under the imperial régime, the Grand Preceptor of the emperor and his assistant occupied a high rank among the state-officials, for it was the duty of these functionaries to inculcate in the heir-apparent the maxims of good government and conduct. Being called *t'ai shi* ("grand preceptor") and *shao shi* ("small preceptor"), a pun was easily suggested by the designation of the lion, which likewise is *shi*. Thus the representation of a large and a small lion intimates the wish, "May you obtain the position of the first and second dignitary at the imperial Court!" Again, the lions engaged in playing ball symbolize the peace and prosperity of the empire. In this case, the lions represent military officers, who are not obliged to go to war, but who indulge at home in the harmless sport of ball-playing.

The inscription, consisting of eight gilded characters (four on each side) and set off from a diapered background, reads,—

"Your merits shine like sun and moon,
"Your good deeds vie in extent with streams and mountains."

This panel is bordered by a pair of dragons on the upper and lower sides and by the figure of a phœnix to the right and left.

The principal and most conspicuous portion of any Chinese structure is the roof. In a private mansion, it is the index of the owner's taste and social standing. In a public building, it indicates at once its peculiar character; and, according to the simplicity, grandeur, massiveness, vigor, or elegance of construction, as the case may be, foreshadows its scope and importance. Color symbolism expressed by the hues of the glazed tiles is another means of identification. In the gateway, each passage is surmounted by a roof of its own, the three roofs being so organically connected that the impression is conveyed of a double roof. The graceful, picturesque curves on the corners ("flying eaves," as they are styled by the Chinese) are each surmounted by four lions, believed to be faithful guardians and to ward off evil influences from the monument.

The roof is supported by eight struts (four on each side), carved into the appearance of bamboo stems with exuberant foliage. Rafters and tiling are so skilfully brought out in the woodwork as to inspire a perfect illusion. Sixteen carvings of figures in the round are displayed over the roof: they represent military officers, eight on horseback, and eight on foot, each in full armor, equipped with spears, halberds, clubs, and other weapons, and attended by footmen holding flags.

The ridge-pole is adorned with a panel carved in open work with two dragons struggling for the flamed pearl, and surmounted by a calabash. The latter is an emblem of the creative power of nature, of fertility and abundance (corresponding to our horn of plenty). The Taoist adepts used calabashes to store the elixir of immortality, and druggists preserve their medicines in them or in gourd-shaped vials. The ridge-poles are flanked by dolphins, head downward and tail upward, the belief being entertained that they safeguard the structure from conflagration. The wooden balls supported by wires are intended for water bubbles rising from the clouds by which the dragons are enveloped.

In the dragon-columns the favorite personages of Taoist mythology are represented. On the two inner columns we note the so-called Eight Immortals, bestowing on mankind old age and all sorts of blessings: Chang Kuo with his magic wand capable of fulfilling his every wish; Ts'ao Kuo-k'iu with his castanets, by means of which he performs magical feats; Chung-li Küan with his fan; Han Siang-tse blowing his flute; Lü Tung-pin with a magic fly-whisk and sword on his back; the beggar Li T'ie-k'uai with a calabash full of blessings; Lan Ts'ai-ho with a flower-basket; and the fairy Ho Sien-ku with the stem of a lotus, the sacred emblem of purity. Compare the two sets of bronze images of the Eight Immortals in Blackstone Chinese Collection, Case 21.

On the two outer columns are portrayed two Taoist sages or hermits, one shouldering a branch with peaches, which ripen but once in three thousand years in the paradise of the goddess Si Wang Mu, and which confer immortality on the adepts of Taoism, and two fairies with miraculous fly-brushes. It was the belief of the Taoists that contact with nature, a contemplative life in the solitude of mountains, is conducive to purification, old age, and speedy salvation; in this manner the soul ultimately is capable of soaring heavenward on the wings of a crane. Thus we see two recluses astride cranes carried upward at the end of their earthly career. In the lower portions of the outer pillars are figured the gay twin genii of Union and Harmony; one, holding a covered jar from which emanates a tree covered with money; the other, a lotus. These emblems are suggested by punning, both the lotus and a jar being called *ho*, and two other words *ho* being expressive of the notions "union" and "harmony."

Other Taoist symbols are illustrated in the panels on the pedestals. One of these shows a crane soaring in clouds above a pine-tree,—both being symbols of endurance and longevity—and a deer, which is emblematic of high official dignity and good income. Another represents a phœnix alighting upon the beautiful Wu-t'ung tree (*Sterculia platanifolia*), famed in legend and art, and the unicorn Kilin, the emblem of perfect good, that appears only at the birth of a virtuous ruler.

The high-relief carvings that decorate the horizontal lintels above the passages carry us back to realistic scenes of human life. They are arranged so that warlike scenes are assembled on one side of the gateway, while representations of peaceful pastimes occupy the opposite side. The main themes of the artist are tournaments of ancient paladins, thrusting halberds or spears at one another in front of a city-wall, from the rampart of which other grandees eagerly watch the spectacle. Or a cavalier turning backward on his galloping steed sends an arrow at his adversary, whose helmet is pierced by it, while tents surrounded by standards lend color to the background of the military action. The genre-scenes depict the tribunal of a high official, old men enjoying themselves in a grove, a lady travelling in a push-cart and escorted by mounted lancers, or a monk conducting a dignitary to the gate of his temple, which bears the name "Temple of Sweet Dew" (*Kan-lu-se*).

There are altogether twenty-two corbels, sixteen being decorated with designs of a phœnix, and six with interesting scenes describing the pastimes of cultivated gentlemen of leisure, as follows:—Feeding ducks, enjoying a cup of wine in a grove of pine-trees, writing a poem on a rock, painting a bamboo sketch on a scroll, reading in the woods at a table formed by a bowlder, playing the lute, dancing around a rock, taking a stroll in the company of a youth, who carries a pot of peonies, playing checkers on a stone board, planting flowers in a bed, examining the growth of plants, going a-fishing with a long rod over the shoulder.

It will thus be seen that the art displayed on this gateway is a marvelous embodiment of Chinese life and thought, a record of cosmogony and mythology, of heaven and earth.

The gateway is carved from teakwood, being 19 feet in height and 16 feet 9 inches in width. It was made in the Chinese Orphanage of Sikkawei, a Jesuit institution, near Shanghai, and was first on exhibition in the Palace of Education at the Panama Pacific International Exposition of San Francisco.

B. LAUFER

功昭日月

163

商、周、汉代的古代中国青铜器

ARCHAIC CHINESE BRONZES

OF THE SHANG, CHOU AND HAN PERIODS

IN THE COLLECTIONS OF

MR. PARISH-WATSON

ACCOMPANIED BY NOTES OF

BERTHOLD LAUFER

NEW YORK

M CM XXII

FOREWORD

ALTHOUGH the collection of archaic Chinese bronzes illustrated and briefly described on these pages is comparatively small, not exceeding ten in number, it easily takes the lead and foremost rank in quality among any gatherings of bronzes that have ever been permitted to pass the borders of China. The fundamental value of this collection, secured for Mr. Parish-Watson by Mr. H. A. E. Jaehne on his recent expedition to China, rests on the fact that here for the first time we are permitted to study at close range well-authenticated bronze relics of the Shang period (1783–1123 B.C.), which marks the climax in the development of the art of Chinese metal-casting. The collection includes five superb examples of this early art (Plates I–V), all rare and precious, and equally distinguished both by artistic merits and antiquarian interest, one of these (Plate III) being unique. The four specimens of Chou bronze art (Plates VI–IX) are no less beautiful and perfect, and here again we encounter a novel type not represented in any other collection of China or Japan, not to speak of Europe and America. There is one Chou vessel with cover (Plate IX) the exact date of which is authenticated by a lengthy inscription, and may be determined with certainty at 999 B.C. All of these features make this collection one of absorbing interest and paramount importance, and all lovers of Chinese antiquity are indebted to Mr. Parish-Watson for placing these extraordinary bronzes on permanent record.

Work in bronze may justly be regarded as the oldest of the national arts in ancient China, and as belonging to that province of art in which the national soul of the Chinese is most typically and felicitously crystallized. *Art,* I say advisedly, not *artcraft;* the archaic bronzes virtually belong to the realm of art, and their makers were full-fledged artists, not artisans. Only the epigones of the T'ang, Sung, and more recent periods degraded the art of bronze to the level of an industrial process; theirs was the technique, yet not the spirit. It is the spirit which makes art and imbues it with life, and it is the peculiar spirit developed in the early religious concepts of the Chinese which created the admirable casts of those metal founders, almost at the threshold of civilization. This was the creative epoch of forms, types, designs, symbols, and expressions of religious sentiments.

Copper and bronze of various alloys were employed in casting from the earliest dawn of Chinese history, not only for sacrificial vessels inspired by the worship of ancestors and nature-gods, but also for every-day implements, such as swords, daggers, spears, axe-blades, knives, lamps, clepsydras, hooks, buckles, and so forth; for the smelting and forging of iron was as yet unknown. Like the nations of western Asia and the prehistoric peoples of Europe, the Chinese passed through a bronze age of long duration, while iron but gradually came into use from about 500 B.C. Implements were cast in copper or clay moulds, but the process of casting, as far as the large vessels are concerned, was that *à cire perdue.* It is amazing that vessels, many of them of great complexity, were in ancient times produced in a single cast, inclusive of the bottom and handle or handles. The bronze experts are inclined to look upon this point as a characteristic feature of an archaic bronze, and in their examination first inspect the bottom of a vessel; if it turns out that the latter is cast separately and soldered in, the piece must lose any claim to ranking in the San Tai (the three dynasties Hia, Shang, and Chou, as the archaic period is styled). In most of the Sung and later bronze vases and jars, bottom and even handles are moulded separately. A strikingly large variety of metal alloys was utilized, different alloys being employed for different classes of objects. A special investigation of the composition of Chinese bronze is prepared in the Field Museum of Chicago, where chemical analyses of the bronze of several hundred specimens have been made, the results of which promise to be very interesting.

The great stimulus to the development of early art was ancestor worship with all of the virtues resulting from it, including the unceasing care for the departed, who were constantly alive and awake in the minds of the people. This deep-rooted reverence for the dead culminated in a minutely developed ritualistic cult which required vases for meat, grain, and wine offerings, and accordingly inspired the bronze-founders to supreme efforts.

Chinese archæology is still in its incipient stages, and, as a matter of fact, we know little about the earliest bronzes. The scholars of China were chiefly interested in their inscriptions, and from the style and technique of the characters have drawn conclusions as to the chronological setting, and classified the material according to historical periods. The results, in general, are sound and acceptable.

The study of bronzes was inaugurated as late as the beginning of the twelfth century at the instigation of the art-loving emperor Hui Tsung of the Sung dynasty, but this was at a time when continuity of historical contact and tradition was broken up. The Chinese scholars struggled to correlate the numerous types of vessels with the succinct and sometimes enigmatic references in the ancient texts and to offer more or less plausible interpretations of the bewildering variety of design displayed in the decorations.

In conformity with the impersonal, sacrosanct, and hierarchic character of this primitive art, all trace of naturalism or realism is conspicuously absent, but this subconscious art was strictly national, untouched by outside currents, and is refreshing in its groping for naïve expression of ideas. The human figure rarely appears in decorative art, all principal designs being of strictly geometric character, and receiving a symbolic interpretation evolved from the minds of agriculturists, who formed the bulwark of Chinese society. Being keenly interested in weather and wind and all natural phenomena exerting an influence on fields and crops, their attention turned toward the observation of the sky and stars, which resulted at an early date in a notable advance in the knowledge of astronomy. Hence, we encounter such interpretations as thunder and lightning, clouds, winds, and mountains, for purely geometric forms of design. Among animals we meet the tiger, always dreaded and revered as one of the great forces of nature, the elephant, the rhinoceros, the domesticated sheep and ox, fantastic birds, and a variety of reptiles. Of insects we find represented with predilection the cicada, whose wonderful life-history excited admiration, and which developed into an emblem of resurrection. The familiar dragon conception of later times is absent from early art, and makes its appearance only in the stone sculpture of the Han epoch; the alligator, however, is felicitously portrayed in the marvelous bone carvings of the Shang period. What is termed "dragon" in ancient bronzes by Chinese art-critics is a different species from an artistic viewpoint.

In many cases the surface of the vessels is covered with a diapering of frets, scrolls, and spirals, of very fine and delicate execution in good pieces, and this may serve as the background for a superimposed design in high or flat relief. In these relief pictures we frequently encounter anthropomorphic or zoömorphic heads or faces, of extremely conventionalized character, sometimes grotesquely distorted or disintegrated into separate parts of geometric pattern. The nose is usually unmistakable, and the eye is always prominently accentuated in the shape of a boss with a slit in the center to indicate the pupil, sometimes inlaid with gold. It has been customary among Chinese scholars since the days of the Sung emperors to honor these monsters with the general appellation *t'ao-t'ie,* which designates a *glutton,* and which was the nickname of a greedy man who is said to have lived in the good old days of the legendary emperor *Yao,* and to have been banished by *Shun.* His portrayal on ancient bronze vessels, then, is pleasantly accounted for as a warning against gluttony and avarice. This rather banal explanation strongly savors of an afterthought which, in my estimation, was alien to the people of the Shang and Chou periods.

For want of something better, European books on Chinese art constantly treat us to *t'ao-t'ie,* "men-devouring ogres" (and such like vague nomenclature), which are easily detected everywhere. But whoever has taken the trouble to delve a little into the mythical and religious lore of the ancient Chinese cannot believe that their power of imagination could have been so dwarfish and arid. It has been a painless operation for me to cast the soulless *t'ao-t'ie* overboard and to reload the ship with better goods. I am disposed to regard all of these manifold variations of monstrous heads as manifestations of the mythological spirit, as the iconography of the pantheon of ancient Chinese religion. Here also are the gods, and through a close study of mythology, in combination with the ancient bronzes, we may hope, in the course of time, to unravel their characteristic traits and to restore to life what has been dormant and misunderstood so long under the cover of thoughtless phrases.

It has been observed that, "apart from their archæological interest and from their beauty of form, Chinese bronzes have a quality of substance which no other bronzes exhibit. The beautiful green patina which we see on Greek and Roman statues, and the more elaborate colored patinas discovered by the ingenuity of the Japanese, are dull compared with the brilliant and jewel-like incrustation with which fine specimens of Chinese bronze are adorned." In regard to patinas, we have many interesting and suggestive theories formulated by Chinese antiquarians, and a germ of truth is doubtless contained in them. But here we are in need of a thorough chemical investigation of all the diverse kinds of patina from bronzes of different periods to obtain a solid basis for the discussion of this problem. In this respect Mr. Parish-Watson's remarkable collection is of utmost importance and utility, for on no other bronzes of antiquity have I ever seen finer and richer patinas. The reproductions of these objects, excellent as they may be, do not render justice to their striking beauty and exquisite coloring.

ARCHAIC
CHINESE BRONZES

1. Bronze Libation-Cup

Type *tsio* 爵.
Shang Period (1783–1123 B.C.).

Dimensions: 9¾ inches in height; opening, 9 inches in length, 4½ inches in width.

This superb specimen excels in all of the essential characteristics associated with the *tsio* of the Shang dynasty. It may be confidently asserted that it is not a *tsio* of the Chou period. The Chou examples present somewhat debased forms, as they are much lower in height, and generally distorted in their proportions. They are also plain, or have merely a narrow band of spiral design around the waist. In the specimen here illustrated the proportions are evenly balanced, and in its bold outlines the vessel stands with the perfect majesty and convincing force of a masterpiece.

The three feet rise in elegant curves, and are slightly turned outward. The body is divided into four sections, formed by three projecting and denticulated ribs, or ridges, and the single loop-handle, that springs from a conventionalized zoömorphic head. Both symmetry and a studied asymmetry, simultaneously applied, have always been the great principle underlying Chinese art. If we were to imagine another loop-handle attached opposite the present one, the impression would be decidedly unfavorable, while a fourth ridge in the place of the single handle would carry the principle of symmetry to an extreme, and the vessel would forfeit much of its present charm. The loop-handle unexpectedly breaks the symmetry of arrangement, adding a pleasing effect to the whole work; nor is it incidental that it has found its place just above one of the three legs. The designs are chased with wonderful clarity, being compositions of plain and convoluted spirals, the projecting eyes in the center hinting at some hidden anthropomorphic or mythological concept. To the artist of that archaic period, at any rate, the production of a sacred vase was a religious duty, and his creation was a reality imbued with the power of life and vision. The triangular patterns in the upper panel are interpreted by the Chinese archæologists as mountains (compare Plate 7).

The type *tsio* is explained by Chinese archæologists as being derived from an inverted helmet to which three feet are added; and, with some stretch of imagination, we might be disposed to argue that the hero of ancient days, desirous of celebrating a victory, doffed his helmet on the battle-field, offering in it a libation to the gods, and that subsequently the helmet was chosen as the model for a libation-cup. On second thought, however, this explanation is hardly convincing; the Chinese never were so warlike that a military headgear would have commended itself as an emblem worthy of being introduced into the ritualistic cult, nor is the alleged coincidence perfect. Above all, the three slender feet are so organically connected with the vessel that the two form an inseparable unit. Another interpretation seems more probable. The character *tsio,* used for writing the name of this type of vessel, properly denotes "small birds, like sparrows," etc., and it is not unlikely that the form of this vessel has grown out of the figure of a bird, undergoing, of course, numerous stages of gradual conventionalization. This theory is confirmed by the fact that there are specimens provided with a cover terminating in an animal's head. Animalized forms in vessels are frequent in both ancient and mediæval China (compare, for instance, the cooking-kettles of the Han dynasty with dragon and serpent heads). The three feet indicate plainly that the vessel was put over a fire; and as it is repeatedly referred to as a wine-libation cup in the ancient texts, it is obvious that the rice-wine was heated in the vessel itself. As is well known, wine, in China, is usually taken hot. That part of the vessel forming the bird's head is chamfered into a spout. The two spikes surmounted by knobs (explained as *chu,* "posts, supports"), and set on the edges, were probably made for the purpose of lifting the hot cup from the charcoal fire. The Chinese also give a symbolic interpretation of these spikes, comparing them with the stalks of cereals—evidently in allusion to millet and other grain from which the sacrificial wine was prepared.

The service of this libation-cup was particularly required in the worship of the great cosmogonic deities, Heaven and Earth, the interaction of which was believed to have created all things in nature. During the Chou period, when, each spring, the Son of Heaven performed the ceremony of ploughing the fields, he was accompanied by the three great ministers of state, the nine other ministers, all princes present at court, and the grand prefects. The Son of Heaven, himself, ploughed three furrows, the great ministers, five, the other ministers and the princes, nine. Upon returning to the palace, the Son of Heaven assembled his companions in his chief apartment, and, taking the cup *tsio,* addressed them as follows: "I offer you this wine in compensation for your trouble." The *tsio* was also employed on ceremonial visits during the ceremonies held in ancestral temples, when the master of the house offered wine from this cup to the representative of the dead ancestor. The *tsio* contained one pint (*sheng*), and was regarded as more honorable and dignified than larger vessels containing three, four, and five pints. The *tsio* were also carved from jade.

Plate I

Under the Chou, the *tsio* were regarded as valuable presents exchanged by vassal kings. Under the T'ang (A.D. 618–906), they were still used by the emperors in solemn ceremonies addressed to the deities, Heaven and Earth, on the summit and foot of Mount T'ai in Shan-tung. Under the Ming (A.D. 1368–1643), the *tsio* was a favorite type, but was degraded into profane purposes; during the marital ceremony bride and groom alternately drank wine from a cup of this shape. I have also seen bronze *tsio* of the Ming period with date-marks inscribed on them. In the age of the Manchu, *tsio* were frequently imitated in plain and decorated porcelain, also in silver; partly for ornamental purposes, partly for the nuptial ceremony.

2. Bronze Libation-Goblet

Type *kia* 斝.
Shang Period (1783–1123 B.C.).

Dimensions: 20⅛ inches in height; diameter of opening, 9¼ inches.

This type of vessel originated under the Shang dynasty, and subsequently it was ordained in the Book of Rites (*Li ki,* VII, II, 4) that this goblet should be exclusively owned and used by the princes of K'i, descendants of the house of Hia (2205–1784 B.C.), the princes of Sung, descendants of the house of Shang, and the princes of Lu in Shan-tung owing to their connection with Confucius, while any other prince or noble usurping this privilege was guilty of infringing the law.

This specimen is distinguished by its dimensions, which considerably exceed those adopted under the Chou; but, above all, surpasses the Chou examples in elegance and refinement. It served the same purposes as the *tsio,* and, like the latter, figured also in ceremonial drinking-bouts arranged by the kings. As to its composition, this type is constructed on the same plan as the *tsio,* save that, instead of the bird-shape of the cup, it is a drum-shaped vessel with flaring rim, and the bottom is flat, while that of the *tsio* is globular. The decoration is complete, covering the three spear-shaped legs, as well as the entire surface of the body and the two spikes surmounted by knobs. A deeply grooved band divides the surface into two horizontal panels, displaying the head of a mythical monster set out in high relief from a background of cloud designs in spiral form. The vigorous monster-head from which the single loop-handle springs is powerfully designed, and, like all these mythological motives, is calculated to impress and inspire the minds of those engaged in the performance of the ritual, and to inculcate the principles of right conduct.

This piece is said to have been excavated from a tomb situated at Lung-hu Shan ("Dragon and Tiger Hill") in the province of Kiang-si.

Plate II

3. Square Bronze Goblet

Type *fung kia* 鳳斝.
Shang Period (1783–1123 B.C.).

Dimensions: 13 inches in height; opening, 10¼ inches in length, 5 inches in width.

This vessel, known as "phœnix goblet" (*fung kia*), is unique. It is probably the only one of its class in existence. Even the emperor K'ien-lung, in all his glory, did not possess a piece like this in his museum collection. In its structure this goblet is a *tsio* "made square," the squaring-up process affecting not only the form of the vessel, but even extending to the spiral designs. As we have round *ting* with three feet and rectangular *ting* with four feet, so we are confronted here with four spear-shaped legs in conformity with the square form of the bowl. Likewise, each leg is quadrilateral, while in the *tsio* it is trilateral. The two external sides are ornamented with a conventionalized human or animal head dissolved into geometrical combinations of spirals of square shape. The eyes are plainly indicated by small strokes in quadrangular enclosures, and the nose is forcibly brought out. Toward the point, the artist was compelled to adapt his composition to a given space and to follow the outlines of the leg. Eight tooth-shaped ridges dissect the four surfaces of the vessel into eight panels. Each side is divided by a groove into an upper and lower section. The elements of the decoration are based on what the Chinese archæologists term the motive of the reclining or sleeping silkworm cocoon. These elemental forms, again, are so combined on each side as to form a face, which, if a conjecture is permitted, may represent the Silk Goddess. These designs, wrought in undercut relief, are set off from a background of delicately traced spirals, symbolic of thunder and atmospheric conditions. On the exterior of the spout we encounter the strongly conventionalized figure of a phœnix (*fung*), "dancing in the clouds" (a well-known motive), the clouds being expressed by the spirals. The term "phœnix," it must be understood, is merely a convenient word used by us, but, as a matter of fact, bears no relation to the phœnix of the Occident.

What is the significance of this vessel? Neither form nor design nor composition is meaningless to the Chinese artist. Numbers play a conspicuous rôle in ancient cosmogony, and everything in the old rituals was reduced to a fixed standard of numbers and categories reflected in celestial phenomena. It will be remembered that the deity Heaven was worshiped by the primeval Chinese under the emblem of a perforated jade disk, and that the deity Earth was revered and symbolized by a jade tube rounded in the interior but square on the outside, and provided with projecting teeth along the corners. I therefore venture to suggest that the circular *tsio,* as shown in Plate I, was in its origin employed for pouring out wine-offerings to the cosmic power of Heaven, and that, correspondingly, the square variation of this type, as represented by this unique and memorable bronze goblet, at the outset served for libations to the deity Earth, which was conceived to be square, and which was regarded next in importance to Heaven. In fact, the lower square section of the bowl, rising above the four legs, bears a most striking resemblance to the jade tubes *tsung,* which symbolize the deity Earth. As the perforated jade disk served as an emblem to the emperor, who was believed to receive his mandate from Heaven, and who ruled by his command as the Son of Heaven, so the jade tube, signifying female power, was the sovereign emblem of the empress. In the same manner, we meet, on this goblet, symbols alluding to silk and possibly even the Silk Goddess. Now, silk was looked upon as one of the precious gifts of Mother Earth, and the rearing of silkworms, as well as the spinning of silk, is ascribed by tradition to a woman's initiative. The empress took a profound interest in the welfare and promotion of the silk industry. When silk cocoons were offered to the empress, she used a jade image of Earth as a weight-stone, in order to weigh the quantity of silk. In the imperial worship performed by the Manchu emperors, silk was still offered in the sacrifice to Earth, being buried in the ground. Finally, the "phœnix dancing in the clouds" is an attribute of love and veneration, and occurs as such on the ancient jade girdle-ornaments worn by women and interred with them in the grave as a symbol of resurrection. All of these facts taken together prompt us to the conviction that this vessel had an extraordinary place assigned to it and a specific function in the rituals performed by the empress in her homage to Earth and Silk.

The beauty of this bronze is enhanced by a rich patina of the brown of autumn-leaves, interspersed with specks of malachite blue-green.

Plate III

4. Bronze Beaker

Type *ku* 觚.
Shang Period (1783–1123 B.C.).

Dimensions: 10½ inches in height; diameter of opening, 6 inches; diameter of foot, 3½ inches.

It is recorded that this vase was discovered in an ancient well at Wu-ch'ang on the Yang-tse, capital of Hupeh Province. It is equally beautiful for its well-balanced proportions, its noble simplicity, its purity of form and design, and the exquisite quality of its patina. This type was first produced under the Shang, and was subsequently adopted by their successors, the Chou. Judging from a famous passage in the Confucian Analects (*Lun yü,* VI, 23), it appears that this vessel underwent some changes in the age of the great sage, but, nevertheless, retained its old name. Confucius denounced the government of his time, which indulged in high-sounding phrases without applying the wise principles of the ancients, and illustrates the folly of using words that do not express the reality underlying them by an allusion to the vessel *ku,* Confucius maintaining that the word *ku* meant essentially that which had corners, while the vessel so named had none. However, the four slightly projecting, dentated ribs around the stem and foot might be so considered in this example.

The spiral composition is chased with unequaled vigor and firmness, and the asymmetry in the arrangement of the designs is a noteworthy feature. The two raised knobs in the middle portion and on the foot are intended for eyes, hinting at the fact that the artist meant to bring out the head of some mythical creature in the seemingly arbitrary combination of these scroll designs. As the spirals symbolize clouds, and the peculiar lanceolate designs, in combinations of four or six, are explained as representing the winds, we shall not err in regarding this head as that of the Storm God moving over a clouded sky.

Some of these *ku* are entirely bare of ornamentation, while others are decorated from top to bottom. Others, again, like the specimen here illustrated, are ornamented in the middle and lower portions; a few, also, in the middle portion only. All, however, are built in three sections, plainly set off by grooved zones, and have the same slender, graceful body and flaring trumpet-shaped opening. The specimen here illustrated is a production decidedly characteristic of the Shang period (compare the analogous Shang pieces illustrated in the *Po ku t'u lu,* Chapter XV, pp. 23, 32, 34, and one in the Catalogue of the Collection of the late Viceroy Tuan Fang, Vol. IX, p. 27).

No other nation can boast of having conceived a vase that could rival this type in grace and beauty of form and sense of pleasing proportions. In regard to its use, nothing definite is known beyond that it served as a wine-vessel. It does not seem to be mentioned in the ancient Rituals.

The entire specimen is coated with a very fine, lustrous, deep olive-green patina.

Plate IV

5. Sacrificial Bronze Jar

Type *i* 彝.

Shang Period (1783–1123 B.C.).

Dimensions: 7½ inches in height; diameter of opening, 9⅛ inches.

A chapter of early mythology, the record of which is unfortunately lost, is pictographically embodied in this bronze. A fierce tiger-head, moulded in high relief, looks down from the center of the upper zone, and is repeated on the opposite side, being surrounded by four creatures of strongly conventionalized forms (possibly birds), the same being repeated in the lower zone along the foot. In the middle zone is vigorously traced the colossal face of some powerful deity, the eyes and nose being prominently delineated, the other parts being filled in by geometrical designs. There are four curious geometric ornaments to the right and left of the handles, which spring from a demon's head,—very much like those connected with the creatures in the upper and lower zones. The style and content of the whole ornamentation are thoroughly characteristic of the Shang period. It is very instructive to compare this object with a similar one in the collection of the emperor K'ien-lung (illustrated in the *Si ts'ing ku kien,* Chapter XIII, fol. 17), which, however, comes down from the Chou period, and which lacks the spontaneous force and direct expressiveness of the Shang production (compare also the fine Shang specimens in the Catalogue of the Collection of the late Viceroy Tuan Fang, Vol. I, pp. 47, *et seq.*). Special attention should also be called to the fine proportions of the vessel and the graceful curves of the loop-handles.

This specimen, it is reported, was exhumed from a grave at Wu-kang chou, in the province of Hu-nan. It is covered in the interior and exterior with thick layers of a patina brilliant with red, gold, blue, green, and brown tinges. The Chinese archæologists term this a five-colored patina.

Plate V

6. Bronze Bell

So-called Dragon Bell (*lung chung* 龍 鐘).
Chou Period (1122–247 B.C.).

Dimensions: 16¼ inches in height; shoulders, 9 inches in width; base, 11½ inches in length, 7½ inches in width.

Bells occupy a prominent place in Chinese antiquity, and belong to the noblest and most admirable achievements which the Chou artists have created in bronze. Elaborate rules for the making of bells are formulated in the *Chou li,* the old State Handbook of the Chou dynasty, which with minute detail sets forth the court ceremonial, the functions of the officers and regulations for their guidance, as well as the productions of the imperial workshops. Bells were invented in China independently of the Occident; the ancient Chinese bell is a type of its own, and also differs considerably from the spherical bell subsequently introduced with Buddhism from India. The independence of the Chinese type is demonstrated by its peculiar form and the absence of a clapper, the instrument being struck outside near the lower rim by a wooden mallet. It was chiefly used in the ancestral hall to summon the spirits of the departed, in order to partake of offerings of meat and wine. A bell was also suspended in front of the banquet hall, and was sounded to call the guests. It likewise had an orchestral function in accompaniment with other musical instruments, for music, as in Plato's republic, formed an integral part of Chinese education and ceremony. Most of the early bells have the two coats set with bosses, arranged according to a fixed scheme, in groups of three, distributed over three rows, three times three being enclosed in a rectangle, so that eighteen appear on each face, making a total of thirty-six. There are many bells, however, without any bosses, and a few have twenty-four of them. There has been much speculation among Chinese and other archæologists as to the function of the bosses. Wang Fu, author of the *Po ku tʻu lu,* published in A.D. 1107, has compared them with nipples, which he takes as an emblem of nutrition, arguing that nipples are represented on bells, because "the sound of music means nutrition to the ear." The simile with nipples, however, does not occur in any ancient text, above all, is absent in the *Chou li,* which speaks merely of knobs. It can hardly be imagined that these bosses—of which, by the way, there is a large variety of different shapes, many of which show no resemblance whatever to nipples—should have served a purely ornamental or esthetic purpose. They were doubtless made with a practical end in view, and, as supposed by some Chinese authors on music, were originally used for regulating and harmonizing the sounds of bells, while later generations forgot this practice, and merely applied the bosses ornamentally (compare *Po ku t'u lu,* Chapter XXV, p. 31). This subject is deserving of a close investigation, for which, naturally, a number of ancient, authentic bells would be required to carry on practical experiments.

The bell here reproduced is remarkable for its imposing simplicity and grandeur of conception. It is a truly classical example of Chou art, inspiring a feeling of reverence and admiration, such as we receive from the lofty arches of an old Gothic cathedral. The *Po ku t'u lu* (Chapter XXIII, p. 14) illustrates a Chou bell very similar to our specimen, except that it is adorned with eight dragons (or perhaps lizards) instead of four, two being added on the right and left sides. The thirty-six nipple-shaped bosses (eighteen on each face) are perfectly modeled, and the five vertical lines of the central zone, as well as the raised meander bands, are delineated with unsurpassed precision and firmness. It is very much to the point that, as an Arabic writer wittily remarked, "Allah, when distributing his gifts among mankind, placed them in the heads of the Greeks, in the tongues of the Arabs, and in the hands of the Chinese." But the Chinese also had the right spirit and knew how to embody the spirit in their art.

The entire bell is coated with a beautiful blue-green patina speckled with gold and brown, which was produced by chemical action underground.

Acquired for the Miss L. M. Buckingham Collection,
The Art Institute of Chicago.

ɪte VI

7. Square Bronze Vase

Type *fung tsun* 鳳尊 ("Phœnix Tsun").
Chou Period (1122–247 B.C.).

Dimensions: 13¾ inches in height; opening, 10⅞ inches in width; shoulders, 8¼ inches in width; base, 6¾ inches in width.

This majestic piece is constructed in three sections clearly set off from one another, although the whole piece is cast in one mould. As in the case of the *ku* (Plate IV), the corners are provided with projecting ribs, and each of the four sides is divided into two panels by a similar rib running through the center. The composition of each zone, however, presents a unit, the same subject being repeated on each of the four sides. The upper panel is occupied by eight triangular fillets, which are intended to symbolize mountains (compare Plate I); for this reason, they always have their place on the neck of a vase, the point or summit reaching its edge. Being suggestive of a towering mountain scene, they lend the vase a feeling of loftiness and sublimity, and strike our imagination. As the triangular bands are filled in by cloud and thunder patterns, we have a symbolic representation of mountains overcast with clouds, ready to pour down fertilizing rain on the fields. Such was the wish of the farmer, and in this simple, impressionistic manner he conveyed his thought. In the lower segment of the upper zone we note a pair of conventionalized animals in strong relief, facing each other, their bodies being formed of spiral designs, the eyes being indicated by ovals. In the rectangles forming the base is brought out a pair of similar or identical creatures. The two birds confronting each other in the middle zone exhibit a certain tendency to realism, especially in the bold outlines of their tail-feathers, while again circles, half-circles, spirals, and curves are resorted to in order to make up the composition. It is possible that this bird is intended for the fabulous *fung* (so-called phœnix) for which this vase is named, but this remains a conjecture for the present.

The vase is finely incrusted with a deep greenish-brown olive patina on three sides, the fourth exhibiting a light green tinge.

Plate VII

8. Rectangular Bronze Vessel

Type *shuang fu* 雙簠 ("Double Fu").
Chou Period (1122–247 B.C.).

Dimensions: 11½ inches in length; between the handles, 13¼ inches; 8⅜ inches in width; 7¼ inches in height.

This object is perfectly unique, and none like it is traceable in any Chinese catalogue of bronzes. It is composed of two equal parts, each in the shape of a rectangle, posed on a hollow base with sides slanting outward. Each single part would form a vessel in itself, and this type of vessel was frequently used in ancient times, being known under the name *fu*. In the origin, the *fu* was a basket, defined by the ancient dictionaries as "square outside and round inside, used to hold boiled millet in State worship." To every student of basketry, baskets which consist of two equal halves, perfectly fitting one over the other (for instance, globular baskets composed of two hemispherical pieces), are well known; and the supposition seems to be well justified that the caster of this bronze derived his inspiration from such a double basket; hence I have proposed for this novel type the name *shuang fu* ("double basket"). The *fu* were also carved from wood or moulded from clay, and a few specimens of Han pottery *fu* have survived; but the favorite material for them was bronze. In the museum of the emperor K'ien-lung there were sixteen bronze *fu*, figured in the *Si ts'ing ku kien* (Chapter XXIX), but he had no double *fu* like this one.

Conventionalized animal-heads, cast in prominent relief on the narrow sides of the upper and lower vessels, take the place of handles, and six smaller zoömorphic faces (two on each long side and one on each narrow side) are so fitted as to hold the two parts closely together. The slanting sides of the upper and lower bases have gracefully cut-out arched openings, making four feet in the corners. The long, massive bands of meander patterns laid around the body in close combination are so delicately traced that the reproduction hardly renders them justice.

The patina which covers the entire object on the exterior and interior is very extraordinary in its delightful shades of light blue and green.

Acquired for the Miss L. M. Buckingham Collection,
The Art Institute of Chicago.

Plate VIII

9. Honorific Bronze Jar

Type *tui* 敦.

Chou Period: Inscription yielding date 999 B.C.

Dimensions: 11¾ inches in height; diameter of opening, 11½ inches; entire length between handles, 17 inches.

This highly artistic bronze vessel has a lengthy, clear-cut inscription, consisting of 152 characters made in the cast, in the archaic script of the Chou period, so much at variance with the modern form of writing. The text of the inscription is given twice,—on the bottom of the vessel and on the inside of the cover. This inscription has been reproduced and transliterated in modern characters by the celebrated scholar Juan Yüan (1764–1849) in a work which contains an extensive collection of inscriptions to be found on ancient bronze objects. It is therefore reasonable to infer that the *tui* here in question once belonged to Juan Yüan's collection, or at any rate was well known to him (compare also a remarkable *tui* figured in the *Kin shi so,* Vol. I; and *Si ts'ing ku kien,* Chapter XXVII, pp. 8 *et seq.*). To give a complete translation of this inscription would require a lengthy philological dissertation to be accompanied by numerous explanatory footnotes, and I hope to make it the subject of a special study in the near future, to be published elsewhere. The essential points of the story are that King Mu (1001–946 B.C.) of the Chou dynasty, when he dwelt in the ancestral temple of K'ang and Chao, accompanied by the chief minister Hung, ordered the annalist Kuo Sheng to issue a diploma in favor of a certain Sung, who was to receive the appointment to a new office; a black robe, a girdle with a buckle, jade ornaments, a standard with small bells, and bridles adorned with bells, were conferred upon him. Sung prostrated himself before the Son of Heaven, expressing his thanks and extolling the glory and benevolence of his majesty. On this occasion, he had this precious bronze vessel cast in commemoration of his venerable departed father, Kung-shu, and his venerable departed mother, Kung-se, in order to cultivate filial piety and to solicit their constant and powerful protection. The most important point is that the inscription opens with a date, "the sixth day of the fifth month of the third year"; and as it appears from the context that the mention is of King Mu, we are confronted with the date 999 B.C. This must also be the date for the casting of the vessel, which thus looks back to the respectable age of 2922 years. The inscription shows also that this vessel was made to commemorate an important event in the life of a high official.

It rests on three low feet, each being surmounted by the head of a monster. The two massive handles spring from well-modeled elephant-heads, a motive not infrequently employed in the art of the Chou. The body is adorned with six deeply-grooved bands, and correspondingly we meet such bands on the top of the cover. The same composition is brought out in the zone laid around the rim of the vessel and along the border of the cover, the same motive being repeated eight times. Despite the purely geometric character of these spirals, the plastic eye in the center is suggestive of a watchful or all-seeing deity. The cover is surmounted by a bowl-shaped flange, on the bottom of which are engraved very exquisite designs.

The *tui* were originally carved from wood and served as receptacles for millet offered to the ancestral spirits. This was also the intention of Sung, whose thoughts, on the memorable day of his promotion, were first of all directed toward his parents, to whom he felt obliged to attribute his success, and whose blessings he invoked for himself and his descendants. Such vessels were naturally transmitted as an heirloom from father to son.

Plate IX

10. Gilt Bronze Vase

Type *hu* 壺.
Han Period (206–22 B.C.).

Dimensions: 16½ inches in height; 13¾ inches in diameter.

This exquisite vase was exhumed from a grave located in the prefecture of Chang-te in the province of Ho-nan. As a type it is well known, and, wrought in bronze, goes back to the culture of the Chou period. It was subsequently adopted by the Han, and developed into one of the most favorite vases of that period, not only for every-day use, but also for the equipment of burial chambers. We have numerous specimens of bronze, pottery, and cast iron. The remarkable feature of the present specimen is its heavy coating of gold foil. This is the only gilt vase of this type that has ever come under my notice. It is partially covered with a thick green patina, which, in combination, or contrast, with the luster of the gold, produces an extraordinary effect. The interesting question arises here as to how this patina was produced. Gold making an air-tight coating, the copper under its surface could certainly not oxidize. It seems rather plausible that this vessel, while in the grave, came in close contact with plain copper or bronze objects the oxidation of which was gradually transplanted to it. This, however, like all other problems bearing on the development of patinas, merits a profound investigation to be supported by chemical analyses and other researches.

Acquired for the Miss L. M. Buckingham Collection,
The Art Institute of Chicago.

Plate X

164

西藏地区利用头盖骨和人骨的情况

Use of Human Skulls and Bones in Tibet

BY

BERTHOLD LAUFER

Curator of Anthropology

FIELD MUSEUM OF NATURAL HISTORY

CHICAGO

1923

FIELD MUSEUM OF NATURAL HISTORY
DEPARTMENT OF ANTHROPOLOGY
CHICAGO, 1923

LEAFLET NUMBER 10

Use of Human Skulls and Bones in Tibet

Among the many customs of Tibet none has attracted wider attention than the use of human skulls and other bones both for practical purposes and in religious ceremonies. Weird stories to this effect were brought to the notice of the occidental world by mediæval travellers who visited Cathay or the court of the Great Khan during the thirteenth and fourteenth centuries. In the Tibetan collections (Hall 32, West Gallery) obtained by the Blackstone Expedition in 1908-10 may be viewed (Case 70) bowls consisting of a human cranium, from which libations of liquor in honor of the gods are poured out on the altars of the Lama temples. Some of these skull-bowls are elaborately mounted and decorated, lined with brass or gilded copper and covered with a convex, oval lid that is finely chased and surmounted by a knob in the shape of a thunderbolt (Sanskrit *vajra,* Tibetan *dorje*), the symbol of Indra which is in constant use in nearly all Lamaist ceremonies. The skull itself rests on a triangular stand, cut out with a design of flames, at each corner of which is a human head. These settings are frequently very costly, being in gold or silver, and studded with turquois and coral.

In the case (68) showing musical instruments which are used for worship in the Lama temples are on view small tambourines made of two human skull-caps cemented together by means of a wooden disk.

These drums are shaken while reciting prayers, to mark the intervals between different incantations. There are trumpets made of human thigh-bones, the bones of criminals or those who have died a violent death being preferred for this purpose. These trumpets are consecrated by the priests with elaborate incantations and ceremonies. In the course of this ritual the officiating priest bites off a portion of the bone-skin; otherwise the blast of the trumpet would not be sufficiently powerful to summon, or to terrify the demons. On one side the trumpet has two apertures styled "nostrils of the horse." This is a mythical horse believed to carry the faithful after their death into Paradise; and the sound of this trumpet reminds the people of the neighing of this horse.

A most interesting addition was recently made to this group of objects by the exhibition (in Case 74) of a very valuable bone apron composed of forty-one large plaques exquisitely carved from supposedly human femora and connected by double chains of round or square bone beads. Such aprons are used by magicians in the Lama temples during the performance of mystic, sacred ceremonies accompanied by shamanistic dances, chiefly for the purpose of propitiating evil spirits and exorcising devils. The plaques are decorated with figures of Çivaitic and Tantric deities, some of which are represented in dancing postures.

At the outset, these relics of an age of savagery and a barbarous cult leave no small surprise in a land whose faith is avowedly Buddhistic, and whose people have made such signal advances in literature, poetry, painting, sculpture, and art industries. Buddha was an apostle of peace and universal love, averse to bloodshed, and forbidding the taking of human and animal life. He repudiated all outward ceremonies and offerings, preaching salvation through the efforts of the mind and the perfection of the heart. But there is room for many extremes in both nations and individuals.

Friar Odoric of Pordenone, who travelled from 1316 to 1330, dwells at some length on the burial customs of the Tibetans, and tells the story of how the corpses are cut to pieces by the priests and devoured by eagles and vultures coming down from the mountains; then all the company shout aloud, saying, "Behold, the man is a saint! For the angels of God come and carry him to Paradise." And in this way the son deems himself to be honored in no small degree, seeing that his father is borne off in this creditable manner by the angels. And so he takes his father's head, and straightway cooks it and eats it; and of the skull he makes a goblet, from which he and all of the family always drink devoutly to the memory of the deceased father. And they say that by acting in this way they show their great respect for their father. It must be added, however, that this account is not based on personal observation, but on hearsay. William of Rubruk, a Flemish Franciscan, who visited the court of the Mongol Khan in 1253, mentions the same Tibetan practice and admits that he received his information from an eye-witness. The peculiar burial customs were a characteristic trait of the Tibetans by which their neighbors were deeply struck, and the story of this ceremonial freely circulated among the Mongols who were doubtless inclined to exaggerate also some of its features.

This Tibetan custom reveals a striking parallel to a record of Herodotus. In his account of Scythia, Herodotus (IV, 23) speaks among many other nations also of the Issedonians, who are located east of the Bald-Heads and were the farthest nation of which the Greek historian had any knowledge. "The Issedonians," Herodotus relates, "have the following customs. When a man's father dies, all the near relatives bring sheep to the house; these are sacrificed, and their flesh cut into pieces, while at the same time the deceased man's body undergoes the like treatment. The two kinds of flesh are mixed together, and the whole is served at a banquet. The head of the dead man is dealt with in another way; it is stripped bare, cleansed, and set in gold. It then becomes an ornament on which they pride themselves, and is brought out year by year at the great festival which is observed by sons in honor of their father's death. In other respects the Issedonians are reputed to be observers of justice, and their women have equal authority with the men." Some scholars have assumed that the Issedonians represent a tribe akin to the present-day Tibetans or could even be their ancestors. Be this as it may, the coincidence of the fact of skull-worship among the two tribes would not constitute sufficient evidence for this theory, as the same or similar practice is encountered among widely different peoples.

The preceding case presents a peculiar form of ancestral worship, the son being intent on preserving the most enduring part of his father's body as a constant reminder, and drinking from his skull in his memory on the day of his anniversary. This, without any doubt, has been an indigenous practice in Tibet of considerable antiquity. Aside from this we meet in that country the use of human bones for purposes which move along an entirely diverse line of thought.

The Jesuit Father Andrada, who visited western Tibet in 1625, observed that the Lamas, when engaged in prayer, were in the habit of sounding trumpets made of metal or the bones of the dead, and that the bones of human legs and arms served for making these instruments. "They also have rosaries consisting of beads made from human skulls," he writes. "When I inquired why they employed bones for such purposes, the Lama who was a brother of the king, replied, 'The people, at the hearing of such trumpets, cannot fail to be mindful of death. For the same reason we avail ourselves of the bones of the dead for rosary beads. Finally, in order to be still more imbued with this melancholy and sad remembrance, we drink from a cranium." According to the same Lama, the idea of death, no less than prayers, contributes to restrain our passions and to regulate our conduct. "These cups of the dead," he remarked, "prevent the people from becoming too much addicted to worldly pleasures, which are uncertain and fugitive, so that the drink develops into a spiritual antidote for passions and vices." This manner of reasoning is not Tibetan, but is decidedly Buddhistic and, as everything else pertaining to Buddhism, has filtered into Tibetan thought from India.

At the present time, as far as observations reach, it is not known that Tibetans preserve the skulls of deceased relatives as drinking vessels, although it may still happen that bones of relatives are kept in houses from motives of religious piety. There are ascetics, however, who make use of human skulls as eating bowls, in the same manner as they make beads for rosaries out of bits of bones. But this custom bears no relation to the ancient family cult of skulls which, as we have seen, presents a form of ancestral worship. The leaning of the Buddhist hermit toward skulls moves along quite a different line, and is prompted by customs adopted by the Tibetans with the Çivaitic worship from India. In this debased form of religion we find in Tibet numerous terrifying deities who wear wreaths of human skulls as necklaces, are clad with human skins, or hold a bowl consisting of a cranium filled with blood. Such a bowl, for instance, is seen in the hand of Padmasambhava, who is still worshipped as the founder of Buddhism in Tibet (eighth century A.D.), and who besides the doctrine of Buddha introduced a system of wild magic and devil-dances connected with incantations and exorcisms (see image of Padmasamhava in Case 71, second shelf, east end).

In India, skulls were chiefly used by the Aghori or Augar, a Çivaitic sect of Fakirs and religious mendicants, which has now dwindled down to a very few members. They used human calvaria as bowls for eating and drinking. This was done as a part of their practice of self-abasement, and was associated with the cannibalistic habits permitted and encouraged by those ascetics.

The Chinese pilgrim Hüan Tsang, who visited India in the seventh century, mentions "naked ascetics and others who cover themselves with ashes, and some who make chaplets of bone which they wear as crowns on their heads."

Amitābha, the Buddha of Endless Light (also called Amitāyus, "Endless Life"), who presides over the Paradise in the west (Sukhāvati), where every one of his devout adherents yearns to be reborn, was originally a deity of purely Buddhistic character, being represented with a bowl holding a sort of nectar which confers immortality upon his devotees. In course of time, this bowl was replaced by a cranium, and it became customary to offer the god a cranium with an invocation of divine blessing for the donor; thus, another custom came into vogue, to utilize human crania as receptacles for the wine or other liquid offered to the temple-statues of the gods. For the purpose of selecting proper skulls, the Lamas have developed a system of craniology which imparts instruction as to the distinctive symptoms of good and bad skulls and the way to obtain prosperity when once the characteristics of a skull have been determined. It is essential that a skull designed for an offering to the gods should be that of a person known to have been profoundly religious, or to have possessed other high qualifications, such as rank, nobility, wisdom, or learning. Failing such a skull, others may serve as substitutes, and elaborate rules have been laid down to determine those suitable for sacrificial bowls or as offerings to the gods. Skulls of women and children born out of wedlock are unsuitable for sacred purposes. Among the very best are skulls of a clear white color like a brilliant shell, or of a glistening yellow like gold, or like a jewel without unevenness, or of equal thickness and of small cubic capacity, or with a sharp ridge stretching far into the interior like a bird's beak or a tiger's claw, or hard and heavy as stone, or smooth to the touch and polished, or with no line on it save clearly defined sutures. These and similar instructions are contained in a small Tibetan book, which teaches the method of discriminating between good and bad skulls and how, by offering a skull (Sanskrit *kapāla*) to Amitābha, prosperity and worldly goods may be secured. This lore is not Tibetan, but has emanated from mediæval India. The background of the treatise in question is Indian: the Indian caste-system is in evidence, for the skulls of Kshatriyas, Brahmans, and Vaiçyas are good, while those of common people and Chandālas are bad. There are indications from which a good skull may be told in a live person: if he has soft and smooth hair of lustrous black, if his forehead is broad and his eyebrows thick, if on his forehead there is a mark, if he has most teeth in his upper jaw, if the tip of his tongue can touch his nose (this is a peculiarity possessed in even a greater degree by all Buddhas), if his voice is high-pitched and his complexion as fresh as that of a youth, if in walking he throws his left hand and left foot out first.

The ceremony of offering a skull to Amitābha is a complex and elaborate procedure, accompanied by a fixed ritual and many offerings of food arranged on the altar in a prescribed order. A thunderbolt wrapped around with strings of various colors is placed inside of the skull, the underlying idea being that a colored light will radiate from the heart of the officiating Lama, and conducted by the strings binding the thunderbolt will penetrate into the light emanating from the heart of the Buddha of Endless Light (Amitābha) whose statue is assumed to be alive. Through this optical contact and spiritual union, the god's soul will be aroused and communicate to the Lama innumerable blessings. From their united hearts will proceed a light which will remove the sorrows of the poor and fulfill all their wishes, and from the extreme end of this light will pour down a rain of jewels which will replenish all the regions of the world and the devotee's own abode. Holding the strings wrapped around the thunderbolt and raising the skull with both hands to his head, the officiating Lama proceeds to recite a prayer, the beginning of which is thus: "Descending from the wide expanse of heaven, Amitābha who art wise, who art the lord of wealth, whose body is as voluminous as the sun, who art full of precious sayings, thou with ornaments and garments of jewels, grant me thy blessing! Mighty one, grant me might! Bless me, thou powerful one! Thou glorious one, grant me blessings! Lord of life, give me life! Lord of riches, give me wealth, confer on me in endless amount all desirable worldly blessings!" Having thus implored the divine blessings, the countenances of the gods in the temple-hall will show their pleasure by melting into light, which reaches to the heart of Amitābha and to the skull to be offered. The Lama then realizes that all his wishes have been fulfilled, and after an offering to the guardian and local deities, will wrap up the skull in silk coverings and hide it away in the store-house of the temple. The skull must be carefully concealed, and no one must be allowed to touch it; for in this case it would lose some of the qualities which it possesses, and the owner's luck would be impaired or perhaps even utterly destroyed.

A peculiar case has been recorded by the late W. W. Rockhill (Land of the Lamas, p. 273). In an uprising instigated by the Lamas in 1887 against the Catholic missionaries along the borders of eastern Tibet, the bones of Father Brieux killed in 1881 were taken from his grave, and his skull was made into a drinking-cup.

Whereas the use of enemies' skulls is extinct in Tibet, the idea itself is slumbering in the pictures and statues of Lamaist deities. A special class of these have been singled out to act as defenders of the faith and to destroy all enemies of the Buddhist religion. The main attribute of these militant demons is a wreath of human skulls surmounted by a thunderbolt. These skulls are naturally supposed to have been captured from enemies; they accordingly represent trophies and simultaneously convey a warning to others to avoid the same fate. Numerous examples of this kind may be seen in Tibetan paintings and statuary (cf. also the masks employed in the Tibetan mystery-plays, Hall I). In Case 80, at the north end of Hall 32, are on view several Tibetan sculptures on stone slabs. One of these, carved in black slate, represents a Dākini, a female sprite akin to our witches, who holds in her left hand a skull-bowl filled with human blood; she has lifted the cover from the bowl, which she carries in her right hand. Her necklace consists of a row of human skulls.

In the Vinaya, the ancient code of monastic discipline of the Buddhists, monks are forbidden using skulls as alms-bowls, as was then customary among devil-worshipping sects.

The customs of a people may be better understood and evaluated by checking and correlating them with similar or identical usages of other nations.

The typical skull-bowl drinkers in times of antiquity were the ancient Scythians, Iranian tribes of roaming horsemen inhabiting southern Russia. Like the Malayans and other peoples, the equestrian Scythians, as described by Herodotus (IV, 64), were head-hunters. The Scythian soldier drank the blood of the first man he overcame in battle. The heads of all slain enemies were cut off and triumphantly carried to the king; in this case only was he entitled to a share of the booty, whereas he forfeited all claim, did he not produce a head. The scalps were likewise captured and suspended from the horse's bridle; the more scalps a man was able to show, the more highly he was esteemed. Cloaks were made by many from a number of scalps sewed together. The skulls of their most hated enemies were turned into drinking-cups, the outside being covered with leather, the inside being lined with gold by the rich. They did the same with the skulls of their own kith and kin if they had been at feud with them and vanquished them in the king's presence. When strangers of any account came to visit them, they handed these skulls around, the host telling how these were his relations who made war upon him, and how he defeated them; all this was regarded as proof of bravery. The practice of the Scythians in capturing and preserving the skulls of slain enemies was doubtless inspired by the widely prevalent belief in the transference of the powers of the deceased to the victor, who, in accordance with this conception, was enabled to add the skill, prowess and courage of his dead enemy to his own.

Livy relates that the Boii, a Celtic tribe in upper Italy, in 216 B.C., carried the head of the Roman consul Lucius Posthumius into their most venerated sanctuary and, according to their custom, adorned the cranium with gold; it was used as a sacred vessel in offering libations on the occasion of festivals, and served as a drinking-cup to the priest and overseers of the temple.

Paulus Diaconus, in his History of the Langobards, writes that Albion, king of the Langobards, used the skull of Kunimund, king of the Gepids, as a drinking-cup, after defeating him in battle in A.D. 566 and taking his daughter, Rosmunda, for his wife. On the occasion of a merry banquet at Verona he ordered wine to be served to the queen in this bowl and enjoined her to drink gleefully with her father. This brutal act led to the king's assassination in 573 by an agent of his wife. In the mediæval poetry of the Germanic peoples (Edda) there are several allusions to the use of cranial drinking-cups.

Krumus, prince of the Bulgars, defeated in three campaigns the Byzantine emperor Nikephoros, who was slain in A.D. 811. The Bulgar had a fine, silver-lined drinking-cup made from his enemy's cranium. In A.D. 972 the Russian grand-duke, Svatoslav, succumbed in a battle against a Turkish tribe, the Pecheneg. It is recorded in the Russian chronicle of Nestor that Kurya, the ruler of the Pecheneg, had Svatoslav's skull prepared as a goblet trimmed with gold.

The fact that this was an ancient Turkish usage becomes evident also from the Chinese annals which have the following incident on record. When the ruler of the Hiung-nu (Huns), Lao-shang, who reigned from 175 to 160 B.C., had defeated the king of the Ta Yüe-chi (Indo-Scythians), he made a drinking-cup out of the latter's cranium. At a somewhat later date, when two Chinese envoys were sent to the Hiung-nu to conclude a treaty, they drank blood with the Turkish chiefs out of the same skull-bowl, in order to solemnize their vows. The sacrificial animal in this case was a white horse. Blood, as is well known, was of great significance with many peoples in affirming sacred agreements and keeping faith. According to the philosopher Huai-nan-tse, the ancient Chinese in such cases rubbed their lips with blood, while the inhabitants of Yüe (in southern China) made an incision in their arms.

The ceremonial use of human crania, consequently, must have been widely diffused in ancient times among Tibetan, Turkish, Scythian (that is, Iranian), Slavic, Celtic, and Germanic tribes. The custom is not restricted to the Old World, however; there are examples to be found among the natives of America as well.

Oviedo relates in his "Historia General y Natural de las Indias" that the Inca king Atabalida possessed a precious drinking-vessel made from his brother's skull. Along its edge it was mounted with gold, the skin with the smooth and black hair having been retained. The king would drink from this bowl on the occasion of festivals, and is was regarded as one of his greatest treasures and most highly esteemed. Why it was just the skull of his brother is not explained by the Spanish chronicler; nor is, as far as I know, any other instance of such a practice on record from ancient

Peru. Molina, in his "Historia de Chile" (1795), states with reference to the Araucanians that, after torturing their captives to death, they made war flutes out of their bones and used the skulls for drinking-vessels.

M. Dobrizhoffer, who worked as a missionary among the Abipones of Paraguay in the eighteenth century, gives the following account: "As soon as the Abipones see any one fall in battle under their hands, their first care is to cut off the head of the dying man, which they perform with such celerity that they would win the palm from the most experienced anatomists. They lay the knife not to the throat, but to the back of the neck, with a sure and speedy blow. When they were destitute of iron, a shell, the jaw of the palometa, a split reed, or a stone carefully sharpened, served them for a knife. Now with a very small knife they can lop off a man's head, like that of a poppy, more dexterously than European executioners can with an axe. Long use and daily practice give the savages this dexterity. For they cut off the heads of all the enemies they kill, and bring them home tied to their saddles or girths by the hair. When apprehension of approaching hostilities obliges them to remove to places of greater security, they strip the heads of the skin, cutting it from ear to ear beneath the nose, and dexterously pulling it off along with the hair. The skin thus drawn from the skull, and stuffed with grass, after being dried a little in the air, looks like a wig and is preserved as a trophy. That Abipon who has most of these skins at home, excels the rest in military renown. The skull too is sometimes kept to be used as a cup at their festive drinking-parties. Though you cannot fail to execrate the barbarity of the Abipones, in cutting off and flaying the heads of their enemies, yet I think you will judge these ignorant savages worthy of a little excuse, on reflecting that they do it from the example of their ancestors, and that of very many nations throughout the world, which, whenever they have an opportunity of venting their rage upon their enemies, seem to cast away all sense of humanity, and to think that the victors have a right to practice any outrage upon the vanquished. Innumerable are the forms of cruelty which the other savages throughout America exercise towards their slain and captive enemies."

G. F. Angas (Savage Life and Scenes in Australia, London, 1847) writes that the natives around Lake Albert and the adjoining portions of the Coorong in Australia used the skulls of their friends as drinking-vessels. After detaching the lower jaw, they fastened a bundle of bulrush fibre to them, and carried them, whenever they travelled, filled with water; always putting in a twist of dry grass to prevent the contents from upsetting. In another passage of his book he speaks of a girl who carried a human skull in her hand; it was her mother's skull, and from it she drank her daily draught of water.

It is assumed by some archæologists also that skulls were used as drinking-bowls by prehistoric man during the palæolithic and neolithic periods of Europe, merely for practical purposes. There was a time when primitive man did not yet understand how to fashion clay into pots and to bake clay into a hardened mass. Wherever nature offered gourds or calabashes or shells, he took advantage of such means; or vessels for holding and carrying water were made, as, for instance, by the aborigines of Australia, of the gnarls of trees, the bark covering the gnarls, or of a portion of the limb of a tree, or finally of animal-skins. Certain it seems that prehistoric man availed himself of human crania for scooping and drinking water. Such brain-pans wrought symmetrically by means of stone chisels have been discovered in the pile-dwellings of Switzerland, as well as in the Magdalenian and Solutrean stations of the French palæolithicon.

It would be erroneous to believe that this "barbarous" practice was limited to prehistoric times and the "savage" tribes of ancient Europe, Asia, America, and Australia. Like so many other pagan customs, it has persisted until recently among Christian, civilized nations. Even within the pale of Christianity, the skulls of saints have been preserved and worshipped. The village of Ebersberg east of Munich, Bavaria, for instance, boasts of possessing for a thousand years the skull of St. Sebastian. It is kept in a special chapel erected in 1670; there, a silver bust of the saint which hides the relic is placed on an altar. On his name-day, the 20th of January, pilgrimages were made to this chapel, and the pilgrims received consecrated wine from the saint's skull, believing they would be cured from any disease. This is but one example out of many; it was an ancient usage of the church to have the faithful drink out of bowls which formerly were in the possession of saints, and particularly out of their skulls. The same ancient belief in the magical power of bones is seen in the veneration of bodily relics of martyrs and saints. One of the earliest and best known examples is that of Lucilla of Carthage, who habitually kissed a martyr's bone before partaking of the Eucharist.

In Buddhism the worship of relics plays alike a conspicuous role. Particularly the teeth of the Buddha and an excrescence or protuberance of his skull-bone are prominent as objects of adoration among its devotees. The high skull-bone was regarded as one of the characteristic signs of beauty of a Buddha, and a relic of this kind is described as early as the fifth century by the Chinese pilgrim Fa Hien on his visit to the city Hidda in north-western India. It was kept in a shrine covered with gold-leaf and the seven precious jewels, and was jealously guarded by eight prominent men. The king made offerings of flowers and incense to the bone. Such bones were also shown in other temples, e. g., in a temple at Fuchow, China. Hüan Tsang even mentions Buddha's skull as being kept in a temple of India and enclosed in a precious casket; he says it was in shape like a lotus-leaf and yellowish-white in color.

Finally, there is a visible survival of the ancient custom still preserved in our language. German *kopf* ("head") corresponds to English *cup* (Anglo-Saxon *cuppe*), both being derived from Latin *cuppa* ("cup"). In Italian, *coppa* means a "cup;" but in Provençal, the same word in the form *cobs* means a "skull." Latin *testa* refers to a pottery vessel or sherd, as well as to the brain-pan and head. In Provençal, *testa* signifies a "nut-shell;" in Spanish, *testa* denotes "head" and "bottom of a barrel." In Sanskrit, *kapāla* means both a "skull" and a "bowl." This correlation is still extant in many other Indo-European languages.

B. LAUFER.

BIBLIOGRAPHICAL REFERENCES

ANDREE, R.—Menschenschädel als Trinkgefässe. Zeitschrift des Vereins für Volskskunde, Vol. XXII, 1912, pp. 1-33.

BALFOUR, H.—Life History of an Aghori Fakir. Journal of the Anthropological Institute, Vol. XXVI, 1897, pp. 340-357, 2 plates.

BREUIL, H. and OBERMAIER, H.—Cranes paléolithiques façonnés en coupe. L'Anthropologie, Vol. XX, 1909, pp. 523-530.

ROCKHILL, W. W.—On the Use of Skulls in Lamaist Ceremonies. Proceedings of the American Oriental Society, 1888, pp. XXIV-XXXI.

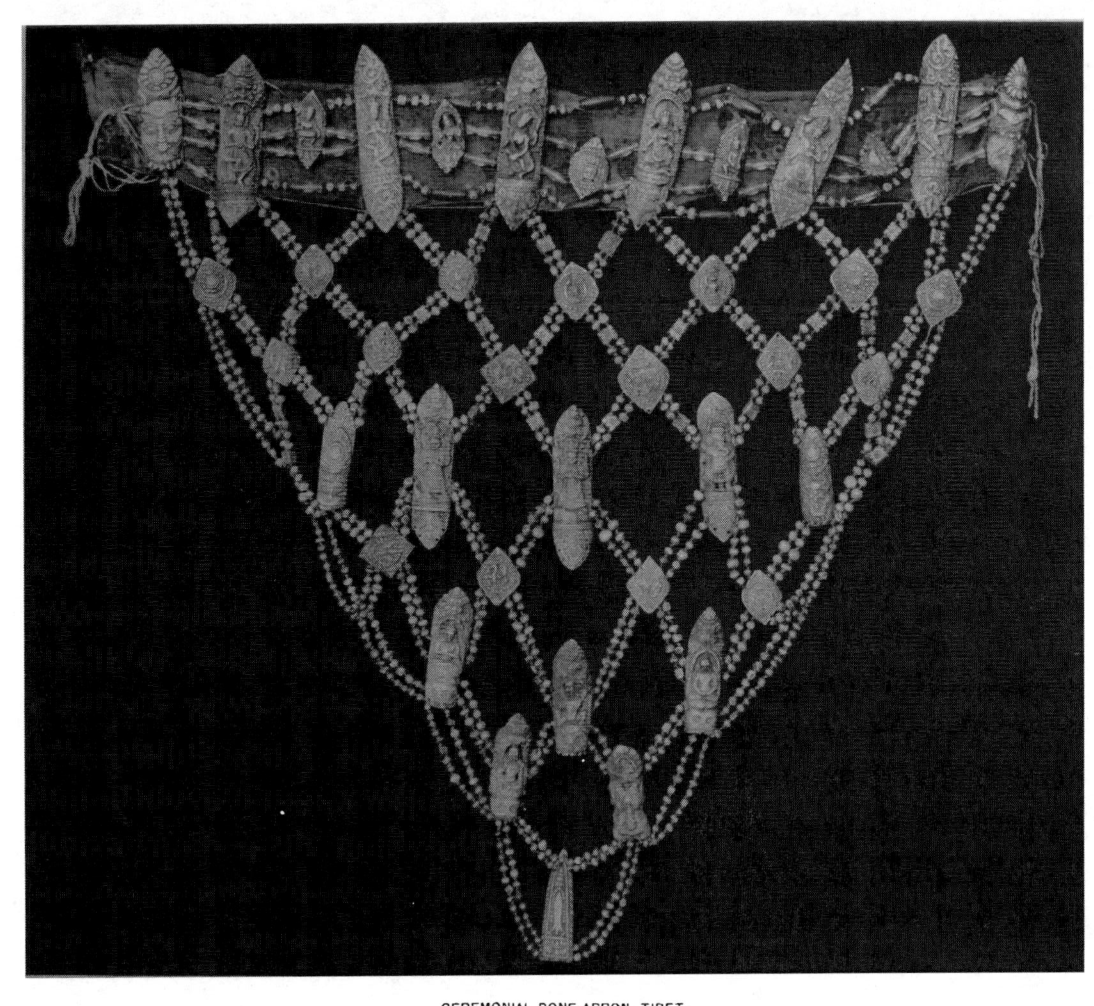

CEREMONIAL BONE APRON, TIBET.
GIFT OF ARTHUR B. JONES, 1922.

165

东方戏剧

FIELD MUSEUM OF NATURAL HISTORY

DEPARTMENT OF ANTHROPOLOGY

GUIDE

PART 1

ORIENTAL THEATRICALS

BY

BERTHOLD LAUFER
Curator of Anthropology

CHICAGO, U. S. A.

1923

SCENE FROM CHINESE RELIGIOUS DRAMA, AS DISPLAYED IN FIELD MUSEUM (Hall I, Case 1):
THE JUDGE OF PURGATORY AND HIS ASSISTANTS PASSING SENTENCE ON THE SHADES OF THE DEAD.

CONTENTS

ORIENTAL THEATRICALS

(Hall I, Ground Floor)

In this Hall the popular entertainments and theatrical performances, as practised in Oriental countries, are grouped together for the purpose of illustrating, as far as the collections at present in the Museum's possession permit, the development and the present state of dramatic art in the Orient. It is proposed to increase and extend these exhibits considerably in the near future.

The exhibits are arranged in geographical order, proceeding from west to east, as follows:—

1. Chinese religious drama, showing the ten purgatories (Cases 1-4).
2. The Lion-Dance, China (Case 5).
3. Chinese masks from an imperial play, illustrating gods and heroes of the Taoist religion (Cases 6-7).
4. Chinese shadow-play figures (Cases 8-9).
5. Masks and masked figures of Tibetan mystery-plays (Cases 10-17).
6. Puppets or marionettes and orchestra accompanying performances, Java (Cases 18-21).
7. Masks, actors' head-dresses and costumes, Java (Cases 22-24).
8. Singhalese masks, Ceylon (Cases 25-28).

THE RELIGIOUS DRAMA OF THE CHINESE

In formulating the ideas of the future life, the Brahmans of India built up a system which has become the high model of all nations of the East. It is based on the fundamental belief in perpetual rebirth and the effects of Karma; that is, retribution in another state of existence for every good or evil action performed in the present. Thus, every individual is self-responsible for his present condition of life, as it is believed to be the result of his conduct in a previous form of existence. To the Indian, pinning his faith on the dogma of metempsychosis, the sojourn in Purgatory is merely transitory, an episode in the constant flight of numerous, successive existences, which may be interrupted at a time when, by saintly conduct, the state of self-destruction, or cessation of all consciousness of existence, the Nirvana, will be reached. There is, accordingly, in Buddhism no belief in everlasting punishment; there is always hope of a final redemption. The torment limited in time is a temporary stage which effects the purification of the soul; and the culprit, running through countless evolutions, at length has an opportunity to improve his moral and physical condition when he will be reborn anew. The descent into Purgatory, consequently, is an incident in the great cycle of transmigrations, and the path which may finally lead to salvation. Nor is Purgatory itself an eternal institution; it is as destructible as this system of worlds (in the Buddhist sense), and will perish in the great conflagration befalling the present world-cycle at the end of a Kalpa (a period of a hundred thousand years). The ethical influence of this belief on the conduct of his votaries must not be underrated; it has acted as a moral stimulus on the lives of millions in the East to whom the esoteric teachings of religion are too lofty to grasp, and who are in need of a tangible memento mori.

In the pre-Buddhist age of China there was a belief in a future life, but the conceptions of the future abodes were extremely vague. Confucius, though reluctant to discuss any questions concerning the dead, recognized a continuation of this life by upholding ancestor-worship. The numerous objects of jade, bronze, pottery, and clay interred in the graves of ancient China demonstrate vividly the people's yearning for the bliss of a greater beyond. Taoism promised immortality as the reward of merit. A clear conception of a future state of happiness for the good and of torment for the wicked took root in China only with the transplantation of Buddhism from India, when, as a great educational means, it deeply agitated and uplifted the masses. The artists seconded the priests by depicting the torments of Purgatory on the walls of temples, or in setting them up in life-size clay models in niches along the temple court-yards. In China, as elsewhere, this was the grossly sensual means used to deter the people from crime and to hold them on the path of righteousness. Wu Tao-tse, the great artist of the T'ang period, is said to have painted scenes of Purgatory, the mere sight of which, it is recorded, made the beholders sweat, and their hair to stand on end. It inspired the butchers and fishmongers at the capital with such a degree of horror that many of them felt compelled to abandon the trades against which the anathema of Buddhism was hurled, and to seek a livelihood in other directions. An art-critic who saw this work and describes its effects on the contemporaries, concludes his discourse by saying, "It has caused them to seek after virtue and give up evil practices; after which, who can say that painting is only a small art?"

"But the Chinese," observes Rudyard Kipling on his visit to the Temple of Horrors at Canton (From Sea to Sea, Ch. X), "are merciful even in their tortures. When a man is ground in a mill, he is, according to the models, popped in head first. This is hard on the crowd who are waiting to see the fun, but it saves trouble to the executioners. A half-ground man has to be carefully watched, or else he wriggles out of his place." It must not be supposed, of course, that the Chinese have ever in reality practised the tortures demonstrated in the ten courts of Purgatory. This lore is not their own, they adopted it from India. It is the visual illustration of what is minutely described in the sacred books of the Buddhists. On the stage, moreover, for which these models are designed, everything is mitigated and permeated by a willful, grotesque humor which makes it difficult for the spectator to take these punishments too seriously. Sceptical and rationalistic as many of the Chinese are, they will be moved to smile at this performance or to entertain doubt as to its reality. The baroque features and semi-comic gestures of the devils contribute to the relief and exhilaration of the audience. The visitor should bear in mind that he is witnessing a fine piece of scenic illusion, which, while moralistic at its root and ethical in its tendency, is far from being calculated to shock the nerves or frighten the conscience, but which, on the contrary, will encourage and elevate by pointing the way to ultimate salvation. The keynote of this drama is not misery and despair, but hope and the possibility of self-perfection.

MASK OF KIANG TSE-YA, CHINA (p. 32).

Salvation is the dominant factor, the sole great goal of religious life in the East. If, as has been said, the average Chinese lives merely for the purpose of being buried, if he lavishes a fortune on his coffin and his funeral, this indicates that he cares more for the future than for the present life and that consequently he makes his preparations thoroughly and in earnest faith. The chief personification of the idea of salvation in popular belief is the goddess of compassion, Kuan-yin, the Madonna of the Buddhists. Her statue with eight arms occupies the centre of the great stage scene, smiling forgiveness and holding out mercy and succor even to the hardened sinner. The contrast between this personification of tender leniency and the stern and immovable god of the lower region, who is also styled the king of the law, is striking; and a more potent and artistic antithesis could not have been conceived by the playwright.

This religious drama is an apotheosis of that most popular deity. According to Chinese tradition, she was brought up as the daughter of a powerful monarch. She took no interest in court life, but turned her thoughts toward religion. She declined to be married in accordance with her father's will, and determined to enter a monastery as a nun. The enraged king had her immured in a convent, where she was assigned degrading duties in the kitchen. Ill treatment did not break her spirit. The sword, by the blow of which she was to be executed at her father's command, splintered into pieces without harming her body. Thereupon she was strangled, when her pure soul wandered away and descended into the infernal regions. The appearance of her beneficent spirit wrought a miracle and converted hell into paradise: the shades of the tormented culprits felt a sudden relief from their agonies. Yama, the god of the inferno, stood aghast at this interference with what he considered law and justice, fearing that his authority might be jeopardized. In order to restore his reputation, he revived the body of the princess and set her afloat on a lotus, which drifted to the Island of P'u-t'o in the Chusan Archipelago, where she lived for nine years, healing disease and rescuing mariners from shipwreck. Subsequently when her father fell ill, she sacrificed her hands and eyes, and made of them a medicine which saved his life. Finally she converted him to Buddhism and obtained the rank of a Buddha. After her death she was canonized and glorified as goddess of mercy, divine grace, and child-blessing, and as such she is still adored all over China and Japan, being frequently represented with a child in her arms. The climax of the drama is reached in her descent to the nether world and her vision of the fate of the damned.

The painted background of the stage is intended to represent the Island of P'u-t'o. The many birds spread over the rocks serve as lanterns, a candle being stuck into a spike, so that during the performance the whole background is illuminated, which lends to the scene a magical effect. P'u-t'o is a solitary rocky isle and a most beautiful spot, where nature and architecture have vied with each other, and are felicitously blended into an harmonious unit. Only the devotion and fervor of Buddhist priests, combined with their innate sense and love of nature, could have overcome the difficulties in making a little paradise of these rocks constantly lashed by the fury of the ocean. It is no wonder that this isle was chosen as the seat of Kuan-yin, the goddess of universal love, and that, with its temples, it is entirely devoted to her worship. It is a curious fact that despite the feminine character of this deity, who is a favorite with women, the monastic rules of the place forbid women to settle there; they may certainly come for brief visits to offer incense and worship, or to make donations to the temples, but they are not allowed to remain. Even the laymen in the employment of the priesthood are forbidden to contract marriages.

In Buddhism, the topography of the lower regions is a complicated system. Eight hot hells are assumed to be situated underneath the earth; but, as each of these establishments has four gates, and outside each gate are four ante-chambers, there are thus altogether 128 hot hells. In addition to these there are cold hells, also eight in number, arranged shaft-like, one beneath the other. This shaft is so shaped that it gradually widens down to the fourth hell, and then narrows again, the first and last hells having the smallest; and the fourth, the greatest diameter. There are, further, eight dark hells. Any individual dying in the first of these hells is at once reborn in the second, and so forth, the period spent in each of them lasting five hundred years. Finally, there are 84,000 small hells, situate on the edge of the universe, which are divided into three classes, according to their location on mountains, on water, or in deserts. There are different torments in the various hells, and the length of detention also varies in each class. People are reborn in one or the other according to their previous merits or demerits. The decision rests with Yama, the king of the infernal regions, who, assisted by eighteen judges and hosts of demons, orders what place and torture are appropriate in each instance. The Taoists adopted this system of retribution from the Buddhists, but simplified it, contenting themselves with ten purgatories, the presiding regents of which are partially associated with the gods of the sacred mountains. It is this conception of purgatory which, with a certain humorous flavor, is shown on the stage in this play. The first purgatory and its paraphernalia are displayed in Case 1, the nine others are installed in the adjoining Cases 2-4.

The visitor should bear in mind that only the figures in the center and foreground of Case 1 represent the real actors, those in the background and on the sides being merely supers. The scene is laid in a temple on the Isle of P'u-t'o, and the statue of the goddess Kuan-yin dominates the altar. She is surrounded by two male and two female attendants. The "Golden Monk" in front of the altar is intended to represent a bronze temple-statue of a Buddhist monk, which is placed in all temples as a sort of guardian.

The chief actor in front is King Ts'in Kuang, absolute sovereign and chief judge in the First Court of Purgatory. His expressive mask is thoroughly characteristic of his infernal majesty, firm, stern, solemn, and bent on meting out justice according to the immutable and inexorable law of his kingdom. In harmony with his exalted position and rank, he is clothed with the imperial robe richly embroidered with dragons and ornamental forms of the character meaning "longevity." Euphemistically he is not regarded as the god of death, but as the ruler of life. His headdress corresponds to that of the ancient emperors of China: sun, moon, the dipper, and earth are represented on the flat board.

In front of the King is the mask of "General Bull," who has the function of a beadle. He is bull-faced, because the bull is sacred to Yama, the ancient Indian god of death, who is usually represented as standing on the back of a bull, or even with a bull's head (cf. Chinese clay figures in Case 18, East Gallery). Opposite are the chief assistant judges. One is a civil officer who takes part in the judicial proceedings. On his official robe are embroidered in gold two blazons with the design of a phœnix looking toward the sun. The other is a military officer in charge of the proper execution of punishments, whose costume is embroidered with golden dragons rising from the sea, with rocky islands below.

The infernal regions are believed to be organized on much the same model as the world above, except in the matter of light. A chancery is established in each of the Ten Courts, and business is conducted on the same lines as in a Yamen, with a staff of secretaries. The three devil-lictors, behind the tortured, represent the infernal police, and have executive power. They arrest the souls of dying persons, being armed with warrants signed and sealed in due form as in the official world above. They drag the culprits who have been duly sentenced to the tortures assigned to them, and see to the execution of the punishment according to the letter of the law. They are divided into various ranks, with a major classification of "big" and "small" devils. The resemblance of their types with our mediæval devils is unmistakable; the same horns and fangs, the same dishevelled hair, the same black or colored faces, and the same diabolic expressions. Their hair is red, yellow, blue, or black.

The "small devil" with a single horn on his head and protruding eye-balls and tusks, standing to the left of the King, is placed under the latter's immediate command, and is the chief executioner. His vest is lined with a tiger-skin emblematic of his "tiger-heart."

As the Judge is often obliged to arrest culprits still living on earth, he has created the office of bailiff or sheriff. The infernal police is not eligible for this dignity, unable as they are to stand the light of the day. The bailiff, therefore, must be of human origin, and it is the ghost of a man who has committed suicide by hanging himself who is appointed; hence he is represented with a long, distorted face and protruding tongue. The four characters inscribed on his tall, white cap signify, "Great luck at one glance!" This means that those whose virtuous actions he has once observed will enjoy felicity, although arraigned before the Judge. In his capacity as lictor to the Tutelary Guardian of every town he has occasion to spy all good and evil deeds of people, which are duly recorded by him for transmission to the office of the lower region. He may also arrest, by order of the Judge, any evil-doers or witnesses required in his judicial procedures.

The King of this first Court has the custody of the book of life and death, keeping a record of all living persons. Every man and woman appears after death before this Court, and if their tale of good and evil works is equally balanced, is thence returned back to life. Male may become female; female, male; rich, poor; and poor, rich, according to the merits acquired during their lifetime. Those, whose good deeds are outnumbered by their bad ones, are sent to a terrace on the right of the Court, called the Terrace of the Mirror of Sin, ten feet in height. In this mirror the wicked souls are able to behold the crimes committed by them on earth. After they have been dragged by devils to the Terrace, they are sent forward into the Second

Court, where they are tortured and dismissed to the purgatory appropriate to their crime. On the stage, an exception is made to this rule, and in order to lend more color to this scene, two tortures are introduced into the First Court. One consists of a mill between the stones of which a woman is ground, as a punishment for adultery; a dog is licking her blood. The other represents a rice-pounder in which culprits are crushed by the pestle. The person under the pestle has just been questioned and tortured, but a lotus sprouting forth from his chest bears witness to his innocence. He will be discharged as not guilty.

Among the supers is the group of Eighteen Arhat, the most advanced disciples of Buddha, who have reached the highest degree or saintship. In this drama they merely illustrate the statues of these saints, as grouped in two rows along the walls of many temples. The other supers represent casts derived from the great mythological romance entitled *Fung shen yen i*, a famed Taoist book of wonders and miracles. In front of the altar, on the right-hand side, is the mask of Erh Lang, worshiped as "the Veritable Prince of the Marvelous Tao" (the way to righteousness). He is the patron and guardian of the dogs. He owns a supernatural dog whose mask is beside him. He can carry this dog in the sleeves of his coat, and the dog devours any enemy at whom he is set. There is a temple at Peking in honor of this god, where a clay dog is placed at the foot of the altar. He who has a sick dog takes a dose of ashes from the censer on this altar and administers it as a remedy to the sick animal. If the cure is efficient, the grateful owner of the dog will offer one of clay to the deified healer.

In front of the Arhat, on the right-hand side, are

posted the Four Guardians of the World, each protecting one side of the world-mountain Sumeru, as defenders of the Buddhist faith. In front of the altar, on the left, appears the mask of Li T'ien-wang, a Taoist conception of one of these Four Guardians and styled "the Heavenly King supporting the Pagoda." The pagoda carved from wood is carried by the actor in his left hand. In popular belief, it is of pure gold, seven inches high and capable of flight; it can lengthen itself to the height of eighty feet, whereupon it will return to its original proportions.

On the left-hand side, between the Great King and the Arhat, are four Taoist saints: In the foreground is Chao Kung-ming, who resides on the sacred Mount O-mei in Se-ch'uan, and who took part in the national war which ended the rule of the Shang dynasty and set up the house of Chou in its stead (1122 B.C.). He is followed by Po Kien, who was a hero of the mythical age and found a place in the lower region for the following reason. After his death, his soul wandered about for a millennium, searching in vain for a body into which it might enter. Finally he was rescued by a recluse and built the terrace where the souls of the heroes slain in battle are sent, in order to receive appointments as spirits or gods. The duty thus devolved upon him to escort the souls to this terrace, and he ultimately received an appointment as the head of all spirits. The third is Huang, the "Flying Tiger" (Fei-hu), a hero, who at first fought on the side of the last ruler of the Shang dynasty, but disgusted with his evil doings rebelled against him and joined the forces of his adversary Wen Wang of the house of Chou. He was killed by Chang K'uei, the fourth in this row, who owns a magical horse, "Black

Smoke" by name. "Black Smoke" is capable of covering five hundred miles a day and slaying seven men at a blow. Hence he is also honored with the epithet, the "Seven-Killer."

Attached to the background, on the right-hand side, are the masks of Yin Hung, son of the emperor Chousin of the Shang dynasty, who after many adventures was slain and canonized as the Spirit of the Five Kinds of Grain; and Mu Ch'a, who first battled under the standard of the Chou dynasty, but then retired into the solitude of the mountains to live the life of a recluse. His brother No Ch'a, who followed his example, is on the left-hand side of the background. There, also, is Yin Kiao, a brother of Yin Hung, canonized as the god of the year who resides on the planet Jupiter.

Taking leave of the First Court, we turn to the Second Court of Purgatory (Case 2, left of Case 1). Like Court I, it is believed to be situated at the bottom of the ocean. It is under the jurisdiction of King Ch'u Kiang, who rules over a vast realm many miles in extent and subdivided into sixteen wards, as follows:—In the first, nothing but black clouds and constant sand-storms. In the second, mud and filth. In the third, spikes and knives abound. The fourth is the abode of gnawing hunger; the fifth, of burning thirst. In the sixth, people sweat blood. In the seventh, the shades are plunged into a brazen cauldron filled with boiling water. In the eighth, they are tormented by sharp-edged brass tools. In the ninth, they are put into iron clothes. In the tenth, they are stretched on a rack to specified length. In the eleventh, they are pecked by fowls. In the twelfth, they have nothing but rivers of lime to drink. In the thirteenth, they are hacked to pieces. In the fourteenth, the leaves of the trees are as sharp as sword-points. In the fifteenth, they are pursued by foxes and wolves. In the sixteenth, all is ice and snow.

In this Court will be found those who filch letters, pictures and books entrusted to their care, and then pretend to have lost them; those who injure a fellow-creature; those who practise medicine without any knowledge of medical art; those who will not ransom grown-up slave-girls; those who, contracting marriage for the sake of gain, falsely state their ages, and many others.

On the stage is shown the torment of the Fiery Pillar, to which the culprits are tied to be roasted. The beadle, called "General Horse," wears the mask of a horse-head.

In the sixteenth subdivision of Court II there is a huge surface of smooth ice on which the culprits are thrust by the devil-lictors, and pushed and rolled to and fro, till their bodies become flattened out. The two figures represent the shades of a man and a woman whose bodies display the effects of this treatment.

Court III is under the jurisdiction of King Sung Ti. His realm is more than 500 *li* (Chinese miles) in compass, being subdivided into sixteen wards:—The first is the place of brine-wells and salt-pits; everything is salt; above, below, and all around, the eye rests upon salt alone. The shades feed upon it, and suffer horrid torments in consequence. In the second, the culprits are bound with cords and carry heavy pillories. In the third, they are being perpetually pierced through the ribs. In the fourth, their faces are scraped with iron and copper knives. In the fifth, their fat is scraped away from their bodies. In the

hacked to pieces. In the fourteenth, the leaves of the trees are as sharp as sword-points. In the fifteenth, they are pursued by foxes and wolves. In the sixteenth, all is ice and snow.

In this Court will be found those who filch letters, pictures and books entrusted to their care, and then pretend to have lost them; those who injure a fellow-creature; those who practise medicine without any knowledge of medical art; those who will not ransom grown-up slave-girls; those who, contracting marriage for the sake of gain, falsely state their ages, and many others.

On the stage is shown the torment of the Fiery Pillar, to which the culprits are tied to be roasted. The beadle, called "General Horse," wears the mask of a horse-head.

In the sixteenth subdivision of Court II there is a huge surface of smooth ice on which the culprits are thrust by the devil-lictors, and pushed and rolled to and fro, till their bodies become flattened out. The two figures represent the shades of a man and a woman whose bodies display the effects of this treatment.

Court III is under the jurisdiction of King Sung Ti. His realm is more than 500 *li* (Chinese miles) in compass, being subdivided into sixteen wards:—The first is the place of brine-wells and salt-pits; everything is salt; above, below, and all around, the eye rests upon salt alone. The shades feed upon it, and suffer horrid torments in consequence. In the second, the culprits are bound with cords and carry heavy pillories. In the third, they are being perpetually pierced through the ribs. In the fourth, their faces are scraped with iron and copper knives. In the fifth, their fat is scraped away from their bodies. In the

HÜAN TSANG THE WHITE HORSE SUN WU-K'UNG CHU PA-TSIE SHA WU-TSING

SHADOW-PLAY FIGURES, CHINA.
"THE JOURNEY TO INDIA" (p. 41).
Case 9, Bottom.

sixth, their hearts and livers are squeezed with pincers. In the seventh, their eyes are gouged. In the eighth, they are flayed. In the ninth, their feet are chopped off. In the tenth, their finger-nails and toe-nails are pulled out. In the eleventh, their blood is sucked. In the twelfth, they are suspended from their feet, head downward. In the thirteenth, their shoulder-bones are split. In the fourteenth, they are tormented by insects and reptiles. In the fifteenth, their thighs and knees are beaten. In the sixteenth, their hearts are extracted.

In this Court will be found those who obstruct funeral obsequies or the completion of graves; those who lose all record of the site of their family burying-place; those who incite others to commit crimes; those who promote litigation; those who repudiate a betrothal; those who forge deeds and other documents; those who counterfeit signatures and seals; those who alter bills; wives who slight their husbands; sons who fail in their duties; working partners who behave badly to the moneyed partner, and many others. All these, even though they have a credit for good deeds, must pass through the misery of every ward. On the stage, two torments are shown, a man being simmered in an oil-kettle, and another having the intestines pulled out.

Court IV is under the jurisdiction of King Wu Kuan. This Court is 500 *li* (Chinese miles) in compass, being subdivided into sixteen wards, as follows: —The first is a pool beneath a high precipice where the shades of the wicked are hung up, and water is continually poured over them. In the second, they are made to kneel on chains and bamboo splinters. In the third, their hands are scalded with boiling water. In the fourth, their hands swell and stream with perspiration and blood. In the fifth, their sinews are cut, and their bones pulled out. In the sixth, their shoulders are pricked with a trident, and the skin is rubbed with a hard iron brush. In the seventh, holes are bored in their flesh. In the eighth, they are made to sit upon spikes. In the ninth, they wear iron clothes. In the tenth, they are crushed under heavy pieces of wood, stone, earth, or tiles. In the eleventh, their eyes are plucked out. In the twelfth, their mouths are choked with dust. In the thirteenth, they are perpetually dosed with abominable drugs. In the fourteenth, they are slipping on oiled beans and constantly falling down. In the fifteenth, their lips are painfully pricked. In the sixteenth, their bodies are buried under gravel, with the head projecting above it.

To this purgatory are condemned those who cheated the customs and evaded taxes; those who repudiated their rent, used weighted scales, sold sham medicines, made base coin, got deeply in debt, sold silks and satins with a false gloss on them; those who did not make way for the cripples, old and young; those who stole bricks from walls as they passed by, or oil and candles from lamps; those who allowed their mules and ponies to be a nuisance to other people; those who destroyed their neighbor's crops or his walls and fences, and many others.

The tortures shown on the stage are the fiery iron bed on which a woman is being roasted, and a stake to which a female culprit is tied to have her eyes plucked out.

Court V (Case 3) is under the jurisdiction of King Yen-lo; that is, the ancient Indian god of death, Yama. It is 500 *li* (Chinese miles) in compass, being sub-

SHADOW-PLAY FIGURES CHINA.

STORY OF THE WHITE SNAKE.

divided into sixteen wards, as follows:—In the first are sceptics and those who do not worship. In the second, those who have destroyed life or hurt living creatures. The punishment in both is pulling out of the heart. In the third are those who have not fulfilled their vows. In the fourth are believers in heretical doctrines, magicians, sorcerers, and those using evil charms to prolong life. In the fifth are those who deceived the good, tyrannized over the weak, but cringed to the strong, also those who openly wished for another's death. In the sixth are those who tried to thrust their misfortunes on to other people's shoulders. In the seventh are those who committed adultery. In the eighth are those who injured others with a view to personal profit. In the ninth are those who are niggardly and refused to help others in distress. In the tenth, are those who committed theft and involved the innocent. In the eleventh are those who forgot kindness or sought revenge. In the twelfth are those guilty of conspiracy, who stirred up others to quarrel, keeping themselves out of harm's way. In the thirteenth are those who spread false reports or enticed to wickedness. In the fourteenth are those who were fond of brawling and fighting, and implicated others. In the fifteenth are those who envied and hated the virtuous and wise. In the sixteenth are those who are lost in vice, evil-speakers, slanderers, impenitent and stubborn people, and such like persons.

All those shades who come before this Court have already suffered long tortures in the previous four Courts, whence, if they are hardened sinners, they are passed on after seven days to this Court, where they are subjected to a further process of purification. On the stage are shown the tortures of the "knife-hill" (a hill bristling with knives on which the delinquents constantly climbing up and down become impaled) and cutting the tongue. Attention is called especially to the mask of the "Devil with the Big Head," a humorous caricature of the stage.

Court VI (Case 3) is under the jurisdiction of King Pien Ch'eng. It extends 500 *li* (Chinese miles) in circumference, and around it are sixteen wards, as follows:—In the first, the shades of the wicked are made to kneel for long periods on iron shot. In the second, they are immersed up to their necks in filth. In the third, they are pounded till the blood runs out. In the fourth, their mouths are opened with iron pincers and filled with needles. In the fifth they are bitten by rats. In the sixth, they are enclosed in a net of thorns and preyed on by locusts. In the seventh, they are crushed to a jelly in a mortar. In the eighth, their skin is lacerated, and they are beaten on the raw. In the ninth, their throats are choked with fire. In the tenth, their skin is blistered with burning mulberry-wood. In the eleventh, they are subjected to fetid odors. In the twelfth, they are butted by oxen and kicked by horses. In the thirteenth, their hearts are scratched. In the fourteenth, their heads are rubbed till their skulls come off. In the fifteenth, they are chopped in two at the waist. In the sixteenth, they are flayed, their skin being rolled up into spills.

In this Court will be found those discontented ones who rail against Heaven and revile Earth, who are always finding fault with wind, thunder, heat, cold, fine weather, or rain; those who steal the gold from the inside or scrape the gilding from the outside of Buddhist images; those who show no respect for written paper by destroying it, who throw down rubbish near a temple, who use filthy kitchens and stoves for preparing the sacrificial meats, who do not abstain from meat; those who have in their possession blasphemous or obscene books and do not destroy them, who tear books teaching man to be good; those who embroider the Svastika (because it is the symbol on Buddha's heart) on fancy work; those who secretly wear clothes adorned with the dragon and the phœnix (emblems of imperial dignity); those who buy up grain and corner it until the price is exorbitantly high. Persons guilty of the above crimes, if they will abstain from animal food on the third day of the eighth month and register a vow from that date to sin no more, and on four other days practise abstinence, vowing moreover to exert themselves to convert others, shall escape the bitterness and torments of all the wards of the Sixth Court. On the stage is shown the torture of a culprit being hacked in a chaff-cutter.

The majestic God T'ai Shan, the powerful spirit of the sacred mountain of the same name in Shan-tung, reigns at the bottom of the great Ocean, away to the northwest, below the Wu-tsiao rock. His realm represents the seventh Court of Purgatory and measures 500 *li* (Chinese miles) in circumference, being subdivided into sixteen wards, as follows:—In the first, the shades of the wicked are made to swallow their own blood. In the second, their legs are pierced and thrust into a fiery pit. In the third, their chests are cut open. In the fourth, their hair is torn with iron combs. In the fifth, dogs gnaw their bones. In the sixth, large stones are placed on their heads. In the seventh, their skulls are split open. In the eighth, they are scorched by fire and pursued by dogs. In the ninth, they are flayed, carrying off their skins. In the tenth, they are pecked by huge birds. In the eleventh, they are hung up and beaten on the feet. In the twelfth, their tongues are pulled out, and their jaws bored. In the thirteenth, they are disemboweled. In the fourteenth, they are trampled on by mules and bitten by badgers. In the fifteenth, their fingers are burnt with hot iron. In the sixteenth, they are boiled in oil.

To this Court will be condemned those who practise swallowing certain drugs, in order to procure immortality; those who spend more than is wise upon wine; those who kidnap human beings for sale; those who steal clothes and ornaments from coffins; those who break up the bones of a corpse for medicine; those who separate people from their relatives; those who sell the girl brought up in their house to be their son's wife; those who conspire to cheat others in gambling; those who unduly beat and injure their servants; those guilty of extortion, those disobeying their elders, talking at random and going back on their word, and those stirring up others to quarrel and fight. Whoever on the 27th of the third moon, fasting and facing toward the north, will register a vow to pray and repent, and to publish the whole text of the Buddhist Eschatology (a popular tract in which the Courts and torments are illustrated and described) for the enlightenment of mankind, may escape the bitterness of this Seventh Court.

On the stage is shown the torture of sawing a culprit in two. Additional features are two delinquents with hands chained to their backs and tied to a stake, and the figure of a culprit wearing the cangue or pillory around neck and hands. On the cangue, the words are written, "Be it known to all that he is condemned to carry the cangue and wear a lock,—Depart-

THE TIBETAN MYSTERY-PLAY.

MASK OF ONE OF THE SIX MAHĀKĀLA, DEFENDERS OF THE LAMAIST FAITH.

Case 12, Bottom.

ment of the Dark Regions." The wearing of the wooden collar is a frequent punishment in China, particularly for such offences as petty larceny, etc., which are duly inscribed thereon. The time for which it may be imposed, and its weight (up to a limit of thirty pounds), are regulated by law. It is generally taken off at night; during the day while the wearer is exposed to the public eye on the market, he must be fed by friends. The word *cangue* used in the East is derived from Portuguese *canga* ("an ox-yoke," or "porter's yoke for carrying burdens"). The practice was introduced into China by the Tungusian dynasty of Wei in the fifth century.

Court VIII is ruled by King Ping Teng. It is 500 *li* (miles) in compass, and subdivided into sixteen wards:—In the first, the shades of the wicked are rolled down mountains and overrun by carts. In the second, they are shut up and suffocated in iron tanks. In the third, their flesh is minced. In the fourth, their noses, eyes, and mouths are stopped up. In the fifth, their tongues and uvulas are cut off. In the sixth, they are imprisoned in an iron cage. In the seventh, their extremities are broken. In the eighth, their intestines are cut out. In the ninth, the marrow of their bones is cauterized. In the tenth, their bowels are scratched. In the eleventh, they are inwardly burnt with fire. In the twelfth, they are disemboweled. In the thirteenth, their chests are torn open. In the fourteenth, their skulls are split, and their teeth dragged out. In the fifteenth, their bodies are lacerated. In the sixteenth, they are pricked with tridents and pitchforks. Those who are unfilial, who neglect their parents or fail to bury them after their death, who subject their parents to fright, sorrow, or anxiety,

—if they do not quickly repent of their former sins, the God of the Domestic Hearth will report their misdoings and gradually deprive them of what prosperity they may be enjoying. Those who indulge in magic and sorcery will, after death, when they have been tortured in the other Courts, be brought to this Court, and dragged backwards to be tormented, and in time they shall be born again as animals. But those who believe in the doctrine of Buddhist eschatology and make a vow of repentance, repeating it every night and morning to the God of the Hearth, shall receive from him, at their death, one of the three words "obedient, willing, or penitent" traced on their foreheads. They shall escape half the punishment from the First to the Seventh Court, escaping the Eighth Court altogether, to be finally reborn from the Tenth Court among mankind as before.

On the stage is shown the torture of being suspended from a scale.

Court IX is under the jurisdiction of King Tu Shi. This Court corresponds to the last of the eight hot Infernos of Brahmanism (called in Sanskrit, Avici), where the culprits die, and are reborn without interruption, but with the hope of final redemption. It is a vast, circular place, securely enclosed by an iron net, and subdivided into sixteen wards, as follows:—In the first, the shades of the wicked have their bones beaten and their bodies scorched. In the second, their muscles are drawn out, and their bones beaten. In the third, crows peck out their hearts and livers. In the fourth, dogs devour their intestines and lungs. In the fifth, they are hurled into a cauldron of boiling oil. In the sixth, their heads are crushed in a frame, and their tongues and teeth are drawn out. In the seventh, their brains are taken out, and their skulls filled with worms. In the eighth, their heads are steamed, and their brains scraped. In the ninth, they are dragged about by sheep till they drop to pieces. In the tenth, they are squeezed in a wooden press and pricked on the head. In the eleventh, their hearts are ground in a mill. In the twelfth, their entire bodies are scalded by dripping boiling water. In the thirteenth, they are stung by wasps. In the fourteenth, they are tortured by ants and venomous insects; they are then stewed, and finally wrung out like clothes. In the fifteenth, they are stung by scorpions. In the sixteenth, they are tortured by poisonous snakes entering their nostrils.

All who on earth committed one of the ten great crimes, and deserved either the lingering death, decapitation, or strangulation, shall, after passing through the tortures of the previous Courts, be brought to this Court, together with those guilty of arson, of producing indecent pictures or obscene books, of making stupefying drugs, and of other disgraceful acts. Then, if it be found that, hearkening to the words of the Buddhist eschatology, they subsequently destroyed the printing-blocks of these books, burnt their prescriptions, and ceased practising the magical art, they shall escape the punishments of this Court and be passed on to the Tenth Court, thence to be born again among human beings. But those who will not repent shall suffer in intensified form all the tortures from the Second to the Eighth Court. In this Court, they shall be bound to a hollow copper pillar, clasping their hands and feet around it. Then the pillar will be heated and burn their hearts and livers. On the stage, the torture of this fiery copper pillar is shown.

A noteworthy figure in this Court is that of a

hungry shade. It is a man who died without issue, and consequently has nobody to feed him after death with the prescribed ancestral sacrifices. In consequence he is vagabonding in the other world as a lonesome and starving shade, eking out a living by begging for alms. Such figures are on view in the temples as a warning to childless people, and exhorting them not to commit race-suicide.

Court X is under the jurisdiction of King Chuan Lun; that is, "the king turning the wheel of the law." He has charge over the six bridges corresponding to six forms of re-birth, of which the Golden and Silver Bridges are shown here. He examines all shades before they are allowed to pass over, and controls the new form of existence which they are allowed to assume; for at this point they have reached the end of their long migrations. This Court is therefore called "the Wheel of Transformation." Originally a figurative expression, later it was popularly conceived as something real, the belief being entertained that the souls are passing round in a huge Wheel of Fate, where their next form of life, the duration of their existence, and the place of their abode are determined. The place where the Wheel of Fate revolves is many miles in extent, enclosed on all sides by an iron palisade. Within are 81 subdivisions, each of which has its particular officers and magisterial appointments. Beyond the palisade, there is a labyrinth of 108,000 paths leading by direct and circuitous routes back to earth. Inside, it is as dark as pitch, and through it pass the spirits of priest and layman alike. But, to one who looks from the outside, everything is visible as clear as crystal, and all the attendants who guard the place have the faces and features that they had at their birth. These attendants are chosen from virtuous people who in life were noted for filial piety or friendship, and are sent here to look after the working of the wheel. If for a period of five years they make no mistake, they are promoted to a higher office; but if found to be lazy or careless, they are reported to the Judge for punishment.

Those who in life have been unfilial or have destroyed life, after undergoing tortures in the various Courts, are brought here and beaten to death with peach twigs. With changed heads, they are turned out into the labyrinth to proceed by the path which ends in their re-birth as an animal. After myriads of years, animals may resume their original shapes. Whoever will not destroy life during their existences may be born among human beings as a reward.

The main importance of this Court lies in the Terrace of Oblivion to which all shades must proceed. There is a cliff in this region on which the following lines are written: "To be a man is easy, but to act up to one's responsibilities as such is hard. Yet to be a man once again is harder still. For those who would be born again in some happy state there is no great difficulty; all that is required is to keep mouth and heart in harmony."

There are six bridges in the other world, of gold, silver, jade, stone, wood, and planks, over which all souls must pass. The shades of good people, after having been passed on from the First to the Tenth Court, are examined by the Judge of the latter Court, and according to their merits are allowed to cross one of these bridges in order to be sent back to earth. There, they will be reborn as they have deserved by their conduct in a previous existence, as men, women, old, young, high, low, rich, or poor. A list of the names of all people passing the bridges is kept and forwarded monthly by the Judge of the Tenth Court to the Judge of the First Court for transmission to Feng-tu, the capital city of the Infernal Regions. An official and his wife are crossing the bridges shown in the play.

A remarkable figure in this Purgatory is presented by the mask of Mother Mong, whose story is as follows: Mother Mong was born at the time of the Earlier Han dynasty. In her childhood she studied books of the Confucian school; when she grew up, she chanted the liturgies of Buddha. She exhorted mankind to desist from taking life and become vegetarians. At eighty-one years of age her hair was white, and her complexion like a child's. She retired to the hills and lived as a recluse until the Later Han. Because certain evil-doers, under the pretext of a knowledge of past existences, used to beguile women by pretending to have been their husbands in a former life, the Judge of the Tenth Court commissioned her to build the Terrace of Oblivion, and appointed her as guardian. It was arranged that all shades who had been sentenced in the Ten Courts to return in various conditions to earth should first be dosed by her with a decoction of herbs. Thus they forgot everything that had previously happened to them. Good spirits who go back into the world will have their senses of sight, hearing, smell, and taste very much increased in power, and their constitution generally will be improved. Evil spirits will experience the exact contrary of this, in retribution of previous sins. The pot on a red-lacquered stand is the one in which the drink of oblivion is boiled.

THE LION-DANCE, CHINA

The lion-dancers first appeared in China under the T'ang dynasty (A.D. 618-906), and made their début at the court of the kings of Tibet about the same time. The lion-dancer represents a form of the Indian mime or burlesque juggler, who originally exhibited tame lions and trained monkeys, wandering from place to place, entertaining crowds at fairs and religious festivals. Along the high roads of Shen-si Province in northern China are still to be seen numerous square pillars surmounted by figures of lions, monkeys, and stage-fools. In some cases the monkey squats on the lion's back, in others it is the fool stretching out his hands into the animal's jaws, who sits astride the lion's back. He wears the characteristic, spacious, conical fool's cap, the tip of which falls down in front, his countenance expressing a somewhat melancholy, but humorous good nature. The mimes, at first, simply covered their faces with a lion mask (as shown in Case 11), and it is an interesting coincidence that the European harlequin also frequently appears with the mask of the lion; for the lion had developed into the emblem of the buffoon.

Live lions were not obtainable in China and Tibet, and as their transportation from India was a costly matter, the strolling mimes soon hit upon the expedient of representing the lion's body by a covering of cloth and using two men instead of one,—one carrying the head; the other, the hind portion; the trousers of each forming the animal's feet (Case 5). Buddhists recognize the lion as the emblem of Buddha, who in the sacred texts is referred to as "he with the lion's voice." The lion is regarded as the protector of his religion; hence huge statues of lions are generally to be found in front of Buddhistic temples. In this way the lion-dance came to be looked upon also as a demon-expelling ceremony. In Peking companies of acrobats have been organized to cultivate this specialty. The blue and yellow lions perform a contra-dance, displaying an astounding skill and agility; the eyeballs, tongue, jaws, ears, and tail in rapid motion, while the bells of the neck-collars tinkle to the accompaniment of gongs. The lion being credited with a fondness of playing with a ball, the main feature of the performance is the pursuit by the lions of an enormous ball which is thrown in front of them or across their path. They will even leap on to the roof of a one-storied house, and jump down from there into the courtyard.

IMPERIAL PLAY, CHINA

On the occasion of a birthday or wedding, it is customary in China to invite a company of actors to give a performance in the family circle. Such private entertainments are always introduced by a dramatic scene of symbolical and allegorical character in which, for example, all the benevolent genii of Heaven appear on the stage to convey their good wishes to the person honored and the guests, and bestow on them their blessings and super-natural gifts, above all, those of long life, wealth, and progeny. The masks here on view (Cases 6-7) are those of a pageant formerly given in honor of an emperor's birthday, and are of especial interest, as several mythological groups, taken from the popular pantheon, appear on the stage to congratulate the emperor and invoke blessings upon him. The following are worthy of special mention:

The Twenty-Eight Patriarchs are a series of venerable priests, who, as fathers of the church, pursued the mission of propagating the doctrine of Buddha and preserving its original purity. They were all natives of India, some of them contemporaries of Buddha, others living in different periods down to the sixth century A.D.

The Buddhist patriarchs are followed by the Gods of the Twenty-Eight Lunar Mansions. The lunar mansions form a zodiac consisting of twenty-eight constellations near the ecliptic. The number 28 is connected with the period of the moon's revolutions. This system is found, not only among the Chinese, but also among the Indians and Arabs. In fact, the astronomical lore of the great civilized nations of Asia appears to have had a common foundation. The Chinese series of the lunar mansions was well established in the third century B.C. Each mask is characterized by the figure of the animal with which its lunar mansion is associated, and it personifies "the official" presiding over the constellation. On the individual labels the names of the lunar mansions are given in Chinese and Sanskrit, the names of the animal and the element to which they belong, the corresponding constellation, and symbolism connected with it, being added.

Next appear on the stage the Dragon Kings of the Four Seas, corresponding to the Nāga kings of Indian mythology, who are conceived as serpents, and are believed to be the guardian spirits of the waters. In China they were associated with the dragon, a rain-sending deity whose abode is in the ocean and in the clouds. Next follow eight female fairies, viz.; the fairy of the pear blossoms, the fairy of the cinnamon blossoms, the fairy of the lotus, the fairy of the spring breeze, the fairy of the thousand flowers, the fairy of the hundred flowers, the fairy of the ten thousand flowers, and the fairy of the autumn moon; together with the Eight Immortals accompanied by the gods of longevity, luck, and prosperity, as well as the twin genii symbolizing harmony and union.

The procession is closed by a group of twenty-one masks representing the principal casts of the great mythological romance *Fung shen yen i.* This is a kind of wonder book with a hundred long chapters, the plot centering around the adventures of Wu Wang, the founder of the Chou dynasty (1122 B.C.), in his contest with Chou-sin (1154-23 B.C.), the last ruler of the house of Shang, notorious for his tyranny and cruelty. Wu Wang is assisted in this national war by a countless host of saints, magicians, and spirits, and finally overcomes all his enemies. The prominent heroes fighting on both sides are promoted to the ranks of gods by the new ruler (the title *Feng shen* means "investiture with the dignity of a spirit"). The many hundreds of legends and episodes, of which this complex romance is composed, reflect a strange medley of Taoist and Buddhist motives and beliefs, characteristic of that syncretism and process of assimilation prevalent in the popular religion, which is neither pure Buddhism nor pure Taoism, but a blending of these two elements reinterpreted in a fantastic manner. These twenty-one masks are as follows:—

1. Kiang Tse-ya (illustrated). He was the chief counselor to Wu Wang, founder of the Chou dynasty, serving him and his son for twenty years and aiding them in consolidating the empire. The emperor first met him when he was a man of eighty years of age, and discovered him in the act of fishing with a straight

iron rod instead of a hook, thus offering as little inducement as possible to the fishes, which, attracted by his virtue, readily allowed themselves to be caught, in order to satisfy the needs of this wise and contented angler. His reputation among the present-day people rests on the authority which he is believed to have exercised over the spirits of the unseen universe. The mere phrase "Mr. Kiang is here!" written on the door of a house is a sufficient means of frightening away any evil demon.

2. K'iung Siao, a nature goddess residing on Three Fairies Island (San sien tao), who, with the two following fairies, forms a triad.

3. Pi Siao, a nature goddess, inhabiting the same island.

4. Yün Siao, a nature goddess of the same island. She is the owner of a pair of magic scissors which resemble two dragons, contain the essence of heaven and earth, sun and moon, and are capable of cutting gods and men in two.

5. Tao-hing T'ien-tsun, a saint residing in the cave Yü-wu ("Jade Room") on Mount Kin-ting ("Golden Summit"). He is in the possession of powerful charms, by means of which he provides the starving army of Wu Wang with grain, and finally takes an active part in his battles against Chou-sin.

6. P'u-hien Chen-jen, the Bodhisatva Samantabhadra as a Taoist conception, residing in the Cave of the White Crane on the Mountain of Nine Palaces. The Taoist pantheon has borrowed much from Buddhism, and has adopted numerous Buddhist deities with slight alterations of name and form. This saint assisted Wu Wang in his struggle, availing himself of a white elephant as riding-animal.

7. Wen-shu Kuang-fa T'ien-tsun, the Bodhisatva Manjuçri as a Taoist conception, residing in the Cave of Fleecy Clouds (Yün-siao) on the Mountain of Five Dragons (Wu-lung). He belongs to the company of saints aiding Wu Wang in his national war, availing himself of a lion as his riding-animal.

8. Huang T'ien-hua, son of Huang Fei-hu (cf. above, p. 14), general of Chou-sin, and like all characters of this story, is an adept in magic. He revives his slain father, fights battles with the magical "fire-dragon club," is killed, restored to life, suffers death again, and is finally deified.

9. Hung Kin, an officer of the emperor Chou-sin, who takes advantage in battle of a wonderful black flag which is capable of being transformed into a gate, thus deceiving his adversaries. He is vanquished by a goddess, daughter of Si Wang Mu, who renders herself invisible by means of a white flag. Hung Kin changes earth into ocean, the goddess follows him on the back of a whale, and binds him with the dragon-fettering chain. He is then sentenced to marry her.

10. Chao Kung-ming (illustrated), a powerful magician residing in the cave Lo-fou on the sacred Mount O-mei in Se-ch'uan. His riding-animal is a tame tiger capable of carrying him through the air. He defeats his adversaries by means of magical weapons.

11. Ts'ing-su Tao-te T'ien-tsun, a saint residing in the cave Tse-yang on the Mount with Green Summit (Ts'ing fung). He rescues the blinded Yang Jen, and by virtue of his magical power, produces new eyes on his hands (see No. 19). He dispatches Huang T'ien-hua to restore his slain father to life, and to assist him (see No. 8).

12. Ts'e Hang Tao-jen, a saint residing in the cave

Lo-kia on the Isle of P'u-t'o in the Chusan Archipelago. He is a Taoist conception of the Buddhist deity Kuan-yin, his name signifying "the Merciful Barge," because Kuan-yin is believed to convey departed spirits from the ills of mortality to a state of bliss. He aids Wu Wang in his national war.

13. Yün Chung-tse, a magician, who is in possession of a magic mirror capable of capturing demons. He brings to Wu Wang a magic sword to expel a fox spirit, and finally takes part in the battles himself.

14. T'ai-yi Chen-jen, a saint residing in the cave of Gold Lustre (Kin-kuang) on Mount K'ien-yüan, and a supporter of Wu Wang in his holy war against the tyrant Chou-sin.

15. Huang-lung Chen-jen, "The Saint of the Yellow Dragon," residing in the cave Ma-ku on Mount Erh-sien, and fighting in the ranks of Wu Wang.

16. Yü-ting Chen-jen, a saint residing in the cave Kin-hia on Mount Yü-tsüan, and who supports Wu Wang.

17. Chi-tsing-tse, a saint living in the cave of Fleecy Clouds (Yün-siao) on Mount T'ai-hua. He restores a slain hero to life and takes part in Wu Wang's battles.

18. T'u Hing-sun, a dwarf, the son of the goddess Si Wang Mu, who excels in healing wounds.

19. Yang Jen, who exhorted the emperor Chou-sin to abandon his evil ways, and whose eyes were pulled out as punishment. By means of a charm, eyes were produced on his hands, with which he was able to perceive all things in heaven and earth.

20. Kuang Ch'eng-tse is a Taoist priest, residing in the cave T'ao-yüan on the Mountain of Nine Fairies, his weapon being a magic seal which turns heaven

upside down. He is believed to be a former incarnation of the philosopher Lao-tse, who himself is a purely legendary figure in popular Taoism.

21. Kiü Liu-sun, a Taoist saint believed to be an incarnation of Çākyamuni Buddha.

Readers in quest of more information on this subject and other stories alluded to in these pages are referred to a recent work by E. T. C. Werner, "Myths and Legends of China" (London, 1922).

THE SHADOW-PLAY, CHINA

At the present time in China, there are, broadly speaking, two popular theatrical pastimes,—the puppet-play or marionettes, and the shadow-play. The scenarios given in these two representations are now identical, being both derived from the literary drama of the legitimate stage. Marionettes and shadow-play make their appeal to the popular taste by having their plots recited in the living vernacular, while the repertoire of the stage in general adopts the literary language which is intelligible only to the educated classes. Despite this modern uniformity, marionettes and shadow-play have in China a very different history.

The shadow-play is, without doubt, indigenous to China. The first mention is made of it in the Historical Annals of Se-ma Ts'ien and relates to the year 121 B.C. This historian narrates the following anecdote. Wu-ti, an emperor of the Han dynasty, lost one of his favorite wives, and was obsessed by a great desire to see her again. One day a magician appeared at court who was able to throw her shadow on a transparent screen.

This story is symbolic of the general idea underlying these early primitive shadow performances. The shadow figures, indeed, were the shadows or souls of the departed, summoned back into the world by the art of professional magicians. This conception of ancestors as shadow-souls is so characteristically Chinese, that it goes far to prove the priority of this performance in China. Its inception, therefore, is purely religious and traceable to spiritistic seances.

During the middle ages a new element was introduced into the subject of these plays when the people became largely attracted by the romantic stories of the Three Kingdoms. In the third century of our era, the country was divided into three states which were at war with one another. Errant knights roamed over the country who chivalrously espoused the cause of the weak, the innocent, and the oppressed. The history of this romantic age was subsequently woven into a semi-historic romance composed of numerous stories which were recited on the public squares by professional story-tellers. Under the Sung dynasty, these story-tellers illustrated their narratives by means of transparencies in which figures represented the principal heroes of the cycle. Later, under the Mongol dynasty, when representatives of all nations flocked to the capital, Peking, the shadow-play attracted wide attention; and owing to the military and political expansion of the Mongols all over Asia, it was conveyed to the Persians, the Arabs, and the Turks. The famous Persian historian Rashid-eddin, who died in 1318, records a very interesting story. At the Persian court of the Mongol emperor, Ogotai, the son and successor of the conqueror Chinggis Khan, actors from China performed behind a curtain wonderful plays in which the types of various nations were represented, among them an old Mohammedan with a white beard and a turban tied around his head and ending in a horse's tail. In the first part of the fifteenth century we find that the shadow-play had established itself as a favorite pastime at the court of Sultan Saladin in Egypt. The Turkish conqueror, Selim, was the first to take a shadow-player from Cairo to Constantinople. The performance seems to have arrived at Constantinople by two routes,—from Egypt, and from the interior of Asia by way of Turkestan and Persia, as we still find it in vogue at Khokand, Tashkend, Khiwa, Bukhara, and Samarkand. In Turkish literature the shadow-play appears first in the seventeenth century.

It did not penetrate to France until 1767. The French name "ombres chinoises" is a reminder of its Chinese origin, and is traceable to the famous work "Description of the Chinese Empire" by the Jesuit Father Du Halde, who was the first European author to call attention to this pastime in China. The performances given in Paris found their way to London in June, 1776. On Goethe's birthday, 28th of August, 1781, "Minerva's Birth, Life and Deeds," and on the 24th of November "The Judgment of Midas," were given by means of Chinese shadow figures. In "The Fair at Plundersweilen" (1774) Goethe brings a shadow-player on the stage.

While the shadow-play was originally of a religious character, and gradually assumed the function of a mere entertainment, the opposite of such a development may be observed in the history of marionettes. Marionettes cannot be traced in the times of early antiquity in China. They are mentioned for the first time in the records of the T'ang dynasty about the year A.D. 630,

BURIAL CLAY FIGURES OF ACTORS FROM KUCHA (p. 39).
CHINA, T'ANG PERIOD (A.D. 618-906).

when the capital of China was Ch'ang-an. At that time Turkestan was subject to China, and the capital swarmed with jugglers, mimes, and actors, hailing from Kucha in eastern Turkestan. Two clay figures representing such actors and interred in a grave for the entertainment of the dead are here illustrated. It is on record that these actors also brought performances given by means of puppets, and at the same time transmitted the Turkish name for the latter, *kukla* (modern Chinese *k'wei-lei*). This word, however, is not of Turkish origin, but is traced to mediæval Greek. From Byzance it spread to the Slavs, and the Slavs handed it on to the Turkish tribes. It is still the general word designating a puppet in all Slavic and Turkish languages, as well as in Gipsy. This migration of the word bears out the fact, confirmed by other evidence, that the idea of using puppets for dramatic plays, as far as we know, first emanated from Greece. Numerous figures of marionettes have been discovered in ancient Egypt, Greece, and Rome. In their construction and method of manipulation these classic marionettes are identical with those still made in China. The ancient Greek name means "puppets suspended from strings or threads," and we meet exactly the same designation in corresponding translations in Sanskrit, Chinese, and Japanese.

Originally the puppet-play had no religious significance whatever. It was purely a pastime, chiefly for the entertainment of women and children. In Asia it rapidly achieved popularity, because respectable women were barred from the public theatre. Indeed, it seems possible that the puppet-show was suggested by the idea of reproducing a stage in miniature, for the purpose of bringing the theatre into the home for the entertainment of the family. It is still customary in China for the itinerant showmen to respond to calls to give performances in the courtyards of private houses. Curiously enough, the development of marionettes in China was such that they were turned to religious purposes at an early date. There are many accounts to the effect that during the Sung and Ming periods (960-1643) performances were given with marionettes after funerals, in honor of the deceased, and even scenes taken from the career of the dead were introduced to honor their memory.

In the shadow-play the figures are flat and ingeniously cut out of parchment, usually ox or sheep skin evenly colored and varnished on both sides. When held against the light they are transparent. A screen of white gauze (Case 8), lighted by means of oil lamps from behind, is stretched between two poles; and the figures, held by wires stuck into bamboo or reed handles, are skilfully manipulated behind the screen. Head, arms, and legs being cut out separately, great agility of motion is assured. Some aver that the shadow-play is the most realistic and picturesque of all the popular performances, as the regular stage (much on the same level as the Shakespearian theatre) is almost lacking in scenery, while all the stage requisites, as sea, clouds, rivers, gardens, mountains, palaces, temples, courts, sedan chairs, boats, gods, demons, and monsters are well represented in the shadow performance.

The performance is always accompanied by a small orchestra, while the various roles are recited by an operator seated behind the curtain. The plots are taken from Buddhist and Taoist lore or incidents in the history of China. The subjects, however, in which

the entertainment of the family. It is still customary in China for the itinerant showmen to respond to calls to give performances in the courtyards of private houses. Curiously enough, the development of marionettes in China was such that they were turned to religious purposes at an early date. There are many accounts to the effect that during the Sung and Ming periods (960-1643) performances were given with marionettes after funerals, in honor of the deceased, and even scenes taken from the career of the dead were introduced to honor their memory.

In the shadow-play the figures are flat and ingeniously cut out of parchment, usually ox or sheep skin evenly colored and varnished on both sides. When held against the light they are transparent. A screen of white gauze (Case 8), lighted by means of oil lamps from behind, is stretched between two poles; and the figures, held by wires stuck into bamboo or reed handles, are skilfully manipulated behind the screen. Head, arms, and legs being cut out separately, great agility of motion is assured. Some aver that the shadow-play is the most realistic and picturesque of all the popular performances, as the regular stage (much on the same level as the Shakespearian theatre) is almost lacking in scenery, while all the stage requisites, as sea, clouds, rivers, gardens, mountains, palaces, temples, courts, sedan chairs, boats, gods, demons, and monsters are well represented in the shadow performance.

The performance is always accompanied by a small orchestra, while the various roles are recited by an operator seated behind the curtain. The plots are taken from Buddhist and Taoist lore or incidents in the history of China. The subjects, however, in which

SHADOW-PLAY FIGURES, CHINA.
STORY OF THE WHITE SNAKE (p. 41).
Case 9, Bottom.

the shadow-player excels, are the comic or satiric; he aims his wit at human weaknesses or, with merciless denunciation, condemns official corruption and other social and political evils. He thus pursues educational and moral purposes, especially when invited into the privacy of a family circle.

In Case 8 is shown a stage-scene with tree, flower-pots, table with embroidered cover, chair with a tiger-skin spread over it, and the Buddhist patriarch Bodhidharma (Ta-mo). The paper figures on the bottom of the case are used as playthings by boys in Ch'eng-tu, capital of Se-ch'uan. On the other side of the screen are displayed a number of animals, such as horse, ox, swine, dog, elephant, lion, tiger, spotted deer, rooster, peacock, crane, pheasant, tortoise, fishes, oyster-shell, and dragon. In Case 9 are shown the complete casts of several famous plays,—a Buddhist drama in which figure the Eighteen Arhat crossing the sea, the story of the White Snake, a tragic love-play, and the Journey to India (*Si yu ki*). This is a romance based on the journey to India of the celebrated pilgrim Hüan Tsang of the seventh century. He is shown in the act of carrying, suspended from a pole, the sacred manuscripts gathered by him during his peregrinations. He is followed by his white horse and his companions Sun Wu-k'ung, the king of the monkeys (the Indian hero Hanuman), Chu Pa-tsie with a pig's head, and the monk Sha Wu-tsing.

In Case 7 are displayed the masks of the same casts, as they appear on the stage. A word may be added about the character of the Pig-headed Companion, who in this allegoric story symbolizes the animal instincts of human nature. His surname Chu means "Swine," and a swine he is, not only in name and appearance, but also in disposition and conduct. His mouth and throat being very large, he is an omnivorous eater and swallows his food at one gulp, but does not perceive its taste. He carries under his arm a quire of coarse brown paper, as a vain pretense of being a literary man. He wears a long robe, to be considered a respectable character. He wears spectacles, in an unsuccessful attempt to hide his real face. He wears a helmet, on the vain assumption of being a great general: but when the roll is called, he beats his retreat. The only weapon he is able to handle is a rake with nine teeth. He sings ballads, but with so bad a rhythm and a tune that he frightens the life out of his hearers. In short, he behaved so exactly like a real hog, that one day when he was playing with a duck, each admired the other's exterior so much that even the innocent duck was deceived and mistook him for the genuine thing,—a swine. A widow proposed to blindfold him with a handkerchief, and promised to give him one of her three handsome daughters that he should catch. He found himself groping in darkness, embracing pillars and bumping against walls and doors, till at length, with bruised and swollen head, he fell on the floor, panting and exhausted.

THE TIBETAN MYSTERY-PLAY

Pantomimic dances with masquerades were cultivated in Tibet as far back as the seventh century A.D., when the country began to emerge from darkness into the light of history, and are described in the native chronicles dealing with the reigns of the early kings. At that time, these dances were part of the indigenous

THE TIBETAN MYSTERY-PLAY.

PERFORMANCE OF THE LAMAS IN THE COURTYARD OF THE TEMPLE AT CHO-NE, WESTERN KANSU.

shamanistic religion, and were chiefly designated to exorcise malignant demons and to ensure the favors of benign deities. With the introduction of Buddhism from India in about A.D. 630, these religious performances received a new impetus and drew fresh inspiration from the superior dramatic art of India. The fertile lore of the foreign religion supplied novel ideas, and although the ancient devil-dance was dressed up in a Buddhist garb, yet many of its former weird features were retained.

The Lamas, the ordained monks of the Tibetan monasteries, reserve to themselves the exclusive right to act in the mystery-plays and to represent, by means of awe-inspiring masks, the gods and demons in their numerous manifestations. This is consonant with the conviction that only the Lama, by virtue of his knowledge and sanctity, can wield authority over gods and demons; he only can, to modify a phrase of Hume, "perform ghostly offices." There is neither theatre nor stage in Tibet, the performances of the Lamas being enacted in the open air in the temple courtyards without any scenic equipment. No words, save wild outcries or interjections, are uttered; it is a pageant, entirely pantomimic in character, calculated to impress upon the onlooking multitude the omnipotence of the Church, the sway it holds over the demoniac and human enemies, and its ability to save the faithful from the clutches of the devils. This, in the main, is the object of most performances.

As to the deeper significance of the various acts, the laymen know hardly anything, and the Lamas, even though they may know, are reluctant to impart information to the inquisitive foreigner. But in most cases they do not know, and this is not strange, as they merely reproduce fixed and stereotyped forms inherited for centuries and contracted to such a degree of conventionality that their true significance was lost long ago. The Lama actors perform their roles in the same mechanical manner as they chant their litanies. With the flourish of trumpets, the sounding of gongs and shawms, and the clashing of cymbals, the mummery usually opens in the court-yard of a temple, which is crowded by men and women in festive attire. The galleries of the temple-buildings, hung with silken draperies, are reserved for the nobility and high dignitaries. As a rule, three or four Atsara (Case 17), with their tight-fitting one-piece garbs on which skeletons are painted, dash first into the arena to perform a terpsichorean extravaganza, pirouetting, leaping, hopping, bending their bodies backward and forward, waving their arms, and turning somersaults, or relapsing into a solemn movement, at a slow ballet-step accompanied by mystical and rhythmical motions of hands and fingers. There is no intermission, and one bewildering phantasmagoria follows closely upon the heels of the other, accompanied by the merciless strains of the orchestra. Different pantomimes are given in the numerous Lamaist monasteries of Tibet, Mongolia, and China. They vary also according to the season and holidays; thus, there is, for instance, a special performance in honor of the Bodhisatva Maitreya, the Messiah of the Buddhists, and another in commemoration of the renowned hermit and poet Milaraspa. It is therefore impossible to give a typical description that would apply to all the performances.

The costumed figures with masks (Cases 13-17) are those employed on the festival of the New Year in the Great Lama Temple Yung-ho-kung at Peking. This

THE TIBETAN MYSTERY-PLAY.
ONE OF THE HEAVENLY KINGS OR GUARDIANS OF THE WORLD (p. 45).

pageant is designed to symbolize the departure of the old year, and to usher in good luck with the new year. It is opened by the big monk Ho-shang, who enters in the arena first, followed by the entire cast. The masqueraders line up in a row, the monk saluting each of them. As the representative of the clergy, he shows his reverence and makes obeisance to the gods. It is remarkable that he is not, as one might expect, a Tibetan monk or Lama. He is, in fact, a Chinese monk, and his smiling features indicate a genial and good-humored soul. He is accompanied by six younger monks, who poke fun at him. He is said to be introduced into the play in remembrance of an historical Chinese monk, called Ho-shang, who came to Tibet from China in the eighth century to enjoy a religious disputation with his Tibetan confrères; these, by their superior wit, scored heavily against him,—at least, so the Tibetans claim. The Tibetan and Mongol Lamas were strongly established and upheld in the capital by the Manchu dynasty for political reasons, as a means of maintaining control of Tibet and Mongolia. It may be that the choice of the defeated Chinese monk was a piece of political propaganda on the part of the Lamas to emphasize their own superiority over the Chinese monks, whom they despise with all their heart. The principal casts are derived from the mythic lore of India. First come the Red, Black, and Blue Kings (Case 13), terrifying embodiments of the so-called Heavenly Kings or Guardians of the World, who are stationed at the foot of the world-mountain Sumeru, guarding the four quarters and combating demons. Their statues are found at the entrance of every temple. They are provided with the eye of wisdom on their foreheads, which enables them to penetrate the past, present, and future. Flames surround their lips, and their mouths are wide agape to indicate that their roaring voice terrifies the demons. Their costumes, embroidered in gold with dragons and lions, are suggestive of armor which alludes to their war-like profession. Two other Kings (Case 15), being yellow and red, wear a wreath of five skulls and on the top of their heads a thunderbolt (*vajra*) surmounted by three flaming jewels. These are emblematic of Buddha, his law, and the clergy. Their costumes are embroidered with dragons rising from the ocean.

The Green-blue, Brown, and Light-blue Officials (Cases 14-15) belong to the retinue of the Guardians of the World who precede them. The masks of this group are of the same style, and are adorned with the same attributes as those of the Great Kings, save that in the place of the gilded crown they wear a wreath of skulls surmounted by the thunderbolt. As warriors they are clad in a costume, after the style of armor, embroidered in gold thread with combating dragons. Each of these is credited with having ninety-one sons, and is attended by eight generals and twenty-eight classes of demons. In their world (the *devaloka*) life lasts five hundred years, but twenty-four hours there equal fifty years on earth. The two female fairies (Case 14) with head-dresses and costumes of Chinese style belong to the retinue of the goddess Lha-mo.

The mask with the head of a bluish bull (Case 16) represents Yama, the ruler of the dead, in this form known as the King of the Law (Dharmarāja). He is the most powerful of the guardian-deities of Lamaism, and is able to protect his adherents from all adversaries. His appearance in the pageant is calculated to impress upon the spectators thoughts of the hereafter and the dreadful spirits from whom the Lamas alone are able to deliver them. The deer-spirit is Yama's messenger. White spots are brought out on the mask, and plum-blossoms are embroidered on the costume, because this species of deer (*Cervus mandarinus*) is known as "plum-blossom deer."

The Yellow Mule (Case 16) is the animal sacred to Lha-mo. When the goddess made her escape from Yama's kingdom on the back of this divine mule, Yama discharged an arrow at him, but Lha-mo transformed the wound into an eye. More information on this goddess is given below (p. 49).

The Dakini (the Tibetan term means "roaming through the air") are a class of aerial fairies, mainly female sprites, akin to our witches, but not necessarily ugly or deformed. They are possessed of supernatural powers, bestowing blessings on saints and women and battling evil demons. There are numerous groups of them, one being called the Fairies of Wisdom. To these belong the Lion-headed Dakini, who appears in the Lamaist pageant (Case 17), and the Dolphin-headed Dakini (Case 16), who has the head of a fantastic sea monster (*makara*) with elephant-like trunk and tusks and a pair of horns on the head. The masks of the lion and tiger are in Case 17; the latter is explained below (p. 51). At Peking, the procession is closed by the Atsara, the ghoul with skeleton painted on his white cotton garb.

The Tibetan masks are very instructive in that they make us acquainted with the mythology and many religious concepts of the people. Among female deities, Tārā is the greatest favorite. She is adored as the savior from the cycle of transmigration. She appears in twenty-one manifestations (Case 10), each of these being invoked by different prayers and charms, and each having the power of rescuing people from a certain danger of affliction. The most popular two of these are the so-called White and Green Tārā The two wives of the first Tibetan king, who ruled in the first part of the seventh century A.D., are believed to have been incarnations of these goddesses. One of these was a Chinese princess, daughter of the emperor T'ai-tsung of the T'ang dynasty who introduced into Tibet silkworms and silk-weaving as well as refined manners. The other was the daughter of Amçuvarman, king of Nepal. Both princesses were devout followers of Buddha, and assisted their husband in propagating Buddhist teachings and the higher culture of China and India. Because of their merits they were subsequently canonized as saints and worshipped under the images of the two Tārā. It is a curious fact that the empress Catherine II of Russia was looked upon as an incarnation of the White Tārā by the Buryat, a Mongol tribe inhabiting the region of Lake Baikal.

In the complex pantheon of the Lamas there is a special class of deities who are pledged to guard Buddha's religion and defend it by means of arms against demons and human enemies, and each god has a distinct sphere of power assigned to him. They are called defenders of the faith (Sanskrit *dharmapāla*). In the mystery-plays a group of sixteen Dharmapāla (Case 12) make their appearance to impress upon the world the preparedness of the clergy and their readiness to strike terror into the hearts of their opponents. They are represented in so-called mild and wrathful or terrifying forms, more commonly in the latter. In the mild form they defeat the enemies of the faith and triumph victoriously. They spy their adversaries at a great distance, appall them with a blaze of flames which constantly surround them, and smite and annihilate their foes with potent weapons. The Sixteen are headed by Yama, the grim god of death with protruding tusks, flaming head-dress, and five skulls hung over his forehead, and by the blood-thirsty Durga, the consort of Çiva; they end with Telopa, a famed Indian magician, who actively opposed the advance of Islam. Islam, on the one hand, and Christianity, on the other hand, still cause uneasiness and anxiety to the Lamas, who practically are the land-owners and rulers in Tibet.

Another bulwark in the defence of the faith is formed by Lha-mo, which means "the Goddess," who is simply so called because she is a great favorite with both clergy and laity, and her real name is held too much in awe to be pronounced. She is an incarnation of the Hindu goddess Kāli, who still presides over the nationalistic movement of the Bengali, and is the subject of a lengthy cycle of legends. She is the patroness of Lhasa, capital of Tibet, where a great mass is celebrated in her honor on the first day of the first month. In the mystery-plays are shown twenty-four emanations of the goddess intended to illustrate the sequence of her legendary story (Case 10). Thus, in the first mask, she is represented in the act of devouring her son. The story runs that formerly she was the spouse of Yama, god of death, before he was converted to Buddhism, and had from him a son who, it was prophesied, would turn an enemy to Buddha's doctrine. She therefore slew her son, flayed him, and saddled a mule with his skin. She escaped from Yama's realm astride this mule, although Yama shot an arrow at the animal. Finally she enlisted among the defenders of the Buddhist religion. In the following masks she is shown with the face of a sea-monster, devouring a snake, and with the face of a lion. Again, she appears as bestowing blessings and longevity, as the protector of the land, exulting over her victories, and finally as the queen of the four seasons.

An interesting group of masks is that representing various animals (Case 11), many of which personify ancient pre-Buddhistic deities of Tibet which take us back to the threshold of a prehistoric age when the gods were embodied in the form of animals, and when the worshipper, in assuming the name and semblance of the gods, believed that he identified himself with them. Prominent in ancient Tibetan mythology is the Raven. It was a sacred bird of solar character, a bird of augury who functioned as the messenger of the supreme deity. From his voice the will of the god was interpreted, and his predictions appeared as the expression of divine providence. It is curious that the mask designated by the Tibetans as that of the Raven is very unlike this bird. It is dark green in color, with curved and hooked red bill, while the raven's beak is straight. A blue eye of wisdom on the forehead, flaming eyebrows, and gold-painted flames protruding from his jaws complete its characteristics. This coincides with the make-up of the mythical eagle Garuda of India, which serves as vehicle to Indra, and is the sworn enemy of the serpents. In the pantomimic dances of the Lamas the Raven attempts to pilfer the strewing oblation, and is driven away with long sticks by two Atsara, ghouls represented by a skeleton which is outlined on their cotton garbs, and equipped with masks having the appearance of skulls (Cases 10 and

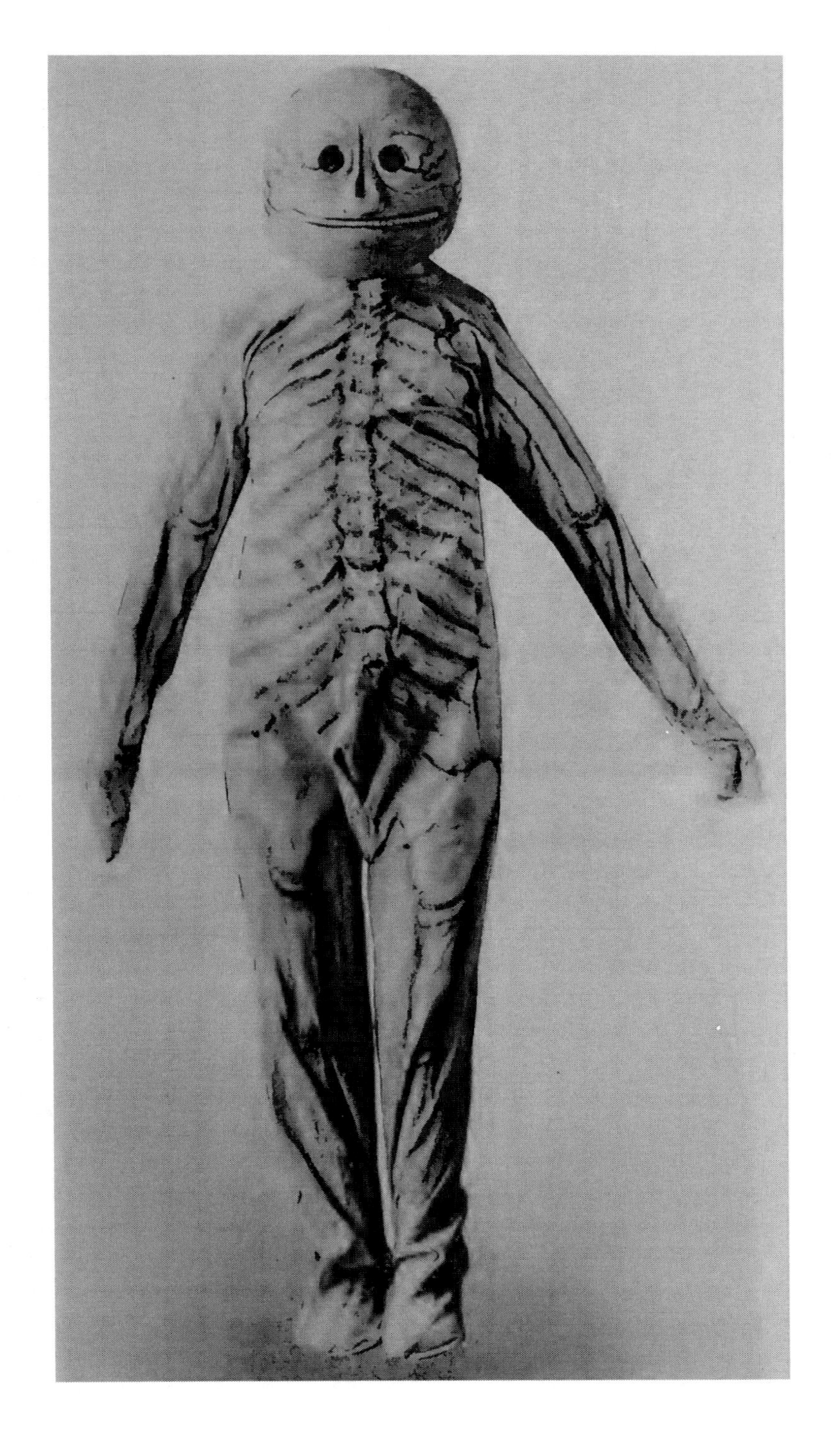

TIBETAN MYSTERY-PLAY.
MASK AND COSTUME OF ATSARA (pp. 44, 47, 50).

16). They play the parts of buffoons. The corresponding Indian tradition is that Garuda captures the *soma* (an intoxicating beverage) or the nectar (*amrita*) destined for Indra. He once even defeated the host of the gods, killed the guardians, extinguished the fire surrounding the nectar, and made away with the latter. Apart from Indian notions, the Tibetan Raven is also imbued with Persian-Mithraic elements, being equipped with six wings and six pinions in ancient texts. In the belief of the Persians, the raven was sacred to the god of light and to the sun; on the Mithraic monuments he perches behind Mithras, who sacrifices a bull. In addition to this, the first of the seven degrees of initiation, which the mystic successively assumed in the Mithraic worship, was styled "Raven."

A very important deity is the tiger, dreaded and worshipped all over eastern Asia. In Tibet he is conceived of as a powerful god living under ground, where he guards great treasures. Many local deities assume, or are represented in, the shape of a tiger. The tiger dance originated in times ante-dating Buddhism, and is connected with the native shamanistic form of religion. It was intended to expel the old year with its demons of ill luck and to propitiate the war-god and the guardian spirits, in order to secure triumph over enemies during the incoming year. The masks of the monkeys likewise are significant. The monkey belonged to the sacred animals of the ancient Tibetans, and was sacrificed together with sheep and dogs once a year, when the high officers assembled for the ceremony of the minor oath of fealty. In their recorded traditions the Tibetans have preserved at great length the story of how they descended from the alliance of a monkey with a female giant (Rākshasi). The Rākshasa are man-devouring demons of Indian mythology, adopted by the Tibetans; their masks are shown on the same screen (Case 11).

In Case 11 is shown a series of ancient masks obtained directly from monasteries in the native Tibetan kingdoms of western Se-ch'uan. These masks are made by means of layers of cloth and paper, and have suffered much from wear and tear. One represents a Brahman of India. Others represent a Hindu layman; Māra, the personification of the evil principle, who tempted Buddha; the famed Indian hermit called Unicorn (cf. Chinese Clay Figures, p. 112 and Plate X); and the dolphin-like marine monster *makara.*

The Tibetans have also a literary drama, the plots of which are all Buddhistic, the style and technique being in imitation of Sanskrit models. There is, further, a popular drama with plots derived from the same source and enacted by professional laymen (not Lamas). Such plays are usually performed in the open air on the occasion of a festival and at the expense of some well-to-do person, the public being admitted free. Few of the actors know their roles by heart, and most of them read the text from a book. It is not a drama in the proper sense of the word, but a story in dramatized form recited, not acted. These itinerant comedians avail themselves of flat cloth masks stuck over with cowries or of painted wood-carved masks (Case 11). The clown, whose buffoonery is constantly in evidence, wields a large stick in his hand. The dancers wear a cap and a plaited girdle of yak-hair and hemp, brass bells being attached to it. This outfit was secured from a party of comedians at Chamdo in eastern Tibet.

THE THEATRE IN JAVA

In Java there are various forms of theatrical representations. The oldest and most important of these is the shadow-play, which is referred to in Javanese poems of the eleventh and twelfth centuries. The figures for the shadow-play are cut out of dried buffalo-skin, and are usually elaborately painted and gilded. There are two varieties with separate names, differing chiefly in the character of the plays presented. In the older one (*wayang purwa*), these are for the most part derived from the old Indian epics, Mahābhārata and Rāmāyana. While the characters and incidents of the stories can usually be easily identified as of Indian origin, the scenes have been transferred to Java, and the people universally regard these stories as relating the adventures and activities of their own ancestral heroes and deities. Among these stories are some, however, that are derived from old Javanese legends and myths, and hence of local origin. As the plays represent the activities of superhuman heroes and deities, they have a religious character, and even to the present day an offering of food is made, and incense burned, before each performance. In the second form (*wayang gedog*) of the shadow-play, the characters and plots are based on Javanese history and legends, and have to do largely with the various adventures of the famous hero Panji. The figures used differ somewhat from those in the *wayang purwa*, the music is in a different scale, and the religious associations are lacking.

A more recent development of the Japanese theatre is the puppet-play (*wayang golek*), in which the characters are represented by wooden puppets with movable arms (Cases 18-19). This is most common in western Java, where it probably originated under Chinese influence, and where it has displaced the older shadow-play to a large extent. In central Java it is less common. The repertoire includes not only all the stories used in the shadow-play, but also many of more recent origin relating to Mohammedan and Malay heroes. The general character of all these stories is much the same, having to do largely with the love adventures and war-like exploits of the principal hero and his attendants. To set off the hero, there is always the opposing group with its leader,—the enemy chieftain or the rival in love. While the hero has every noble quality, his opponents are frequently quite the reverse, and are often represented as in league with demons and evil spirits. These also appear in the plays, and are represented by certain of the puppets.

The puppets representing the two opposing groups are distinguished by certain characteristics, the hero and his associates being lighter in color and more noble in appearance. The prince and princesses on both sides are represented with crowns and special ornaments. Clowns and buffoons are frequently present, and serve to break the monotony of the performances, which are often very long, continuing sometimes for several days. A skilled operator (*dalang*) will also introduce local touches, much to the delight of his audience. If Europeans are present, a sly hit may be made at some of their foibles. These humorous and local touches add greatly to the popularity of the performance.

The puppets are usually shown through an opening in a curtain or screen. The head and body are of separate pieces, the head being fastened to a long stick,

MASK OF CHAO KUNG-MING, CHINA (p. 34).

which passes through the body, having an enlargement or crosspiece on which the body rests. As the sarong or clothing covers the lower part of the puppet, the legs are lacking, and the lower end of the stick is held by the operator (*dalang*), who at the same time moves the sticks attached to the arms with his thumb and forefinger. The puppets not in use are stuck in a banana stem on each side. The *dalang*, of course, recites the story, as well as operates the puppets. He also directs the orchestra, which is placed behind him.

Music is a very important feature in every theatrical performance. The orchestra not only plays before and between acts, but certain instruments often play during parts of the performance, and indicate the nature of the action taking place. All this is controlled by the operator (*dalang*). The orchestra consists of a variable number of musical instruments (Cases 20-21), according to the character of the performance. There are two different musical scales, one having seven notes to the octave, the other only five. All the instruments of any one orchestra would be arranged on the same scale. The one with five notes seems to be the older, and the orchestra having this scale (*gamelon selendro*) is regarded as the best. It is the one used with the old shadow-plays. In both cases, the instruments are the same, except for the number of keys. Most of the instruments are instruments of percussion—xylophones, drums, gongs, and similar instruments,—but there are also two stringed instruments and sometimes a flute. One of the stringed instruments (*rebab*), with two copper strings, is always present, and is played by the leader of the orchestra. The instruments, of course, vary greatly in the care and expense bestowed upon their manufacture. The finest orchestras belong to the princes and nobles of central Java, and their value often runs into thousands of dollars.

The true theatre, in which living actors represent the characters of the story, is probably of independent development. There are two forms,—the *topeng,* in which the actors wear masks (Case 22), and the *wong,* in which they are without masks, and speak their own parts. In both forms rather elaborate head-dresses (Case 23) and costumes (Case 24) are worn. Masked plays (*topeng*) are referred to in old Javanese literature, and in Java are at least as old as the shadow-play, but do not seem to have been so popular, probably owing to the expense. The *wong,* which is much the same as the *topeng,* with the masks and operator (*dalang*) omitted, was first introduced about the middle of the eighteenth century by one of the princes or sultans of central Java, and soon came into common use. Dancing and posturing form a very important part of the performance in either case, and long courses of training are necessary to fit the actors for their parts.

The best performances are to be seen at the courts of the ruling princes of central Java, where the actors and dancers belong to the prince's retinue. Smaller and less elaborate performances are fairly common, however, and there are small traveling troops of players and dancers who give simple performances in the villages. The stories are the same as those in use in the shadow and puppet plays. Masks may, or may not be used.

The masks (Case 22) are made of light wood, painted and decorated to represent the different characters, the differences corresponding to those shown in the wooden puppets. They are kept in place by means of a loop of rattan or a strip of leather held between the teeth. As in the puppet-play, the operator (*dalang*) recites the story, though the actors may occasionally speak their own parts, in which case the mask is held before the face with one hand.

The head-dresses (Case 23) are worn more especially by the actors impersonating kings and princes. The best head-dresses are made of leather or rawhide, but thin beaten brass or copper is also often used. Similar head-dresses may be seen represented on the old stone figures and sculptures of central Java.

The information contained in this chapter has been supplied by Assistant Curator A. B. Lewis.

SINGHALESE MASKS

There are two distinct groups of these masks,—one used in dramatic performances; the other, by sorcerers in the healing of disease.

The masks displayed in Case 25, elaborately carved from wood and painted in bright colors, are employed by itinerant actors in pantomimes and comedies performed in the villages of Ceylon. A slit made under the eye enables the actor to see. The patterns are very ancient, and are always strictly reproduced. Some, like the uncolored carving in the left lower corner of the screen, have artistic merits. Little is known about these Singhalese performances and the significance of the individual masks.

The wood-carved and painted masks displayed in Cases 26-28 represent demons, and are used in devil-dances purported to heal various diseases. Whenever an individual is ill, he summons the sorcerers supposed to have power over the devil causing the complaint. The ceremony and the dance connected therewith are very elaborate. A throne room is built and decorated with fresh leaves, flowers, and plantain-stalks, and is illuminated with hundreds of torches dipped in oil. In front are placed seven floral steps. The patient is made to wear a crown, anklets, armlets, bracelets, belts, etc., made of stems of tender creepers. The performers are dressed in ornamental hats, bodices, and skirts. The chief sorcerer wears the mask and costume of the devil by whom the patient is possessed. After the recital of many invocations, charms, and songs, all decorations are cut down, and the demon of disease is supposed to be exorcised. The underlying idea is that he has taken his abode in the devil-dancer, who is placed on a bier and carried as dead out of the village. The devil, believing that the sorcerer is dead, will voluntarily forsake his body to seek another victim.

In the uppermost row are shown masks representing animals, in the second row turbaned Mohammedans, in the two lower rows masks with high or elaborate head-dresses and sôme without head-gear.

Special attention is called to the mask of Nāgeçvara (Case 27), a serpent-demon or Nāga surmounted by the image of a seated Buddha overshadowed by a cobra, —a favorite theme of Buddhist lore, and to the serpent-demons, male and female, representing Nāgas. These are evil demons believed to be personifications of venomous serpents and to cause diseases, especially leprosy. They have demoniacal faces with bulging eyes and large protruding tusks. Their heads are surmounted by cobras standing erect. In some examples serpents are to be seen stretching along their upper lips, in others they are crawling into their nostrils.

In Case 28, the Rākshasa and Yaksha are grouped together. These demons are man-eaters, feeding on the flesh of corpses and drinking human blood. They are believed to have been the earliest inhabitants of Ceylon. There are many classes of them, and for each there is a different ceremony for its propitiation, which includes offerings and dances accompanied by music.

166

史禄国《满族的社会组织》书评

American Anthropologist

NEW SERIES

ORGAN OF THE AMERICAN ANTHROPOLOGICAL ASSOCIATION, THE ANTHROPOLOGICAL SOCIETY OF WASHINGTON, AND THE AMERICAN ETHNOLOGICAL SOCIETY OF NEW YORK

VOLUME 26

1924

Reprinted by permission of the original publishers

KRAUS REPRINT CORPORATION, NEW YORK 17, N.Y.

1962

has been said to show that Professor Karsten has made a valuable contribution to our knowledge of a very interesting people.

ROBERT H. LOWIE

ASIA

Social Organization of the Manchus, a Study of the Manchu Clan Organization. S. M. SHIROKOGOROFF, anthropologist of the Museum of Anthropology and Ethnography of the Russian Academy of Sciences at Petrograd. Royal Asiatic Society, North China Branch, Extra Vol. III, Shanghai, 1924. 194 pp.

In this learned and critical monograph the social organization of the Manchu is studied and set forth for the first time. The author's researches were made at Aigun in northern Manchuria where the Manchu element is preserved in a purer form and less influenced by Chinese than in southern Manchuria and Peking. He formulates the following definition of the clan: The Manchu clan is a group of persons united by the consciousness of their common origin from a male ancestor and through male ancestors, also united by the recognition of their blood-relationship, having common clan spirits and recognizing a series of taboos, the principal of which is the interdiction of marriage between the members of a clan. The clan also includes the women adopted as wives by the male members of the clan. A clan-meeting at which all members from boys to old men are obliged to be present is called once a year. On the first day a sacrifice is offered to the ancestors and clan spirits, an animal (not connected with totemistic practices) being selected in accordance with the rites of the clan and the sense of the assembly. A master of ceremonies elected by the members directs the sacrifice and the banquet following it. On the second day of the meeting the clan-chief is elected. He is usually a young man, but as a rule at least 25 years of age. Social position and wealth are not considered, but the candidate is required to be well bred, honest, tactful, and able to govern. There is no vote, the candidate is proclaimed *viva voce*. There is no restriction to the duration of office, but in case of inability he may offer his resignation. The clan-chief's power equals that of a governor-general of Manchuria, and there is no appeal from his decisions. He presides over the clan meetings, has the function of a supreme judge, decides all important clan affairs, supervises the clan rites, watches over public and private morality, keeps the clan register, issues permits for marriage and inheritance, and advises the clan members on business matters. Sometimes he maintains his position for twenty years and even longer, and he may be succeeded by his son. In case his actions are contrary to law or meet with the disapproval of the assembled clansmen, the matter is laid officially before the meeting, and the chief is obliged to submit his resignation, which according to the case may be accepted or declined. Frequently the assembly is content to address a reprimand to him and allows him to remain in office. His position is unsalaried, and when rich, he may help poor clan-members. His power within the clan is almost absolute, and all clan members are treated alike within the organization. The women take no part in the men's assemblies, but address their petitions to the clan through the chief or to the latter directly. The women, however, are organized separately, exactly like the men. On the third day of the clan meeting a women's gathering takes place; it embraces all unmarried women born in the clan, as well as the women adopted by the clan through marriage. The women elect a woman clan-chief imbued with the same functions and rights as the male clan-chief. Questions of a similar nature are discussed, the same rules and customs are observed as at the men's meeting. No man save the clan-chief is permitted to attend the women's convention. Affairs concerning both men and women are submitted to a joint meeting. In case of a complaint against a woman, the man in question must state his case to the female clan-chief. It is not surprising that the author assures us that he could not make any direct observations regarding the female organization, but that what he knows he owes to the communications of men who naturally know little about the subject, as they are barred from the women's meetings and know merely what the women please to tell them. The point is interesting, for it bears out the fact that the Manchu women understand the art of keeping their secrets. The author concludes that "obviously the women's organization as a social phenomenon has no such importance as the men's organization; Manchu society like all societies known up to the present time is based on the preponderance of man, not of woman." This is a somewhat hazardous generalization: we know of many primitive societies in which woman wielded as great a power as man or even a greater power; in Asia, the classical example is presented by Tibet. Military service which the author gives as a reason for male hegemony certainly was not the direct cause of it, but was the outcome of historical developments which naturally had the tendency to strengthen man's influence. "Finally," the author remarks, "hunting and agriculture could not be women's trades, either" (p. 56); but on p. 105 (again p. 123) he states that two processes of fishing and searching for edible grass and roots are duties exclusively reserved for women, and that even the care of cattle, swine, and poultry are woman's business.

The clan system also has a religious significance in that the members of a clan are united by ceremonies pertaining to the worship of ancestors and shamanistic practices. Every clan has its own group of spirits peculiar to it. Shamanism is one of the elements forming the basis of the clan organization, and with it the existence of the clan is bound up. Every individual, through the clan, is firmly attached to the system of spirits from whom he receives assistance through the mediation of those familiar with the clan-rites. A list of 42 clans with their Manchu and Chinese names, as well as 10 tables explaining the systems of relationship, is given. The author refers to a Manchu book which contains a complete description of the clans, their names and history, and regrets that he was unable to obtain a copy of it in Peking, 1917-18. A complete copy of this rare and important work is in the reviewer's library, and he would be pleased to place it at the disposal of any one who is qualified to make the proper use of it.

There are two detailed chapters on marriage, exogamy, levirate, sexual taboos, wedding, divorce, division of labor, woman's rights, childbirth, education, and woman's work. As to customs bearing on childbirth and marriage it would have been desirable that the author might have collated his data with those of W. Grube (Zur Pekinger Volkskunde), who has given a very full description of the analogous customs of the Peking Manchu; this work is not cited in the author's bibliography, either. His description of a Manchu cradle, accompanied by a sketch, may interest our Americanists. The discussion of economic conditions, as agriculture and domesticated animals, is very interesting. Swine-breeding is fundamental, and pork is the principal flesh food. Pigs are the unit in calculating the bridal price or a ransom. Some clans do not use the pig as a sacrificial animal; others, like all Mongol-Manchu clans, use the sheep; others again, the badger. Artificial incubation of chickens is understood and practiced. The author observes that in distinction from the Chinese the Manchu do not sow maize in the field, but in the kitchen-gardens. This held good for nearly all parts of Europe and Asia when maize first became known after the discovery of America. It first was merely planted in gardens and at a later stage rose into a field-crop, as I hope to show in detail in a new study of maize; in Fu-kien Province maize is solely known as a garden plant. Tobacco-cultivation and the characteristic long pipes of the Manchu are not mentioned. For the first time the author figures and describes a peculiar Manchu plough in the shape of a sledge (p. 135). There is every reason to be grateful to the author for his thorough investigation of a people that is in a complete state of disintegration and threatens to disappear as an ethnic unit. The Asiatic Society of Shanghai also merits our thanks for having published this volume. Our only regret is that the work has not had the benefit of an English proof-reader. Mr. Shirokogoroff has written English for the last two years only and struggles with the language nobly, but not always successfully; grammatical and stylistic errors are numerous, and there are many unintelligible sentences. Who is able to understand the long sentence on p. 104 (second paragraph) consisting of fourteen printed lines? Since 1915 the author has been carrying on extensive researches in eastern Mongolia, Manchuria, and Siberia, studying the ethnology of the Tungus, Dahur, Manchu, and Chinese. He has made large collections of Tungus folk-lore in texts and translations, and has prepared a comparative dictionary of Tungus dialects on cards. Much important work from his pen may be anticipated in the near future.

B. LAUFER

OCEANIA

The Material Culture of the Marquesas Islands. RALPH LINTON. (Memoirs of the Bernice Pauahi Bishop Museum, Vol. VIII, No. 5, pp. 263-471, plates XL-LXXXIV.) Honolulu, Hawaii. 1923.

Ralph Linton's contribution to Polynesian study is the fifth publication resulting from the Bayard Dominick Polynesian Expedition and is devoted to a survey of the Material Culture and Archaeology of the Marquesas group. A recent renascence of scientific interest in Polynesia renders any exposition of new data from that field especially welcome.

Linton's paper observes an elaborate care for details, such as methods of house and boat sennit lashings, house descriptions, boat descriptions, wood carving, tools and their uses, that is most commendable. The chapter on stone artifacts shows basic knowledge and careful application. It is noted with pleasure that close attention is given to artistic design units and motifs.

167

烟草及其在亚洲的使用

TOBACCO AND ITS USE IN ASIA

BY

BERTHOLD LAUFER

CURATOR OF ANTHROPOLOGY

ANTHROPOLOGY

LEAFLET 18

FIELD MUSEUM OF NATURAL HISTORY
CHICAGO
1924

The Anthropological Leaflets of Field Museum are designed to give brief, non-technical accounts of some of the more interesting beliefs, habits and customs of the races whose life is illustrated in the Museum's exhibits.

LIST OF ANTHROPOLOGY LEAFLETS ISSUED TO DATE

1.	The Chinese Gateway	$ .10
2.	The Philippine Forge Group	.10
3.	The Japanese Collections	.25
4.	New Guinea Masks	.25
5.	The Thunder Ceremony of the Pawnee	.25
6.	The Sacrifice to the Morning Star by the Skidi Pawnee	.10
7.	Purification of the Sacred Bundles, a Ceremony of the Pawnee	.10
8.	Annual Ceremony of the Pawnee Medicine Men	.10
9.	The Use of Sago in New Guinea	.10
10.	Use of Human Skulls and Bones in Tibet	.10
11.	The Japanese New Year's Festival, Games and Pastimes	.25
12.	Japanese Costume	.25
13.	Gods and Heroes of Japan	.25
14.	Japanese Temples and Houses	.25
15.	Use of Tobacco among North American Indians	.25
*16.	Use of Tobacco in Mexico and South America	.25
*17.	Use of Tobacco in New Guinea	.10
18.	Tobacco and Its Use in Asia	.25
19.	Introduction of Tobacco into Europe	.25
*20.	The Japanese Sword and Its Decoration	.25

D. C. DAVIES
DIRECTOR

FIELD MUSEUM OF NATURAL HISTORY
CHICAGO, U. S. A.

*In preparation—November 1924.

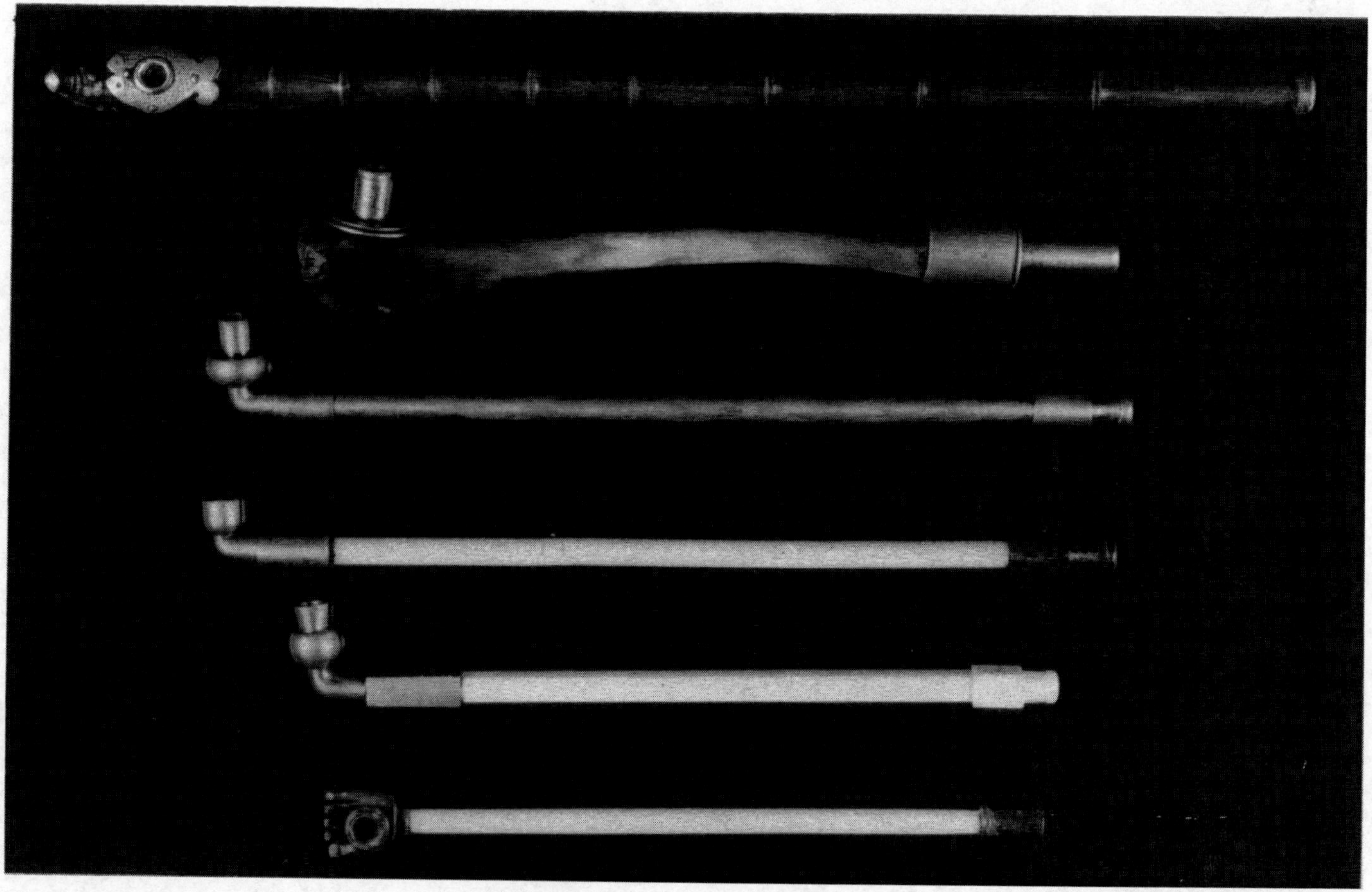

1

2

3

4

5

6

CHINESE DRY TOBACCO PIPES.

FIELD MUSEUM OF NATURAL HISTORY
DEPARTMENT OF ANTHROPOLOGY
CHICAGO, 1924

LEAFLET NUMBER 18

Tobacco and Its Use in Asia

In this sketch conditions, as they prevailed in the seventeenth and eighteenth centuries, and the use of tobacco, as it grew out of native customs, are briefly set forth, but modern conditions, as created by international commerce and colonial enterprise, are disregarded.

The earliest datable reference to the use of tobacco in the Far East occurs in an entry under August 7th, 1615, in the diary of Captain Richard Cocks, who was chief of the English Factory of Hirado in Japan from 1613 to 1621.

"Gonosco Dono came to the English house, and amongst other talk told me that the King [that is, the Daimyo of Hirado] had sent him word to burn all the tobacco, and to suffer none to be drunk in his government, it being the Emperor's pleasure it should be so; and the like order given throughout all Japon. And that he, for to begin, had burned four picülls or hundredweight this day, and had given orders to all others to do the like, and to pluck up all which was planted. It is strange to see how these Japons, men, women, and children, are besotted in drinking that herb; and, not ten years since it was in use first."

"Four tobaka pipes purchased at Kyoto" are entered in the Log-book of William Adams (1614-19). Tobacco, accordingly, was smoked from pipes by the Japanese at that early date.

It follows from Cocks' contemporaneous notice that tobacco was introduced into Japan about the year 1605 and that it was planted and eagerly indulged in by all classes of the population within a decade after its introduction. On the whole, this inference agrees with Japanese records. We are informed by these that the tobacco plant is not a native of Japan and that tobacco-leaves were first traded to the country by the Portuguese (Namban) toward the close of the sixteenth century. About 1605 the first tobacco plantations were established at Nagasaki, the habit of smoking spread rapidly despite prohibitory decrees, and in the latter part of the seventeenth century the cultivation was practised on an extensive scale. Tobacco was also utilized for medical purposes. The word *tabako* used by the Japanese in both their literature and colloquial speech is based on the Spanish-Portuguese form *tabaco* and confirms the correctness of Japanese tradition.

The Portuguese, however, are not responsible for the transmission of the tobacco plant into China. This is outwardly demonstrated by the word *tan-ba-ku* or *tam-ba-ku* under which tobacco first became known in Fu-kien Province in the beginning of the seventeenth century when the Ming dynasty was still in power. The Fukienese were enterprising mariners and maintained regular intercourse with the Philippines, in particular with the island of Luzon, several centuries prior to the Spaniards' conquest and colonization. In the same manner as they obtained on Luzon the peanut and the sweet potato, they also got hold there of tobacco seeds which were transplanted into their native country, first into the province of Fu-kien, whence the novel plant was diffused southward into Kwang-tung and northward into Che-kiang and Kiang-su. The first author who has left an interesting account of tobacco is Chang Kiai-pin, a reputed physician from Shan-yin in the prefecture of Ta-t'ung, Shan-si Province. He carefully studied the physiological effects of smoking and made a number of correct observations. He felt somewhat sceptical when he first came into possession of the plant, but several trial smokes convinced him of its usefulness and superior quality. He highly recommends it as a remedy in expelling colds, for malaria caused by mountain mists, for reducing the swellings brought about by dropsy, and for counteracting cholera. "In times of antiquity," he writes, "this plant was entirely unknown among us; only recently, during the period Wan-li (1573-1620) of our Ming dynasty, it was cultivated in Fu-kien and Kwang-tung, and from there spread into the northern provinces. Wherever it may be planted, it does not come up in quality to that of Fu-kien which is a bit yellow in color and so fine that it has received the name 'gold silk smoke'; it is very strong and of superior aroma. Inquiring for the beginnings of tobacco-smoking, we find that it is connected with the subjugation of Yün-nan Province. When our forces entered this malaria-infested region, almost every one was infected by this disease, with the exception of a single battalion. To the question why they had kept well, these men replied that they all indulged in tobacco. For this reason it was diffused into all parts of the country. Every one in the south-west, old and young without exception, is at present addicted to smoking by day and night."

Therefore, in the same manner as in Europe, tobacco first served as a remedy in China and gained its first adherents among the men of the army. An imperial edict issued in 1638 prohibited the use of tobacco and threatened decapitation to those who would clandestinely hawk it. As everywhere, such decrees remained inefficient, and a contemporaneous author writes that this order was soon rescinded, because there was no better remedy than tobacco for colds in the army. The cultivation has never been discontinued, as the labor was easy and the profit to be made was considerable. The same writer says that in his childhood he was entirely ignorant of what tobacco was, while in the closing years of the period Tsung-cheng (1628-43) there was hardly a boy three feet tall who did not smoke tobacco, so that he concludes that from this period onward tobacco culture was in a flourishing condition. This state of affairs was not altered by another edict promulgated against smoking in 1641, where the pointed paragraph occurs that this practice is a more heinous crime than the neglect of archery, which was regarded as the chief exercise of the army. Addressing the princes and high officers, the emperor laments, "It has become impossible to maintain the prohibition of tobacco-smoking, because you princes and others smoke privately, though not publicly; but the use of the bow must not be neglected."

No species of the genus Nicotiana is a native of China; in fact, none is indigenous in any other part of Asia. Nor can there be any doubt that the species first introduced into China from Luzon was *Nicotiana tabacum*, the typical species of America, the species with large cabbage-like leaves and purple flowers. This becomes perfectly evident from the descriptions of the plant in the early Chinese sources. Moreover it is this species which at present is most commonly cultivated all over China and the adjacent territories. *Nicotiana rustica*, the species with yellow flowers and broader leaves, is cultivated only to a limited extent, chiefly in northern Shen-si and in the mountainous districts of Hupeh and Se-ch'wan, in the latter province up to an altitude of 9,000 feet for

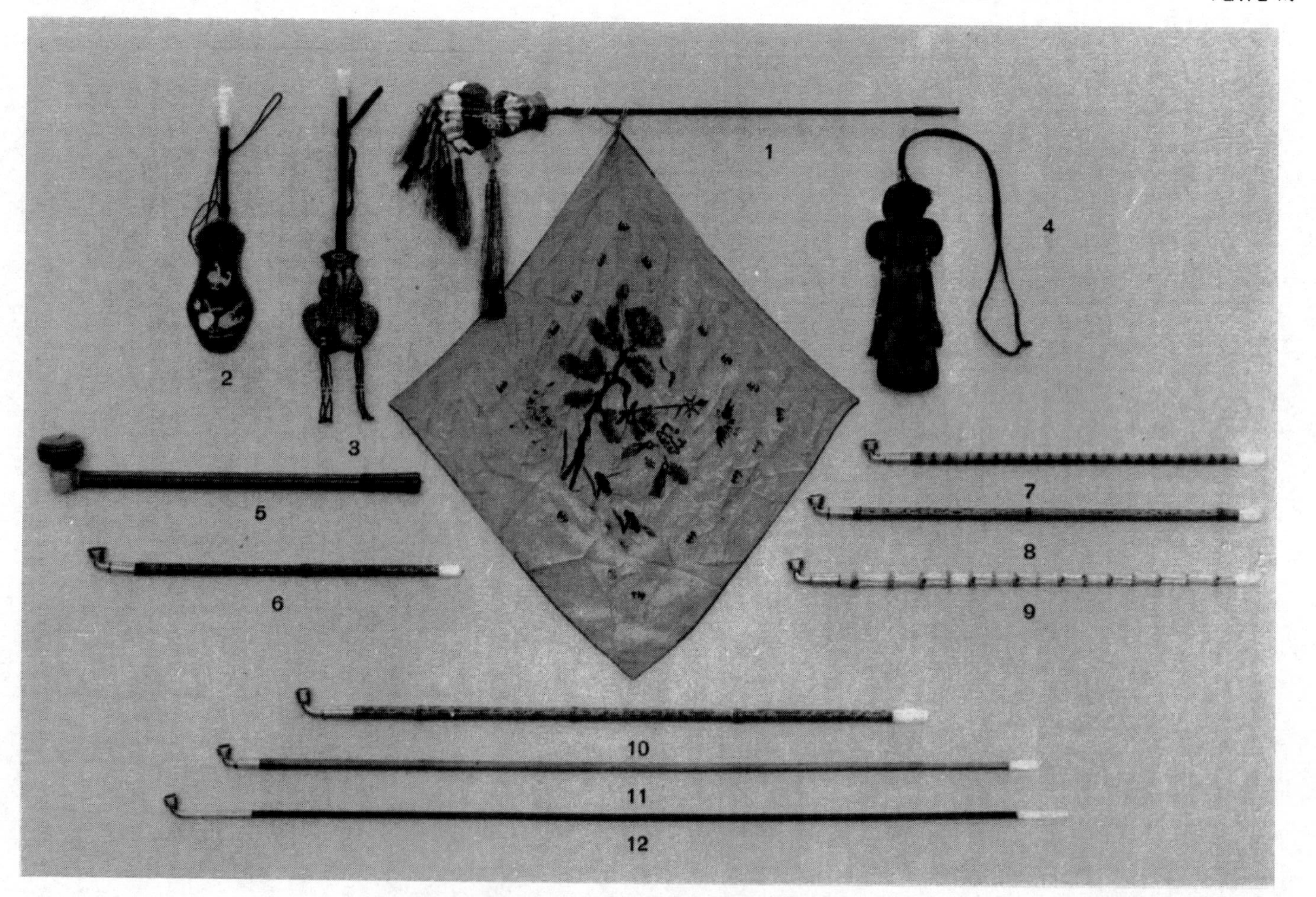

CHINESE DRY TOBACCO PIPES AND POUCHES. 5, OPIUM-PIPE.

About one-tenth actual size.

purely local consumption. At these high elevations the other species would not succeed. The leaves of *N. rustica* receive no preparation beyond being dried in the sun, and the quality of this tobacco is naturally inferior.

In the cultivation of *N. tabacum*, which by the end of the seventeenth century was firmly established throughout the length and breadth of the land, the Chinese have displayed a great deal of natural acumen and aptitude without receiving lessons from foreign nations. Skilful and experienced farmers as they are, they have conceived rational methods comparable to our own and resulting in an excellent leaf. Fertile soil and land beyond the reach of inundations are selected for the successful culture of the plant. To produce a luxurious growth, the farmer will trench his fields deeply, and will manure them with bean-cake. Manure of vegetable origin is preferred to cattle dung which has a tendency to impart to the leaves a rather disagreeable flavor. In the spring the seeds are sown in a well-cultivated bed, and in those provinces where the nights of the vernal season are still cold, the seed-beds are carefully covered with straw or mats. The fields into which the seedlings are to be transplanted are formed into narrow ridges, each about two feet wide and a few inches apart. The seedlings are carefully removed from the seed-bed by means of small spades, great pains being taken not to shake the earth from their roots, and are set in holes previously dug in the field. The plants are arranged in two rows at a distance of sixteen inches from one another. The farmers endeavor to keep the field clean of weeds which would greatly interfere with the growth of the crop. The soil between the plants is loosened at frequent intervals. A few plants are set aside and allowed to blossom for the purpose of gaining seeds. The buds of all other plants are removed, so that the leaves, as the Chinese say, may "gather all strength, grow thick, and improve their flavor." The leaves which occupy the lower parts of the stems are plucked, so that those which cluster around the upper parts may have a chance to expand. The bunch of leaves that grows in the crown of the plant is regarded as especially fine and aromatic, and the tobacco from such leaves is known as *kai-lu* ("covered with dew"). The stems grow to a height of four or five feet, each producing from ten to twenty leaves. In the autumn, the latter assume a very pale green color with a slight tinge of yellow. The appearance of the yellow tinge is the signal for the harvesting of the leaves. The stems are cut very close to the ground, and are left in their places for a few hours to dry. Before the close of the day, however, they are gathered into the garner, as exposure to the night-dew would prove injurious to them.

Now the process of fermentation begins. For the purpose of sweating the leaves are piled up in heaps for four days, and then are placed in light, airy rooms to dry. When dry, they are exposed to another sweating process and laid in heaps upon trays of trellis work, being covered with mats. They are frequently examined in order to prevent the heat becoming too excessive. The leaves are finally plucked from the stems and tied up in bundles for sale. In this manner they are used for the long or dry pipes, while for the water-pipe they are finely shredded or shaved by means of a plane. In this case the leaves have to be completely stripped of their ribs and fibres, and are trampled on by men on a wooden threshing floor. At intervals they are sprinkled with oil or wine, and are finally pressed between two hard boards by means of a huge lever. The cake of tobacco thus formed is then shaved into fine shreds which are parceled out. These packages are dried in charcoal ovens to free the tobacco from the oil.

On the whole, the Chinese medical profession cast its vote in favor of tobacco, but several physicians also recognized and denounced its deleterious effects in undisguised language. In general, the Chinese, as well as the Japanese, are moderate in the use of tobacco. Aside from being administered in malaria, a decoction of tobacco is used for destroying insects and in parasitic skin-diseases. Prepared tobacco is used to staunch the flow of blood wounds. The flower stalk of the plant is considered to be more poisonous than the leaves. It was formerly employed for stupefying fish, being chopped fine and bruised together with green walnut hulls and thrown into a pond. The vapor inhaled of the juice of fresh leaves combined with pine resin is believed to benefit the blood vessels in defective circulation. The bruised leaves were also applied in snake bite, and the dried leaves were sometimes put into beds, or burned under the bed, to drive away bedbugs. The deposit in the interior of an old tobacco-pipe stem was regarded as a sovereign remedy for the bite of venomous snakes. This substance, as well as the water from a water-pipe when sufficiently saturated with nicotine, was believed to be the product of the five elements (water, fire, wood, metal, earth) developed in the process of smoking, and was hence named "pill of the five elements." This "pill" was used to kill insects and to cure skin diseases, snake and centipede bites, and the like.

Chinese terminology relating to tobacco is also of some interest. The foreign word *tam-ba-ku* has always been restricted to the written language, and is now obsolete, but it survives in the form *ma-ku* (abbreviated for *ta-ma-ku*), which is commonly used for cigarette among the Canton and Fu-kien men at the ports. The plant is generally known as the "smoke-herb, smoke-flower, smoke-fire, smoke-leaf," and even "smoke-wine," because, like wine, it is capable of intoxicating people. Tobacco is simply styled "smoke" (*yen*), snuff is "nose smoke;" and the Chinese "eat, sip, or inhale" smoke, while the Japanese and Tibetans "drink" it. There are several poetical names, as "herb of benevolence," "herb of yearning or affection," because he who once tasted it cannot forget it and constantly hankers after it; "herb of amiability," on account of the affectionate feelings entertained toward one another by all classes of mankind since its use became general; and "soul-reviving smoke," because a puff has the power of reviving the energies of the melancholy and wearied. Among the members of the secret sect known as Heaven and Earth Association which has a secret language, tobacco is called "ginger," and "to bite ginger" means to smoke; the pipe is termed by them the "vast bamboo" or the "blue dragon." There is another sect, the Tsai Li, which forbids its members to smoke tobacco and opium and to eat beef.

As a specimen of the Chinese philosophy of tobacco may serve the following extract from an herbal (*Pen ts'ao tung ts'üan*) written in the period Shun-chi (1644-61): "Tobacco has an irritating flavor and warm effect and contains poison. It cures troubles due to cold and moisture, removes the congestion of the thorax, loosens the phlegm on the diaphragm, and also increases the activity of circulation. The human alimentary and muscle systems are aided in their smooth operation as the smoke goes directly from the mouth to the stomach and passes from within to outside, circulating around the four limbs and the hundred bones of the body. There are four principal properties of the smoke: first, it may intoxicate a person, even if he was not drunk before, because the fiery vapors steam the body from both sides, front and back, having the same effect as though he would drink a cup of wine; second, it may

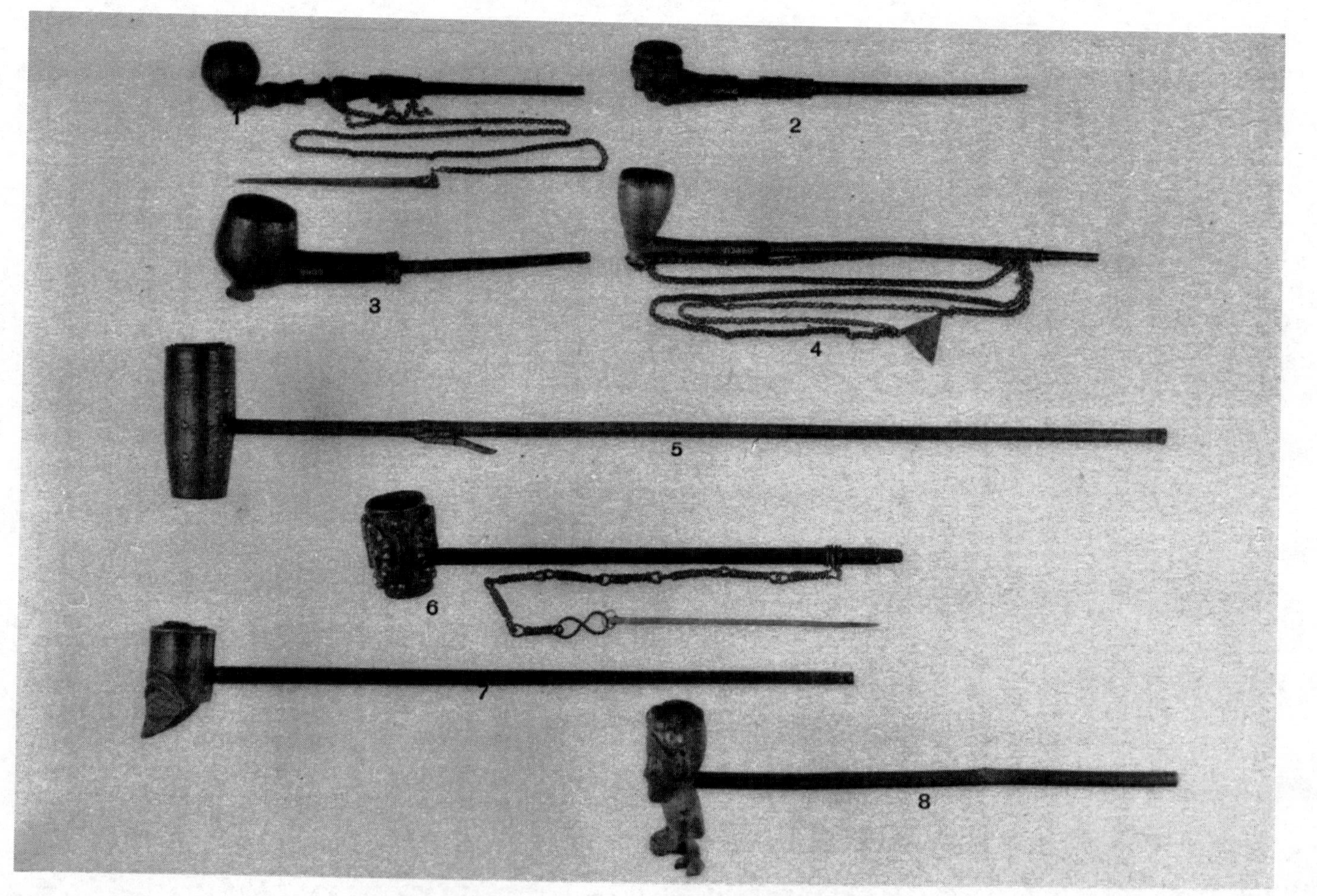

TOBACCO-PIPES. 1-4, FROM THE PHILIPPINES; 5-8, FROM FORMOSA.

About one third actual size.

remove intoxication resulting from wine, because, when used after drinking, it softens the temper, lessens the phlegm secretion, and cures the after-effect of wine; third, it gives man satisfaction whenever he is hungry; fourth, it makes man hungry when he is sated. If a person smokes when he is hungry, he feels as though he has taken plentiful food; and when he smokes after eating sufficiently, it affords good digestion in a most satisfactory manner. For this reason many people use it as a substitute for wine and tea, and never get tired of it, even when smoking all day long.

"We must next consider the matter of respiration, which is closely connected with the circulation of the blood. The blood moves three inches either at one inhalation or one exhalation, with the result that it circulates through the whole body fifty times in the course of a day and a night; during this interval a man makes thirteen thousand and five hundred respirations. Because circulation is fully controlled by the stomach, the smoke which is absorbed by the stomach will pass through the body without interfering with the order of organs, rush around swiftly, and force its way in its progress through the body. However, human energy does not equal the natural fire; there is the only alternative as to the triumph or defeat of one or the other, —but what energy of a human being can stand a wicked fire burning all day long, and depriving him of real power, drying his invisible blood, and shortening his allotment of life without his knowledge? As a rule, the disturbance of the order of the human body will be increased by expelling the cold and wet or deposited phlegm by using the smoke, but this is only good for the person who suffers simply internal impediment or external disorder; but if he has trouble arising from fire inside of his body, smoking will inflame it, so that he should cautiously avoid it."

Our earliest authority for the acquaintance of the Koreans with tobacco and smoking is the Hollander Henry Hamel of Gorcum, who with a party of Dutch sailors travelled from 1653 to 1668, and, being shipwrecked, was held captive in Korea for thirteen years. In his Relation which appeared in 1668 he writes that fifty or sixty years previously the Koreans adopted from the Japanese the cultivation and use of the tobacco plant which had heretofore been unknown to them. The seed, the Koreans were informed on that occasion, came from Nampankouk ("country of the Namban," that is, Southern Barbarians, Portuguese and subsequently Hollanders), and hence they frequently styled the plant Nampankoy. Tobacco, therefore, appears to have been introduced into Korea from Japan early in the seventeenth century, shortly after it had become known in Japan. In the beginning when tobacco was first brought to Korea, Hamel relates, the people bought it for its weight in silver and regarded Nampankouk as one of the best countries in the world. In Hamel's time smoking was general and indulged in by both sexes, even by four or five year old children.

The Koreans tell the following story with reference to the beginnings of tobacco. One of their kings had lost a favorite court-lady of whom he had been very fond, and bewailed her death. The lady appeared to him in a dream and said, "On my grave you will find an herb, called the smoke-herb. Gather it, dry it over a fire, and inhale its smoke! It will stop your grief and make you forget your sorrow." The king obeyed her order, found the herb, and propagated its seed in his country.

Two centuries ago Korean tobacco was a great favorite with the Chinese. Twice a year the Koreans then sent a tribute mission to Peking, and among their gifts presented to the emperor of China on these occasions was as a rule included a finely shredded tobacco which the Chinese preferred to their own product.

According to a Javanese chronicle, tobacco was first introduced into Java in 1601. Probably it was introduced there by the Portuguese, and possibly re-introduced by the Hollanders. G. E. Rumpf, a botanist, who explored the flora of the Malay Archipelago in the latter part of the seventeenth century, writes that old Javanese, according to what they had learned from their parents, told him that the tobacco plant had been well known on Java prior to the arrival of the Portuguese, but solely for medicinal purposes, not for smoking; they stated unanimously that they acquired the custom of smoking from Europeans. Such oral traditions, as a rule, are devoid of historical value. The same Rumpf also learned on Java from an Amoy Chinese that the tobacco plant had from ancient times existed in China, but was rarely cultivated; and this plainly contradicts the Chinese records concerning the recent introduction. No species of Nicotiana is a native of Java; nowhere does it occur there in a wild state, nor do the Javanese have an indigenous name for tobacco. They have only the foreign *tabako* or *tambako*. Moreover, we have many excellent Chinese accounts of Java and her products covering long periods of history, and none of these alludes to a plant that might be interpreted as tobacco.

Tobacco was introduced into India by the Portuguese about 1605, first to the Deccan, and thence it was subsequently diffused to northern India. The first Englishman who mentions it is Edward Terry, who spent two years and a half (1616-19) in Malwa and Gujarat in western India as chaplain to Sir Thomas Roe. He writes in his memoirs, "They sow tobacco in abundance, but know not how to cure and make it strong, as those in the Western India" [West Indies].

We also owe to Terry the first description of the Indian hubble-bubble or hooka.

The following interesting native account of the introduction of tobacco into India is contained in the Wikaya-i Asad Beg, written by Asad Beg of Kazwin, an officer at the court of the emperor Akbar, in 1605:—

"In Bijapur I had found some tobacco. Never having seen the like in India, I brought some with me, and prepared a handsome pipe of jewel work. The stem, the finest to be procured at Achin, was three cubits in length, beautifully dried and colored, both ends being adorned with jewels and enamel. I happened to come across a very handsome mouthpiece of Yaman cornelian, oval-shaped, which I set to the stem; the whole was very handsome. There was also a golden burner for lighting it, as a proper accompaniment. Adil Khan had given me a betel bag of very superior workmanship; this I filled with fine tobacco, such, that if one leaf be lit, the whole will continue burning. I arranged all elegantly on a silver tray. I had a silver tube made to keep the stem in, and that too was covered with purple velvet.

"His Majesty [the emperor Akbar] was enjoying himself, after receiving my presents, and asked me how I had collected so many strange things in so short a time, when his eye fell upon the tray with the pipe and its appurtenances; he expressed great surprise, and examined the tobacco, which was made up in pipefuls; he inquired what it was, and where I had got it. The Nawab Khan-i Azam replied, 'This is tobacco, which is well known in Mecca and Medina, and this doctor has brought it as a medicine for your Majesty.' His Majesty looked at it, and ordered me to prepare and take him a pipeful. He began to smoke it, when his physician approached and forbade his doing so. But his Majesty was graciously pleased to say he must

CHINESE WATER-PIPES OF EARLY TYPE. 1-3, OF BRONZE (K'IEN-LUNG PERIOD, 1736-95); 4. OF TOOTNAGUE, MODERN.
About one-fifth actual size.

smoke a little to gratify me, and taking the mouthpiece into his sacred mouth, drew two or three breaths. The physician was in great trouble, and would not let him do more. He took the pipe from his mouth, and bid the Khan-i Azam try it, who took two or three puffs. He then sent for his druggist, and asked what were its peculiar qualities. He replied that there was no mention of it in his books; but that it was a new invention, and the stems were imported from China, and the European doctors had written much in its praise. The first physician said, 'In fact, this is an untried medicine, about which the doctors have written nothing. How can we describe to your Majesty the qualities of such unknown things? It is not fitting that your Majesty should try it.' I said to the first physician, 'The Europeans are not so foolish as not to know all about it; there are wise men among them who seldom err or commit mistakes. How can you, before you have tried a thing and found out all its qualities, pass a judgment on it that can be depended on by the physicians, kings, great men, and nobles? Things must be judged according to their good or bad qualities, and the decision must be according to the facts of the case.' The physician replied, 'We do not want to follow the Europeans, and adopt a custom, which is not sanctioned by our own wise men, without trial.' I said, 'It is a strange thing, for every custom in the world has been new at one time or other; from the days of Adam till now, they have gradually been invented. When a new thing is introduced among a people, and becomes well known in the world, everyone adopts it; wise men and physicians should determine according to the good or bad qualities of a thing; the good qualities may not appear at once. Thus the China root, not known anciently, has been newly discovered, and is useful in many diseases.' When the emperor heard me dispute and reason with the physician, he was astonished, and being much pleased, gave me his blessing, and then said to Khan-i Azam, 'Did you hear how wisely Asad spoke? Truly, we must not reject a thing that has been adopted by the wise men of other nations, merely because we cannot find it in our books; or how shall we progress?' The physician was going to say more, when his Majesty stopped him and called for the priest. The priest ascribed many good qualities to it, but no one could persuade the physician; nevertheless, he was a good physician.

"As I had brought a large supply of tobacco and pipes, I sent some to several of the nobles, while others sent to ask for some; indeed, all without exception, wanted some, and the practice was introduced. After that the merchants began to sell it, so the custom of smoking spread rapidly. His Majesty, however, did not adopt it."

In 1610 tobacco was grown in Ceylon; smoking became general in India, and English invoices of date 1619 list tobacco as being shipped from India to Red Sea ports. Jahangir, as he himself informs us in his memoirs, issued a prohibition of tobacco in 1617. His own words are as follows: "As the smoking of tobacco has taken very bad effect upon the health and mind of many persons, I ordered that no one should practise the habit. My brother Shah Abbas [king of Persia], also being aware of its evil effects, had issued a command against the use of it in Iran. But Khan-i Alam was so much addicted to smoking, that he could not abstain from it, but often smoked."

J. B. Tavernier, a French gem merchant, who travelled in India, wrote in 1659 that tobacco grew abundantly in the neighborhood of Burhanpur, and that in certain years the people neglected saving it, because they had too much, and allowed half the crop to decay. F. Vincenzo Maria (Viaggio all' Indie Orientali, 1672) even goes so far as to say that tobacco is produced in India in such quantity that both Asia and Europe could be supplied with it.

The Persians first became acquainted with tobacco during a war of Shah Abbas the Great (1586-1628) against the Osmans. Thomas Herbert, who crossed Persia in 1626 on his way to India, is the first traveller who mentions the use of tobacco in the country. A. Olearius, in 1636, found tobacco cultivated in Persia and writes that "there is hardly any Persian, what condition or quality soever he be of, but takes tobacco, and this they do in any place whatsoever, even in their mosques; they highly esteem that which is brought them out of Europe, and call it Inglis Tambaku, because the English are they who bring most of it thither."

From the preceding notes it becomes clear that tobacco appeared in the countries of the East almost simultaneously in the beginning of the seventeenth century and that the literary nations of Asia have preserved records to this effect. The civilized nations who first received it successfully advanced its cultivation and spread it to the surrounding tribes of lower culture.

The Chinese with their mercantile instinct became the most active propagators of tobacco and smoking all over Asia. As distributors of the product they played the same role in Asia as the English in Europe, and covered a larger territory than any modern tobacco trust could ever hope for. Chinese tobacco and smoking utensils are still ubiquitous among all native tribes of the Amur country in eastern Siberia, in Mongolia, Turkestan, and Tibet. As in so many other things, the Chinese set the model for all peoples with whom they came into contact. Wherever the Russians advanced into Siberia in the course of the seventeenth and eighteenth centuries, they found tobacco already cultivated under Chinese influence and the practice of smoking it well established. When Ysbrants Ides, envoy of the Russian czar to the court of China, reached Tsitsikar, a mart of Manchuria, in 1693, he found the Dauri, a tribe of Tungusian stock, in the possession of tobacco cultures. They transmitted it to the tribes of the lower Amur and finally to the Gilyak living at the estuary of the river and on Saghalin Island. The words for tobacco and the pipe in the languages of all these peoples are based on the Chinese prototypes. They smoke, but do not snuff or chew. From the middle of the nineteenth century onward Russian tobacco also reached the Amur tribes through the medium of Cossacks, hunters, and merchants, but Chinese tobacco has always held its ground among them. At the time of my travels in the Amur country in 1898-99 the long Manchurian tobacco-leaf tied up in bundles was the favorite medium of barter.

The Ostyak on the Ob are known to have smoked tobacco in the latter part of the seventeenth century, and did so with a peculiar method of their own. They first filled their mouth with water, and lighting a pipe, swallowed the smoke together with this water. An observer of that time relates that, when they had their first pipe in the morning, they fell to the ground as though attacked by an epileptic fit, as the smoke they had swallowed took their breath away. They were in the habit of smoking only when seated. Their pipes were made of a wretched kind of wood, and when tobacco failed them, they smoked the shavings from the pipe-wood. They preferred Chinese to Russian tobacco.

In 1697 the Russians instituted a tobacco monopoly in Siberia which in the following year was ceded to Sir Thomas Osborne. The English tobacco thus introduced had to struggle with the formidable competition

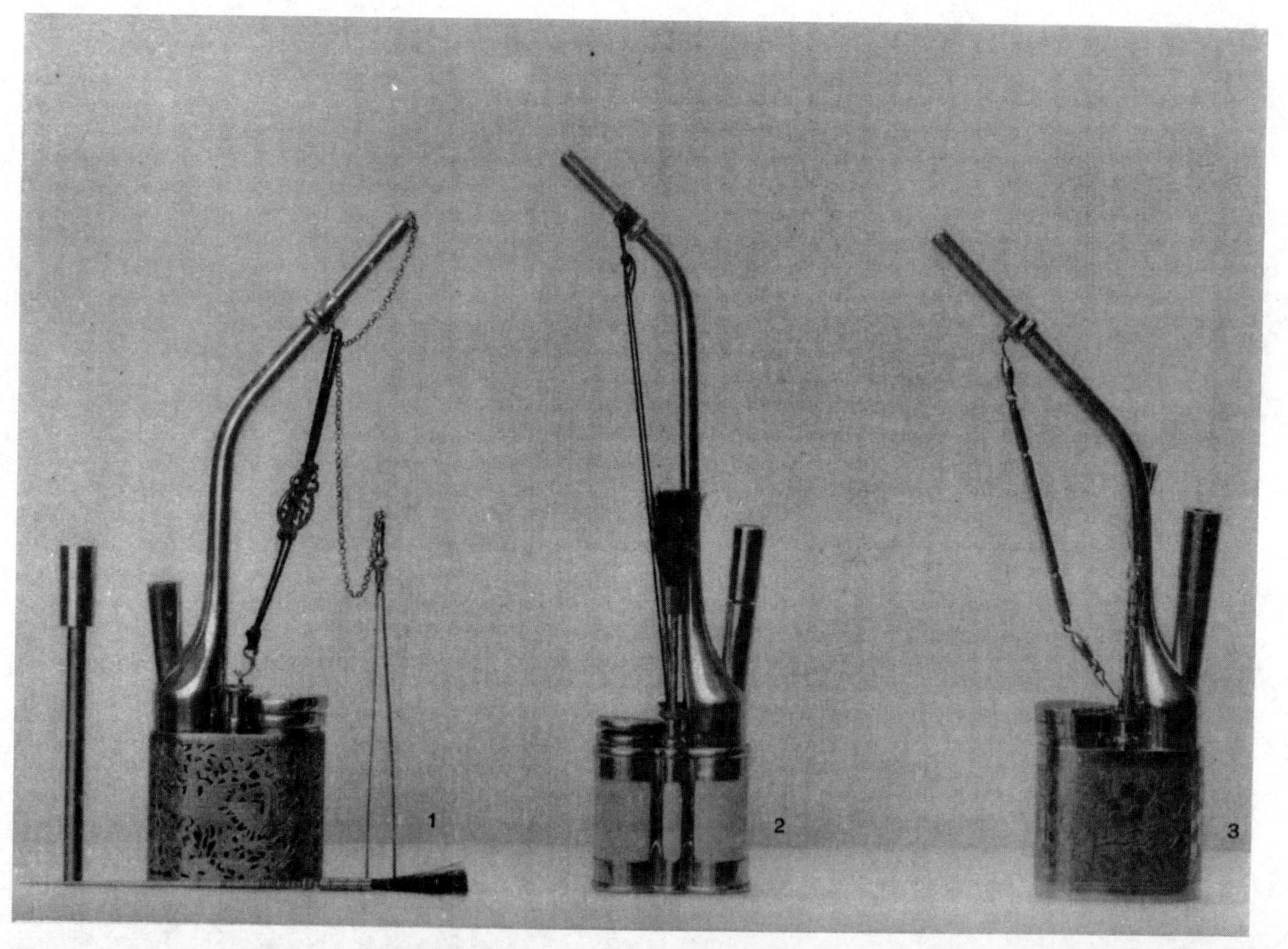

CHINESE WATER-PIPES OF MODERN TYPE

of the Chinese product, so much so that the sale of the latter was finally prohibited in 1701 under penalty of fine and confiscation, to which in 1704 capital punishment for officials was added. The interesting point is that at that moment tobacco had completed its encircling of the globe and that the tobacco having crossed the Atlantic to England and Russia clashed in Siberia with the tobacco having traversed the Pacific to the Philippines and to China, as it were, in a head-on collision.

There are, accordingly, three movements of the tobacco plant into Asia to be distinguished: one from Mexico to the Philippines continued into Formosa and China and from China into the adjacent territories; another from Europe over the maritime route chiefly fostered by the Portuguese, who transmitted the plant to India, Java, and Japan; and a third sponsored by the Russians during their advance into Siberia.

In accordance with its tardy appearance on Asiatic soil, tobacco has not entered into religious ceremonies of Asiatic peoples, as it was customary in America. Among some tribes of Siberia the pipe has endeared itself to his owner to such a degree that it accompanies him into the grave. Thus the Ostyak on the Ob inter a pipe and tobacco with the deceased, and the Tungus of eastern Siberia who bury their dead in trees place their weapons and a handful of tobacco with them. The late Dr. Reinsch tells in his memoirs (An American Diplomat in China) that, when Yüan Shi-kai's funeral took place in Peking in 1916, the usual funeral offerings, as well as the weapons, clothes, and other objects of personal use of the departed were displayed on tables in his residence; long native pipes and foreign smoking sets were included in these paraphernalia. The Ainu of Japan had sometimes valuable tobacco-boxes buried with their owners; before thrown into the grave, the box was smashed to pieces. The Ainu also had a curious tobacco ordeal: a favorite way of trying a woman was to make her smoke several pipes of tobacco; then the ashes were knocked out of the pipe into a cup of water, and she was compelled to drink it. If she passed the ordeal without falling ill, she was regarded as innocent; if she fell ill, she was found guilty.

Finally the curious fact may be pointed out that there is but one people in Asia who does not make use of tobacco in any form, and this is the Yami who inhabit to the number of about 1,700 the small island of Botel Tobago 35 miles east of Formosa. They do not cultivate the plant, nor will they accept tobacco as a gift. Not being acquainted with the preparation of any alcoholic beverage, they are complete prohibitionists.

The distribution of the cigar in Eastern Asia is very curious. Introduced by the early Spaniards from Mexico into the Philippines in the sixteenth century, it is found at present among the native tribes of Luzon and Formosa; in Korea; among the Miao-tse, an aboriginal tribe of Kwei-chou Province in southern China; among the Chinese of Se-ch'wan, the westernmost province of China, and of Shen-si in the north (but not among Chinese of other provinces); among the Karen of Upper Burma, the Burmese, and in southern India. The earliest allusion to Asiatic native cigars I have been able to trace occurs in E. Kaempfer's "Amoenitates exoticae" (published in 1712). In 1688 Kaempfer travelled from Persia to Batavia, visiting on his way the Dutch settlements in Arabia Felix, India, Ceylon, and Sumatra, and observed that the dark-skinned tribes of these regions (Nigritae gentiles) inhale the smoke of tobacco without an instrument, rolling the leaves into a whirl or twist which is lighted at the base, while the upper end is held between the lips and sucked. The natives of Luzon and Formosa, the Miao-tse and Koreans insert the rolled leaves into the pipe-bowl; the Formosans also smoke big cigars without a pipe. Considering the striking resemblance of Luzon and Formosan pipes (Plate III) and the identical method of smoking, there is a high degree of probability in the supposition that tobacco was directly transplanted from Luzon to Formosa, independently of the movement from Luzon to the mainland. Indeed it is affirmed in the earliest Chinese chronicle of Formosa written in 1694 that tobacco was first produced on T'ai-wan (the Chinese name of the island) and that people of Chang-chou in Fu-kien made its acquaintance there and on their return home planted it in Fu-kien. This tradition tends to confirm the conclusion drawn from the above observations. It is even stated by Chinese that the savages of Formosa have a tobacco of a quality superior to their own. In Formosa it is a winter crop which is harvested in the spring.

Cheroot is the name of the truncated cigars, as they were formerly made in southern India and at Manila. The word is derived from Tamil *shuruttu*, Malayalam *churuttu* ("a roll of tobacco"); hence Portuguese *charuto*. In southern India cheroots are chiefly made at Trichinopoly, being known as trichies, and have a straw inserted at the end to be used as a mouthpiece. Those made in the islands of the Godavery Delta are much prized in the Madras Presidency, and are called lunka.

The cigarette, likewise introduced by the Spaniards from America, is of comparatively ancient date among the Malayan tribes; it is described by the botanist G. E. Rumpf (Herbarium amboinense, 1747): the green leaves were dried in the wind, cut into small strips, and wrapped in dried and smoothed banana-leaves, about five or six thumbs long and a finger thick; these rolls were called *bonkos* by the Malays. This word, also spelled *bunco* or *buncus,* is derived from Malay *bungkus* ("wrapper, bundle"); it was also used for a cigar. The cigarette used by the Dusun of British North Borneo is covered with a wrapper made from the flower-spathe of the Nipa palm. These wrappers are sold ready cut at all markets, and are made up into bundles. The Alfur of Ceram also smoke tobacco in the form of cigarettes; dried leaves, the leaves enveloping maize-cobs, or the outer bast of the young leaves of the fan-palm are used as wrappers.

In India cigarettes and cigars under the Malay name (*punka*) are reported as early as the seventeenth century. C. Lockyer writes in his "Account of the Trade in India" (1711), "Tobacco for want of pipes they smoke in Buncos as on the Coromandel coast. A Bunco is a little tobacco wrapt up in the leaf of a tree, about the bigness of one's little finger, they light one end, and draw the smoke thro' the other; these are curiously made up, and sold twenty or thirty in a bundle." In Siam also cigarettes rolled in banana leaves were smoked. It is stated in Japanese documents that the cigarette was used in the early days in Japan, the dried leaf being rolled up in a piece of paper. In certain localities of Japan tobacco is still rolled up in a leaf, generally of the *Camellia japonica,* and smoked like a cigarette.

While Asia owes the tobacco plant to America, it owes nothing to America in regard to smoking utensils, for Asiatics have exerted their own ingenuity and produced their smoking apparatus from resources wholly their own. Whether the early Spanish colonists introduced American pipes into the Philippines, whether the natives of Luzon fashioned their first pipes after models furnished by the Spaniards, and whether the first Chinese who introduced tobacco from Luzon to Fu-kien imitated Luzon pipes, we do not know. All we learn from the early Chinese accounts is that a long tube was held between the lips, and that the tobacco leaves were ignited and the smoke swallowed. Smoking, accordingly, was practised at the moment the plant was introduced and cultivated. Lu Yao, who wrote a small treatise on tobacco in the latter part of the eighteenth century, remarks that bamboo was regarded as the best material for pipe-stems and given preference to ebony and ivory which were apt to crack. For this reason a copper tube was inserted into the ivory stem in Yün-nan Province. In Che-kiang Province, according to Lu Yao, pipe-bowls were carved from wood and the stems made of bamboo, entirely plain for the use of rustic folks, while the gentry could afford the luxury of bowls of gold, silver, copper, or iron with inlaid designs. He also refers to a primitive method of smoking in Fu-kien: the leaves piled upon a heap of old roots in the woods were set fire to, and the rising smoke was inhaled.

A one-piece pipe without a separate bowl and mouthpiece is still found among the poor farmers and workmen of An-hui and Ho-nan Provinces (Plate I, Fig. 1). It is simply a bamboo stem cut off with a part of the root which is naturally thicker than the stem and is hollowed out a little for receiving the tobacco. The specimen figured is old, and is mounted with a metal plaque cut out into the design of a double fish. The hole for the tobacco is lined with white copper. In other provinces this type of pipe is unknown, but a similar one of bamboo is used by the Tibetan and other aboriginal tribes of Se-ch'wan. At Ta-t'ung fu, in the northern part of Shan-si Province, I found last summer a peculiar pipe consisting of a polished sheep-bone, the bowl and mouthpiece being formed of low brass tubes held in place by a brass coin (Plate I, Fig. 2). Among the fishermen of the Luchu Islands a bamboo pipe is still in use with the end scooped out to hold the tobacco. It may be argued that the one-piece pipe had originally a wider distribution in the East, but such a supposition cannot be supported by actual evidence at present.

As a rule, the Chinese long pipe for dry tobacco consists of three separate parts, a round bowl of small capacity, usually of white copper or tootnague (an alloy peculiar to China and composed of copper, zinc, nickel, and iron), more rarely of brass, a stem of bamboo or wood, and a mouthpiece of stone, sometimes jade, ivory, or milk-white glass. The pipe in Fig. 3 of Plate I is entirely of brass. Besides bamboo, ebony, hard black-wood (from *Dalbergia latifolia*), and rattan are esteemed as stems. Mottled bamboo and the square bamboo are highly prized (Plate II, Figs. 6, 8, 11). Pipes are also entirely carved from ivory with round or square stems (Plate I, Figs. 5-6). Mouthpiece and bowl are also made of walrus ivory stained green (Fig. 6); in this specimen the bowl is combined with an ivory hand which serves as a back-scratcher. Pipes for women, as a rule, are much longer than those for men, which holds good also for Japan. The Manchu are fond of very long pipes (Plate II, Fig. 1), and carry them stuck into a tobacco-pouch which is suspended from their girdle. The Manchu women adorn the stem with an embroidered silk kerchief. A new invention was made during the republican era: a metal cigarette-holder is screwed into the pipe-bowl when the owner desires to smoke a cigarette, and is taken out when he wants to smoke tobacco (Plate I, Figs. 3 and 5). Clay bowls, as far as I know, are not used in China for tobacco-pipes, but are the rule in opium-pipes. The bowl is always set vertically on the stem under a right angle, and terminates in a short metal tube made in one piece with the bowl. The stem connects this tube with the mouthpiece, being fitted into the two. This type of pipe is an original invention of the Chinese, and they deserve due credit for it. It is a practical instrument, elegant in shape, light in weight and convenient to handle, pleasant to smoke from and easy to clean.

On Plate II also four tobacco-pouches are illustrated, that in Fig. 1 of embroidered silk, those in Figs. 2 and 3 of silk with appliqué designs, and that in Fig. 4 of plain black leather.

Korean tobacco-pipes are modelled after those of the Chinese, but in distinction from the latter Koreans also make bowls of wood and clay. Good examples may be seen in Case 28, Hall 32. Japanese pipes, though they have a distinct style, consist of three parts in form very similar to the Chinese pipe.

An interesting problem is presented by the interrelation of opium and tobacco smoking. A new investigation of this subject which I made on the basis of Chinese sources has led me to the conclusion that opium-smoking sprang up as a sequel of tobacco-smoking not earlier than the beginning of the eighteenth century. Before tobacco became known in Asia, opium was taken internally, either in the form of pills, or was drunk as a liquid. The Hollanders, who exported large quantities of opium from India to Java, were the first who prepared a mixture of opium with tobacco by diluting opium in water, and who offered this compound for smoking to the natives of Java. This fact is stated in perfect agreement by E. Kaempfer, a physician in the service of the Dutch East India Company, who visited Batavia in 1689, and by contemporaneous Chinese documents. A Chinese author, who wrote a history of the island of Formosa, which was under Dutch rule from 1624 to 1655, even intimates that the inhabitants of Batavia, who were originally excellent fighters and had never lost a battle, were enervated and conquered by the Hollanders by means of opium prepared by the latter for smoking purposes. Be this as it may, the custom was soon imitated by Chinese settlers on Formosa, and smoking-opium was smuggled into that country from Batavia despite prohibitory regulations of the Chinese authorities. Opium was then boiled in copper kettles, and the mass was invariably blended with tobacco; the price for this product was several times greater than that for tobacco alone. It was a much later development to smoke opium in its pure state. The opium-pipe, as it still exists, was invented by Chinese on Formosa in the first part of the eighteenth century. We have several descriptions of the opium-pipe written by authors of that period, which leave no doubt of the fact that in principle the instrument was then identical with the modern one. An old opium-pipe with ivory mouthpiece beautifully stained a deep lustrous brown by an inveterate smoker is reproduced in Plate II, Fig. 5. The stem is lacquered red and ornamented with fine cloud designs. The bowl fashioned from Yi-hing terra cotta and neatly decorated is inserted into a white jade piece carved in the form of a closed hand. This is not the place to go into the details of opium smoking; it is mentioned here merely in order to show that the opium-pipe is based on the tobacco-pipe, and that opium-smoking has grown out of tobacco-smoking.

Four Philippine pipes from Luzon are illustrated on Plate III, Figs. 1-4. Those in Figs. 1-2, made by the Bacun of Igorot stock, have carved wooden bowls of the same rounded form as the Chinese pipes, and have another characteristic in common with the latter

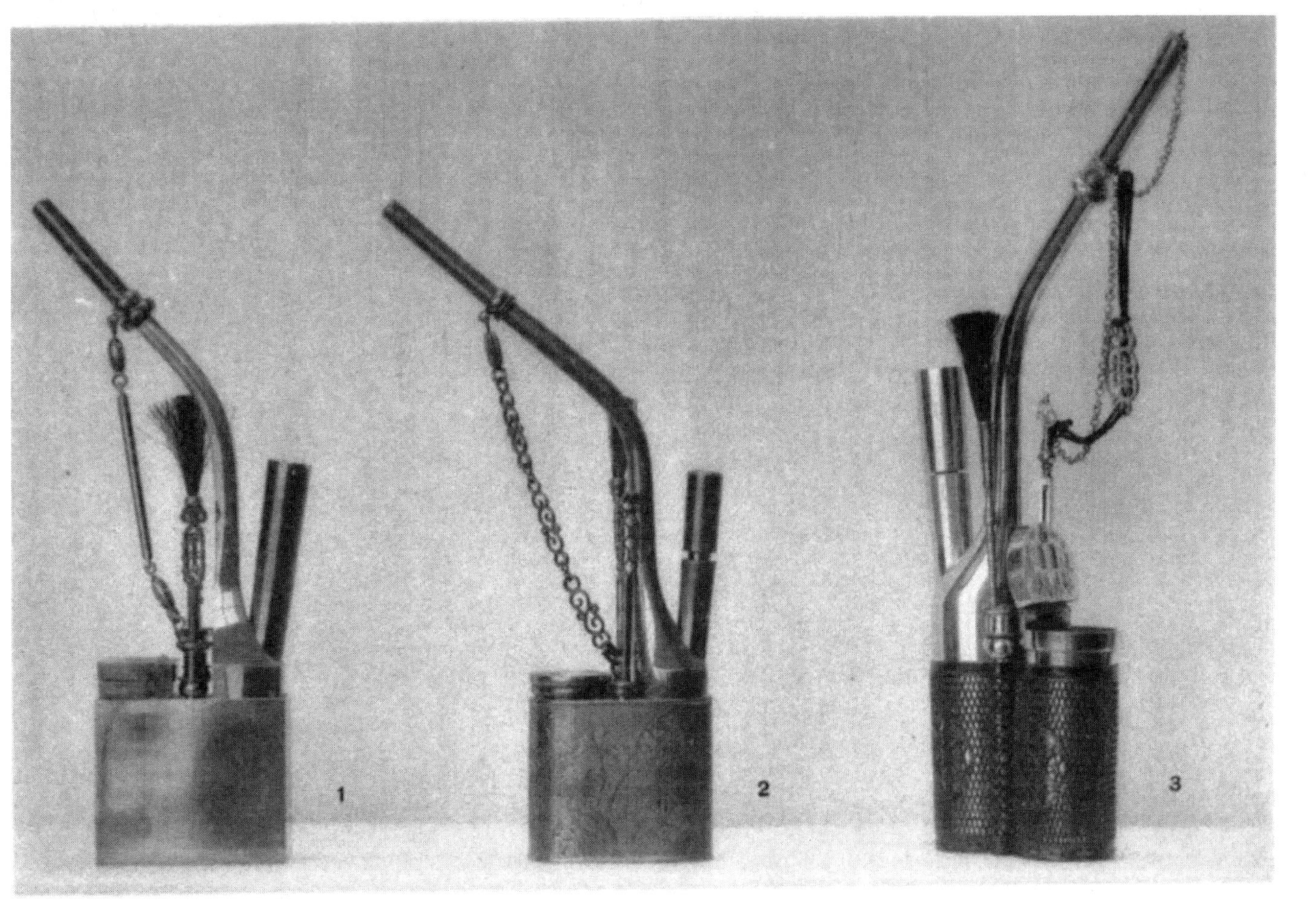

CHINESE WATER-PIPES OF MODERN TYPE. 1-2, OF BRASS, FROM SUCHOW; 3, OF TOOTNAGUE, FROM CANTON.
About one-third actual size.

in that the stem is carved out of the same piece with the bowl; a slender short brass mouthpiece is inserted into the stem. The pipes in Figs. 3-4 are from the Ifugao, likewise of Igorot stock; the former is carved from a hard reddish wood, the latter is entirely of brass, the mouthpiece being wrapped around with brass wire, and a double brass chain being attached to the stem. The pipes in Figs. 1-2 were collected by F. C. Cole, those in Figs. 3-4 by S. C. Simms, on Museum expeditions to the Philippines. Other types of Philippine pipes are illustrated in "The Tinguian" by F. C. Cole (Museum Publication 209, p. 429). The leaf is rolled into thin cigars which are placed in the pipe-bowls,—a practice followed by the aboriginal tribes of Formosa.

The Formosan pipes consist of a cylindrical or barrel-shaped wooden bowl perforated at the side for the insertion of a reed or thin bamboo, which is not provided with a separate mouthpiece. The bowl in Fig. 5 is neatly engraved with geometric and floral designs and decorated with tiny silver studs. That in Fig. 6 has two human faces carved on each side, each face being outlined by rows of silver studs, eyes and mouth being formed by silver pieces; a brass scraper is attached to a wire chain fastened to the stem. On the bamboo bowl in Fig. 7 a single face is carved in front. The pipe in Fig. 8 shows the complete figure of a crouching man of rather naturalistic style, carved from wood, the head forming the bowl; such figure pipes also occur in the Philippines. These four pipes come from the Paiwan tribe, Jamari Village, Formosa, and were obtained by S. Ishii; they were presented to the Museum, together with a representative collection from Formosa, by Dr. Frank W. Gunsaulus in 1919 (exhibited in Case 43, Hall 32).

Besides the dry pipe, a method of wet smoking unknown in America was developed in Asia. Whatever the mode of construction, the principle underlying all water-pipes is the same, and is based on the desire to neutralize, as much as possible, the poisonous properties of tobacco by permitting the fumes, before being inhaled, to pass through water. In this manner a proportion of nicotine is absorbed by the water, and the smoke is purified, cooled, and moderated in strength.

In India this type of pipe is called hooka (also spelled huka), Anglo-Indian hubble-bubble. It consists of a hollow, oval, metal or earthenware vessel or a coconut shell partially filled with water (Case 33, Hall 32). From this vessel arise two tubes—one the mouthpiece, the other being the attachment for the actual pipe, the *chillum*, usually of clay, which contains the tobacco. The fumes pass through the water when the pipe is put to work. The tobacco is cut small or reduced to a powder which is kneaded into a pulp with molasses and a little water. It is thus made into large cakes. It is ignited with a burning piece of specially prepared charcoal, and contact with glowing charcoal is needed to keep it alight.

Early in the seventeenth century the water-pipe was used by the Persians, and it is possible that the instrument was invented in Persia. The first description and illustration of it is found in one of the early books on tobacco, the Tabacologia (written in Latin) by the physician J. Neander and printed at Leiden in 1626. The two pipes figured by Neander correspond exactly to the modern Persian *ghalian*, and are expressly credited by him to the Persians. In view of the fact that tobacco became known in Persia only in the beginning of the seventeenth century, it is somewhat amazing that in the course of a few years, a decade perhaps, the Persians should have conceived the invention of so complex an apparatus as their water-pipe represents; and this state of affairs has induced some authors to advance the opinion that the water-pipe pre-existed in Persia, being formerly used for the smoking of hemp, and was afterwards employed for tobacco. This is a rather attractive hypothesis, but any direct historical evidence is lacking for it; no document has as yet come to light to show that the Persians or Indians really smoked hemp out of an instrument in times prior to the introduction of tobacco, nor is the description of such an instrument on record. All the accounts of hemp-smoking we possess were written after this time, and it is as a rule a mixture of tobacco and hemp that is used for smoking.

Speaking of the coffee-houses of the Persians, John Fryer, who travelled for nine years in India and Persia (1672-81), writes, "They are modelled after the nature of our theatres, that every one may sit around, and suck choice tobacco out of long Malabar canes, fastened to crystal bottles, like the recipients or bolt-heads of the chymists, with a narrow neck, where the bowl or head of the pipe is inserted, a shorter cane reaching to the bottom, where the long pipe meets it, the vessel being filled with water: after this sort they are mightily pleased; for putting fragrant and delightful flowers into the water, upon every attempt to draw tobacco, the water bubbles, and makes them dance in various figures, which both qualifies the heat of the smoke, and creates together a pretty sight."

The Arabs propagated the water-pipe in Egypt and over many tracts of Africa, where it appears in a great variety of forms. As early as 1626 Thomas Herbert found the hooka in use among the inhabitants of Mohilla, one of the four islands forming the Comoro. In 1638 it was noticed on Madagascar by

Peter Mundy, who writes, "Most commonly the men wear about their neckes in a string sundry implements off iron, etc., . . . a mouth peece for a tobacco pipe, having the tobacco growing here, which they draw through the water as in India, their hucka beeing the end off a horne with a short pipe or cane, to the end off which they apply their mouth peece afforesaid." As Mundy had spent several years in India (1628-34) in the service of the East India Company before coming to Madagascar, he is also authority for the early use of the hooka in India. As already mentioned, it is described by Terry as early as 1616. In the latter part of the seventeenth century it advanced to Siam, where La Loubère, a French envoy to Siam, noticed it in the hands of resident Mohammedans. The Mohammedans apparently spread it all over Asia, also to Chinese Turkestan. The Chinese report that the water-pipe made its first appearance at Lan-chou, capital of the province of Kan-su, in the beginning of the eighteenth century, and came from there together with the finely shredded tobacco used for the water-pipe. Lan-chou is still the producer of the best water-tobacco, which in appearance is not unlike the Turkish cigarette tobacco. It is also a centre of the Mohammedan population, and, adjoining Turkestan in the west, Kan-su is likely to have been the home of the Chinese water-pipe. The Chinese, however, received merely the impetus from Persian or Turki Musulmans; for, compared with the clumsy apparatus of Persia and India, their water-pipe is so convenient, simple, graceful, and artistic that it may be put down as an invention wholly their own.

There are two forms of water-pipe in China,—a plain one which originated in the eighteenth century, and a more complicated one, developed in the nineteenth century. The older ones are now very scarce,

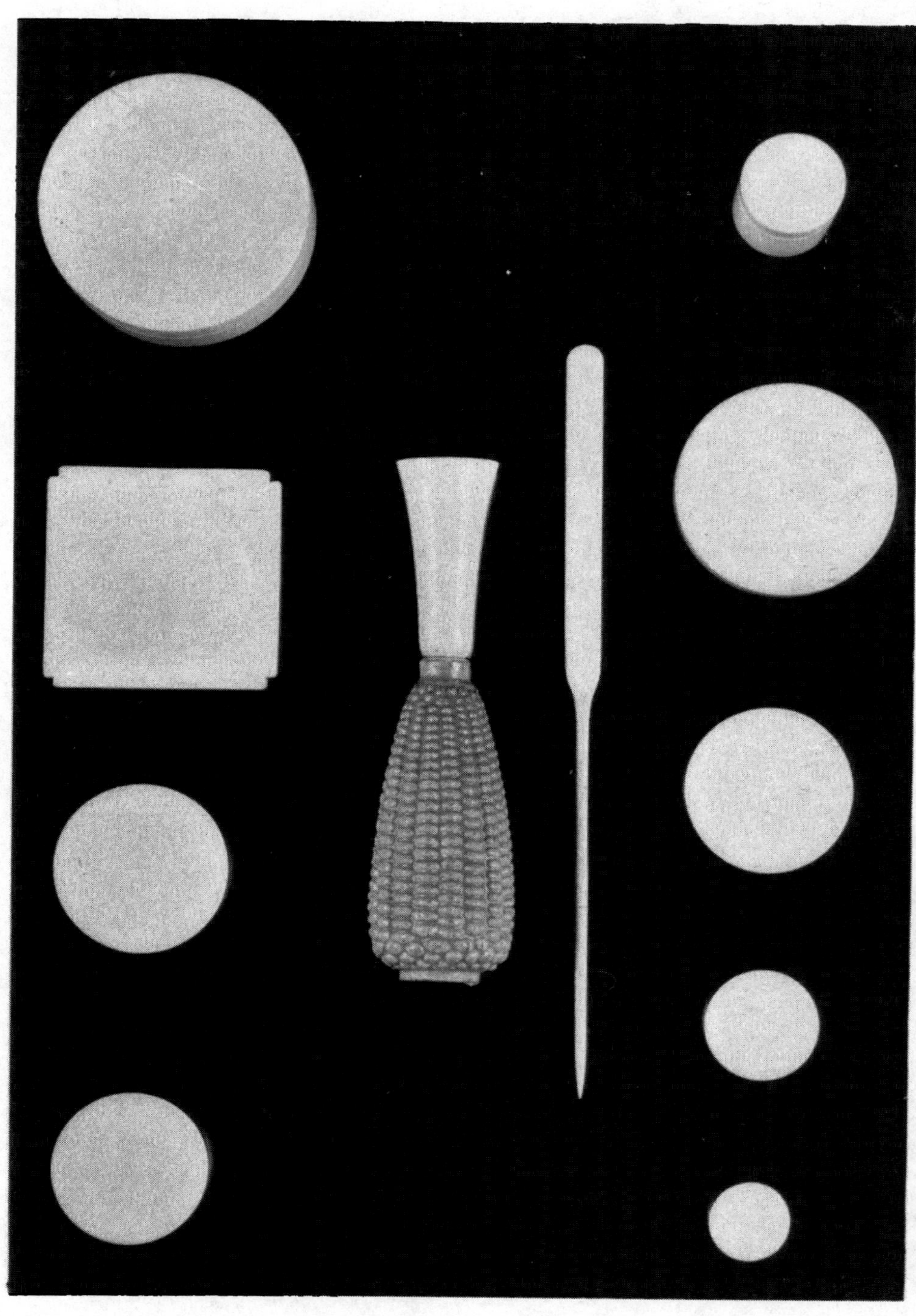

CHINESE IVORY TOBACCO AND OPIUM BOXES (AT TOP); SNUFF-BOTTLE WITH IVORY FUNNEL AND SPATULA (IN CENTER); SET OF IVORY SNUFF-DISHES (ON SIDES).
One-half actual size.

but I have been fortunate enough to secure three good specimens made in the K'ien-lung period (1736-95) and illustrated in Plate IV. The one in Fig. 1 is in shape of a standing crane, the body of the bird forming the water-vessel, and its beak the mouthpiece; the tobacco receiver is lost. The pipe in Fig. 2 represents a crouching elephant caparisoned and carrying on its back a vase, which holds the tobacco; the elephant's trunk forms the smoke-tube. Elephants of this style are frequent in K'ien-lung bronzes, as censers, candlesticks, and flower-vases (examples in Case 24, Hall 24). The water-receptacle of the pipe in Fig. 3 has the shape of an obtuse cone; the stem is worked into the appearance of bamboo, and the whole is covered with very fine engravings of floral and leaf designs. The pipe in Fig. 4, entirely made of tootnague and undecorated, is modern and manufactured by the firm Chang-te-tai at Suchow; it is reproduced after the old style, and is still in vogue among old-fashioned folks.

Six examples of the new type of water-pipe are selected from a large number in the Museum collection and reproduced in Plates V and VI. The principal innovation lies in the fact that a receptacle for storing the tobacco has been added in the form of a cylindrical vessel with hinged cover, which is closely joined to the water-receptacle and together with the latter is encased in an oval box. The pipe proper is loosely stuck into this box and held in position by means of a chain (or sometimes silk cords terminating in tassels). A scraper with brush inserted at the top and a pair of pincers for picking the tobacco are placed in detachable tubes. In a word the tendency is to concentrate and have all the necessary articles conveniently arranged in the instrument, so that it can easily be handled and carried. In Fig. 1 of Plate V, the single parts are shown separately, the smoke tube with a small cavity in the upper end in which a pinch of tobacco is placed (to the left of the pipe), the scraper in front, and the pincers to the right. The surface of the box is treated in open work and decorated with a peacock and a phœnix. The mouthpiece may be closed with a metal cap when the pipe is not in use. This pipe was made at Canton. In Fig. 2 the box is inlaid with ivory. In Fig. 3 the box is finely chased with flowered branches. The two latter pipes come from Hangchow. The pipe in Fig. 1, Plate VI, made at Suchow, is of brass and plain; the box is of rectangular shape. This pipe is regarded by the Chinese as highly artistic. The one in Fig. 2, of brass also, is engraved with a landscape all around the box. That in Fig. 3, of tootnague, is a specialty of Canton; the box is encased in black varnished leather in which designs are neatly cut out.

Water-pipes are also made of pure copper, and there is a great variety of shapes and designs. Those of tootnague are frequently inlaid with ornaments of copper, brass, bone, horn, tortoise-shell, or enamel. It would be easy to collect several hundred different varieties in different parts of the country. A new pattern was recently inaugurated at Shanghai: the water-receptacle is built in the shape of a boot, there is no tobacco receptacle, and the smoke-tube slides into the boot so that it is no longer than eight inches. On account of its reduced dimensions and light weight it can easily be carried in the pocket, but it lacks artistic merit. Water-pipes are used alike by men and women, and are freely offered to guests and visitors; they are lighted by means of paper spills. Formerly water-pipes of enormous size were circulated among the patrons of a theatre, and a cash or two were paid for a puff.

In connection with the water-pipe, a curious custom has developed among some tribes of Assam and Upper

Burma, and this is the use of nicotine water. The women of the Chin smoke the hubble-bubble largely for the benefit of the men. When the water in the water-receptacle is sufficiently saturated with nicotine, it is poured into a gourd. This liquid, however, is not swallowed; the men merely retain it in their mouths for a time and then spit it out. Sir J. George Scott characterizes the process as "merely a lazy form of chewing," though chewing is apparently not involved. The nicotine gourds of the men are often ornamented with ivory stoppers and painted with vermilion. This juice is said to act as a tonic, and travelling Kuki who eat nothing all day keep their strength up by constant sips of this juice which they retain in the mouth not more than three minutes at a time.

Tobacco-chewing is not practised in China, Tibet, Korea, and Japan. It is wholly confined to the zone of the betel-chewers, which includes India, certain portions of Farther India like Siam and Indo-China, and the Malay Archipelago. In this region tobacco leaves are added to the ingredients chewed with the nut of the areca-palm, or tobacco alone is chewed together with lime, while smoking tobacco is reduced to a minimum; cigars and cigarettes prevail in this area over the pipe. Among some tribes, as, for instance, the Karen of Upper Burma, smoking is almost as prevalent as betel-chewing. It is interesting to learn from a Japanese author, who wrote in 1708, that at that time Siamese and other foreigners at Nagasaki were observed to chew tobacco—a practice unknown to the Japanese.

A curious mode of smoking is practised in some localities in the Himalaya, southern Tibet, Kashmir, Baltistan, and Russian Turkistan. This is a stationary earth-pipe. Two holes are dug in the earth of a sloping bank, connected by an underground channel. In one hole is placed the lighted tobacco, and the smoker, crouching over the other opening, sucks out the smoke. A reed is sometimes inserted in the latter as a mouthpiece. It seems that this method is resorted to when a pipe is lacking. An interesting illustrated article on this subject has been written by H. Balfour under the title "Earth Smoking-Pipes from South Africa and Central Asia" (*Man*, 1922, No. 45).

In Asia, snuff is taken by the Chinese, the Japanese, the Tibetans, and the Brahmans of India; it is unknown in Persia. In China snuff-taking has developed into a fine art. The impetus to the practice was doubtless given by the Jesuit missionaries at a time when they wielded a powerful influence at the court of the Manchu emperors. In 1715 the emperor K'ang-hi celebrated his sixtieth birthday, and the festivities held in commemoration of this event and the homages paid to the sovereign are minutely set forth in a voluminous Chinese work. In the list of presents made to the emperor on this occasion figure also two bottles of snuff as the gift of the Jesuits Stumpf, Suarez, Bouvet, and Parrenin. It is no wonder that France, where snuff-taking was an established custom of the elegant world, should have communicated it to China. Snuff was imported from France in packages bearing three lilies as a coat of arms, and this design was adopted by the snuff-dealers in Peking as their emblem. The fleur-de-lis still forms the insignia of a snuff-shop in Peking, and it is even asserted that to this day the chief sellers of snuff are Roman Catholic converts. The largest snuff business in Shanghai, however, where I obtained last summer ten samples of the principal varieties of the article, is in the hands of Mohammedans from Lan-chou.

As early as 1685 snuff occurs in a customs tariff among the foreign imports of Canton. It was be-

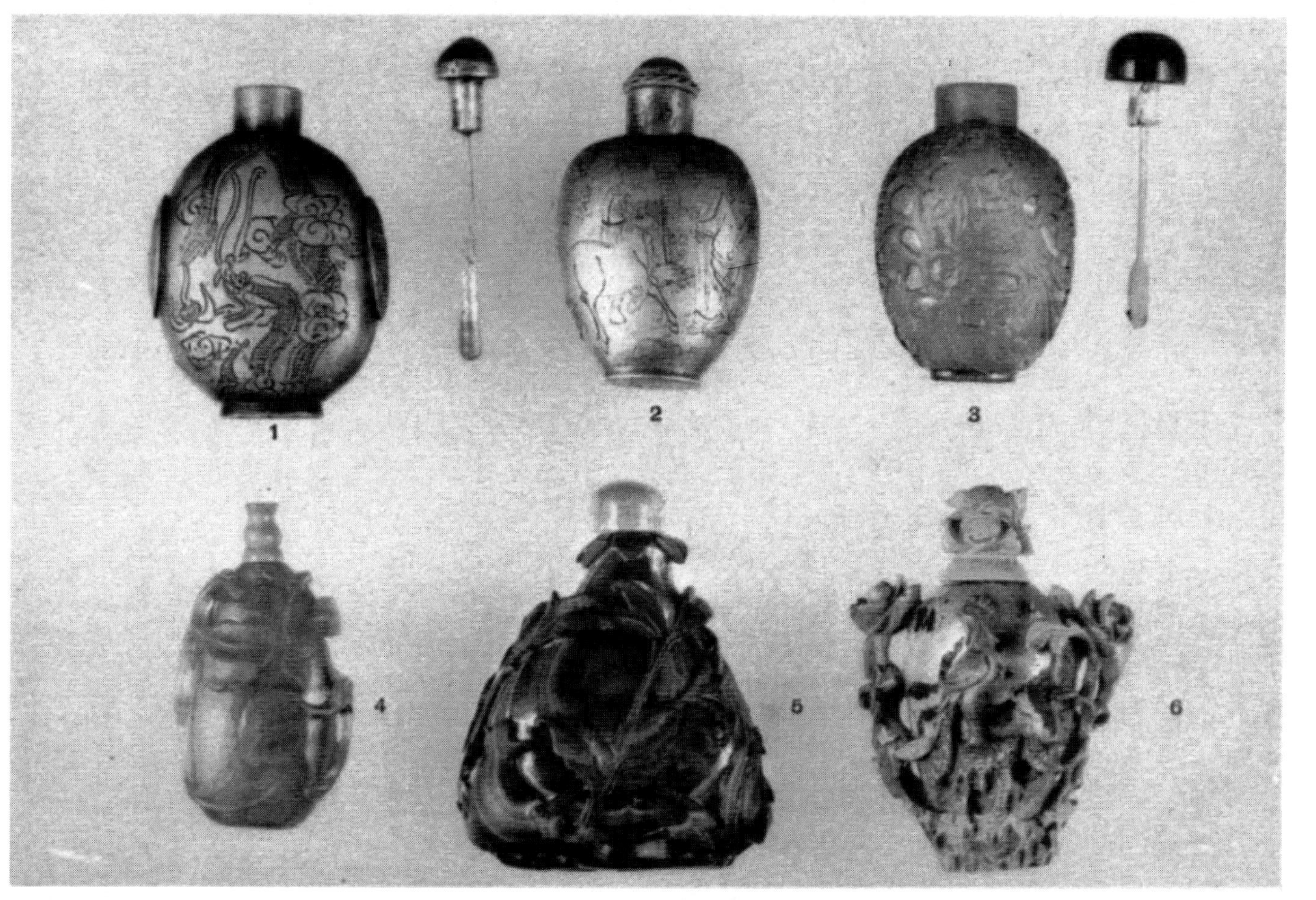

CHINESE SNUFF-BOTTLES. 1, OF BRASS; 2, OF TOOTNAGUE; 3, OF AMBER; 4, OF AGATE; 5, OF MALACHITE; 6, OF TURQUOIS.
One-half actual size.

lieved to dispel colds and act as a sudorific. Soon afterwards it was manufactured in the capital, as we read in the *Hiang tsu pi ki*, a Chinese work written in the early years of the eighteenth century: "Recently they make in Peking a kind of snuff which brightens the eyes and which has the merit of preventing infection. It is put up in glass bottles, and is sniffed into the nostrils with small ivory ladles. This brand is made exclusively for the Palace, not for sale among the populace. There is also a kind of snuff which has recently come from Canton and which surpasses that made for the Palace. It is manufactured in five different colors, that of apple color taking the first rank." Finally we hear that various kinds of snuff are used in the Palace,—snuff imported from abroad, snuff made at Canton, and several other grades made of native tobaccos. That of duck-green color was esteemed most highly, that of rose color ranked next, and that of soy color came third. Mint, camphor, and jasmine were (and still are) the principal aromatic ingrediences; essence of rose was also mixed with it. In the eighteenth century good qualities were sold for their weight in silver, and were a favorite gift among friends. The Portuguese distributed snuff from their settlement at Macao.

On account of its peculiar aroma a certain brand of tobacco growing in Shan-tung Province is given preference in the manufacture of snuff. The dried leaves are carefully freed from the stems and ribs, and are crushed in a mill or mortar to a fine powder which is several times winnowed through sieves until it is as fine as wheaten flour. The tobacco powder is then scented with aromatic substances, and is packed in small tubes of tin. The workmen have to keep their mouths and noses covered during this occupation, in order to prevent perpetual sneezing that the fine dust might provoke. Lu Yao, who wrote a treatise on tobacco in the latter part of the eighteenth century, observes that those who have made a long-continued practice of taking snuff will not sneeze any more.

In Peking, where snuff is more popular than in other cities of China, it is said to be taken chiefly for the benefit of one's nose, protecting it from the plentiful dust of the capital and saving it the offensive street odors. They also attribute to snuff medicinal virtues and beneficial effects, particularly after a heavy dinner, so that it is taken for curative purposes or made the vehicle of conveying other medical agents into the system. It is believed to be good for pain in the eyes, toothache, throat-trouble, asthma, and constipation. Like the Italians, the Chinese have great faith in old snuff, and the Peking dealers in antiques dispose of snuff alleged to be a century old or even older and stored in big glass jars of the same period. It is said that the habit is now on the decline, and it may be doubted that it ever was very general; it seems to have always remained a luxury confined to the well-to-do, especially the class of officials. The fact that it was a popular sport in high society during the eighteenth century is plainly visible from the large number of very artistic snuff-bottles which have come down to us from that period. Almost every substance available for this purpose in the three kingdoms of nature has been utilized for the making of snuff phials: the beak of the hornbill or buceros, ivory, coral, mother-of-pearl, amber, jade, agate, carnelian, chalcedony, rock-crystal, malachite, turquois, lapis lazuli, gold, silver, brass, white copper, porcelain hard and soft, painted enamel, carved lacquer, glass painted or cut in different layers of colors, and even bamboo, nut-shells, and various hard fruits. There is an endless variety of shapes and designs, and many are veritable gems eliciting our admiration for the skill and ingenuity of the lapidary. As to forms they are traceable to older drug-phials, and since snuff was placed in the category of medicine, it is easily understood that a drug-phial did service as a snuff-holder. The old drug-phials, however, were limited to pottery or porcelain as to material, while the manifold varieties of material for the use in snuff-bottles are characteristic of the K'ien-lung period (1736-95).

In former times snuff-bottles were part and parcel of a gentleman's outfit, and people were proud of displaying them. Snuff-bottles are closed by a stopper in the form of a small knob made of jade, coral, turquois, tourmaline, or colored glass. Attached to this stopper is a small ladle of silver, ivory, bone, horn, or bamboo, by means of which the snuff is taken out of the bottle and placed on the thumb-nail (Plate VIII, Figs. 1 and 3). From the thumb-nail the substance is conveyed into the nostrils. In order to fill the bottle with snuff, a slender funnel carved from ivory is inserted into the opening, and the snuff is poured through the funnel by means of an ivory spatula (Plate VII, in centre). The snuff-bottle here illustrated is of yellow glazed porcelain in shape of a maize-cob. Those who wish to serve a variety of snuffs to their friends, especially after dinner, avail themselves of a set of tiny ivory dishes of varying sizes, seven of which are shown in Plate VII; on these the snuffs are arranged and served to guests. An ivory box to hold tobacco and a smaller one for opium are illustrated in the same Plate (at top).

Plates VIII and IX illustrate a small selection of ancient snuff-bottles in the Museum's collections. That in Plate VIII, Fig. 1 (presented by P. J. Bahr of Shanghai), is of brass engraved with a dragon, and is unique in being inscribed on the bottom with the name of the maker and a date of the Shun-chi period (11th year), which corresponds to the year 1653. It was made by Cheng Tsung-chang. A snuff-bottle of tootnague (Fig. 2) is finely engraved with a scene on each side; the one shown represents the Taoist goddess Ma-ku rowing a boat which is a rugged, bare tree and carries her basket filled with gifts of blessing. The next bottle (Fig. 3) is carved from transparent Burmese amber with raised designs of two phœnixes and peonies; that in Fig. 4 is carved from chalcedony with graceful floral and leaf designs in undercut relief, the knob being of turquois. Fig. 5 represents a rare specimen cut from a dark green malachite veined with light green zones and decorated with a pomegranate, leafed branches, and birds; the stopper is of pink tourmaline. The bottle in Fig. 6 is a carving from green and bluish turquois in the matrix decorated with a phœnix perching on a rock and surrounded by flowers and leaves in high undercut relief.

Of agates there is an immense variety: the bottle of triangular shape (Plate IX, Fig. 4) is of moss agate, milk white in color with strata of yellow and black natural designs which look like ferns. The oval-shaped bottle (Fig. 5) is carved from an agate with yellow clouds and dark brown streaks on the sides. In such pieces the lapidary strives at bringing out the peculiar coloration of a stone to its best advantage. The brown agate (Fig. 7) is shaped into a jujube fruit of surprising naturalness, with three peanuts and a bee appearing in high relief. A snuff-bottle of yellowish jade carved into the appearance of a basket, with stopper of tourmaline, is illustrated in Fig. 8.

A snuff-bottle of glass carved in cameo style, two hydras of ruby color standing out in high relief, is shown in Plate IX, Fig. 1. The Chinese were always fond of treating glass like semi-precious stones and cutting and polishing it in its hard state. They know

CHINESE SNUFF-BOTTLES. 1-3, 6, OF GLASS; 4, 5, 7, OF AGATE; 8, OF JADE.
One-half actual size.

how to produce in glass an astounding wealth of colors by means of metal oxides, as iron pyrites, iron oxides, copper oxides, acetate of lead, and others, and try to imitate in glass the tinges of jade, agate, malachite, lapis lazuli, amber, jet, coral, as well as leaves and fruits; thus a gourd in its natural colors of green and white (Fig. 2) and a peach in the process of ripening (Fig. 3). They are skilful also in fusing together glasses of different colors or introducing into the mass spots, veins, or bands with a view of rivaling nature in the imitation of stones which serve as models. Sometimes the color of the glass is brought out to its full perfection by a simple process of polishing; sometimes feet, handle, and neck of a vessel are added with glass of a color differing from that of the body. In many cases there are several layers of glass of various tints placed one above the other, the upper one being cut into scenes or figures that stand out in high relief from the body of the vessel.

An industry characteristically Chinese is the manufacture of glass snuff-bottles with decorations colored by hand in the interior of the glass. The bottles required for this purpose were made at Canton, and were sent there from Peking for painting. As the surface of the glass is too smooth to take pigments, the inside is prepared with pulverized iron oxydul which is mixed with water. This liquid shaken in the bottle for about half a day will form a rough, milk-white coating suitable for receiving paints. In executing the work the artist lies on his back, holding the small bottle up to the light between the thumb and the index finger of his left hand and with a very fine brush in his right hand. The hairy tip of the brush is not straight, as usual, but stands under a right angle against the handle. His eyes are constantly fixed on the outer surface of the glass, thus watching the gradual development of the picture as it emerges from under the glass. He first outlines a skeleton sketch in black ink, starting from below and then passing on to the middle and sides, finally inserting the colors. Half a day is sufficient to complete an ordinary piece, while a whole day and more may be spent on more elaborate work. The subjects include landscapes, genre and battle scenes, as well as flower-pieces. This art-industry commenced in the K'ien-lung period (1736-95), and the little masterpieces turned out at that time are unsurpassed. A good example of this period, dated 1740, is illustrated in Plate IX, Fig. 6. In addition to the figure-painting, the surface of the glass bears etchings of bamboo and plum-blossoms. Specimens of the Tao-kuang period (1821-50) also are usually good, but scarce. The modern output is chiefly intended for the foreign market, and does not stand comparison with the products of bygone days; the bottles are large, coarse, and clumsy, and the paintings are usually crude.

The Portuguese appear to have introduced snuff into India. Gautier Schouten, who travelled in India from 1658 to 1665, writes that the Portuguese women of Goa were in the habit of taking both betel and snuff, and constantly carried a snuff-box in their pockets. Many people appeared in the streets of Goa with a snuff-box in their hands and let it circulate; they seemed to vie with one another in sneezing, and were always seen with lips and noses stained by tobacco.

In cultivating tobacco for the manufacture of snuff, it is customary in India not to irrigate the crop from wells, but to grow the plants by the aid of rain alone. A Brahman may take snuff, but he should not smoke a cheroot or cigar. Once the cheroot has touched his lips, it is defiled by the saliva, and therefore cannot be returned to his mouth. This rule was adopted by the clergy of Tibet; the Lamas must not smoke, and smoking is strictly forbidden to the young clericals. Tibetans smoke but rarely, while snuff is a passion with all classes of the laity and clergy. Snuff is prepared in round wooden boxes, across the interior of which is stretched a fine cloth sieve. The coarse tobacco is put in the top of the box through a hole in the lid, which is closed by a wooden stopper, and by lightly striking the box on the knee the finer parts are sifted through into the lower compartment (Plate X, Fig. 1). By a little aperture in the lower part of the box the snuff is poured out onto the nail of the left thumb held against the index, and is thus inhaled. Tobacco is imported into Tibet from China, Bhutan, India, and Nepal. The leaves of the rhubarb plant are frequently used as a substitute for it, being either mixed with tobacco leaves, or even used pure.

The Tibetan snuff-bottle in Fig. 2 of Plate X is carved from the burl of a maple-tree, and is mounted with brass ornaments; that in Fig. 3 is formed by a bean, the opening in the centre being closed by a wooden plug. The snuff is poured in through this aperture and taken out through the tube of soft stone inserted at the top. Horns of wild sheep, yak, and oxen are largely used as snuff-containers, particularly by the nomadic tribes of eastern Tibet. The horns are well polished, plain or incised with geometric ornaments, or decorated with silver, white copper, or copper bands. They are filled through the lower end which is tightly closed by wooden or metal covers, and the snuff is taken out through the upper pointed tip, closed by a stopper (Fig. 5). The specimen in Fig. 4 is cut out of animal bone, with wooden lid and handle of leather thong. More examples may be viewed in Case 33, Hall 32, where Tibetan tobacco-pipes are also on exhibition.

The preceding brief sketch is based on an extensive manuscript of the writer, which may be published at a later date.

B. LAUFER.

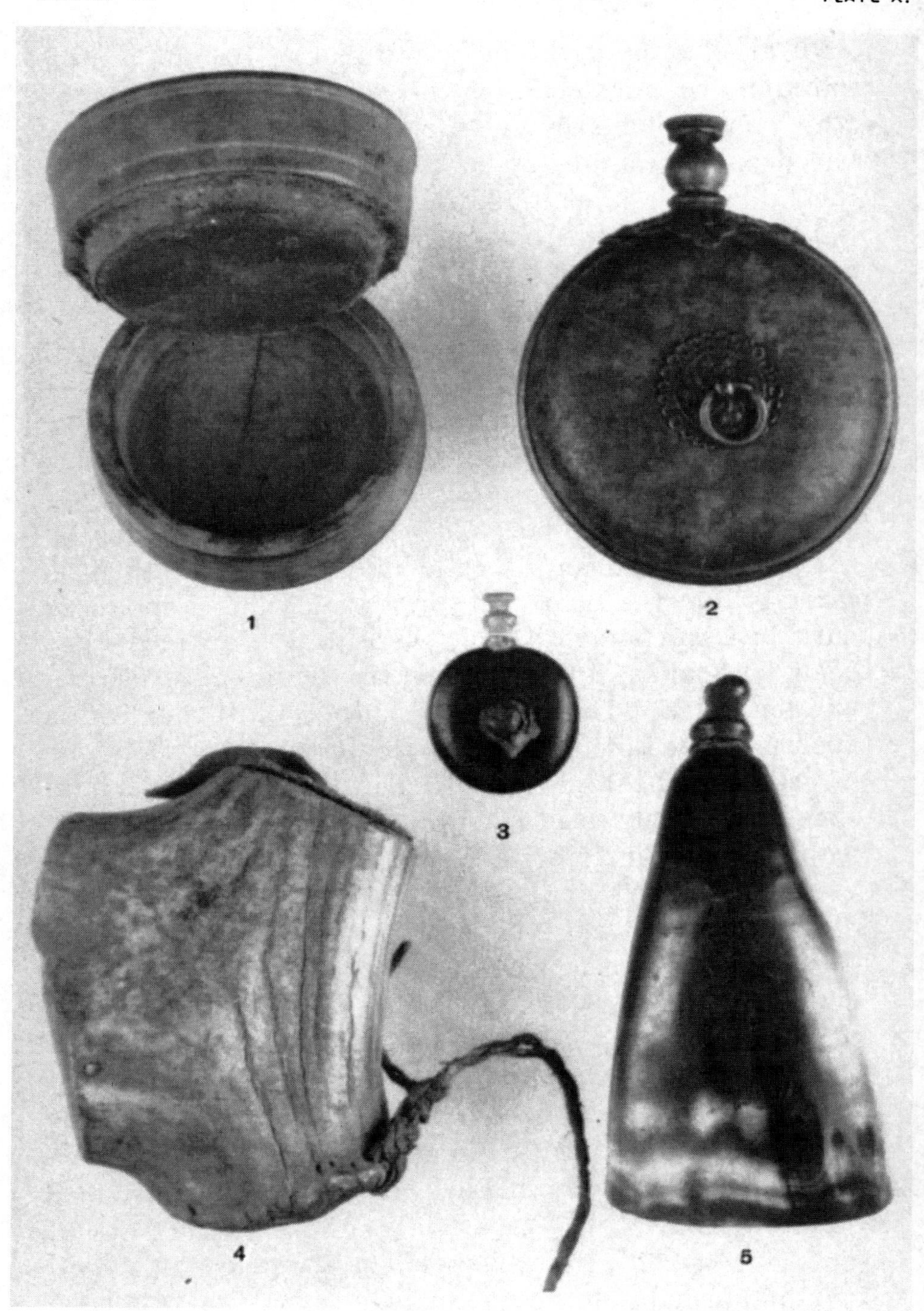

SNUFF-BOXES AND SNUFF-HORN FROM TIBET.
One-half actual size.

168

烟草传入欧洲考

INTRODUCTION OF TOBACCO INTO EUROPE

BY

BERTHOLD LAUFER

CURATOR OF ANTHROPOLOGY

ANTHROPOLOGY

LEAFLET 19

FIELD MUSEUM OF NATURAL HISTORY

CHICAGO

1924

The Anthropological Leaflets of Field Museum are designed to give brief, non-technical accounts of some of the more interesting beliefs, habits and customs of the races whose life is illustrated in the Museum's exhibits.

LIST OF ANTHROPOLOGY LEAFLETS ISSUED TO DATE

No.	Title	Price
1.	The Chinese Gateway	$ .10
2.	The Philippine Forge Group	.10
3.	The Japanese Collections	.25
4.	New Guinea Masks	.25
5.	The Thunder Ceremony of the Pawnee	.25
6.	The Sacrifice to the Morning Star by the Skidi Pawnee	.10
7.	Purification of the Sacred Bundles, a Ceremony of the Pawnee	.10
8.	Annual Ceremony of the Pawnee Medicine Men	.10
9.	The Use of Sago in New Guinea	.10
10.	Use of Human Skulls and Bones in Tibet	.10
11.	The Japanese New Year's Festival, Games and Pastimes	.25
12.	Japanese Costume	.25
13.	Gods and Heroes of Japan	.25
14.	Japanese Temples and Houses	.25
15.	Use of Tobacco among North American Indians	.25
*16.	Use of Tobacco in Mexico and South America	.25
*17.	Use of Tobacco in New Guinea	.10
18.	Tobacco and Its Use in Asia	.25
19.	Introduction of Tobacco into Europe	.25
*20.	The Japanese Sword and Its Decoration	.25

D. C. DAVIES
DIRECTOR

FIELD MUSEUM OF NATURAL HISTORY
CHICAGO, U. S. A.

*In preparation—November 1924.

FIELD MUSEUM OF NATURAL HISTORY
DEPARTMENT OF ANTHROPOLOGY
CHICAGO, 1924

LEAFLET NUMBER 19

The Introduction of Tobacco into Europe

In the four preceding leaflets the history and use of tobacco in the two Americas, in Melanesia, and in Asia have been briefly discussed. It may therefore not be amiss to close this series with a review of the early history of tobacco in Europe, particularly in England,—a subject of general interest.

The white man learned the use of tobacco from the aborigines of America soon after the discovery, and the European colonists who flocked to America rapidly adopted the habit of smoking. Las Casas was already compelled to admit that the Spaniards on Cuba who had contracted the habit could not be weaned from it. Lescarbot applies a similar remark to the French of Canada. "Our Frenchmen who visited the savages are for the most part infatuated with this intoxication of petun [tobacco], so much so that they cannot dispense with it, no more than with eating and drinking, and they spend good money on this, for the good petun which comes from Brazil sometimes costs a dollar (*écu*) the pound." John Hawkins observed in 1564 that the French in Florida used tobacco for the same purposes as the natives. A. Thevet, who visited Brazil in 1555-56, noticed the Christians living there as "marvelously eager for this herb and perfume." Gabriel Soares de Souza (Noticia do Brazil, written in 1587), a Portuguese farmer, who lived in Brazil for seventeen years from about 1570, informs us that tobacco leaves were much esteemed by the

Indians, Negroes (whom he calls Mamelucos), and Portuguese, who "drank" the smoke by placing together many leaves wrapped in a palm-leaf; they used, accordingly, the cigar. The unknown author of the "Treatise of Brazil," written in 1601 and published by Purchas, also describes the mode of cigar-smoking in Brazil and winds up by saying, "The women also doe drinke it, but they are such as are old and sickly, for it is verie medicinable unto them, especially for the cough, the head-ache, and the disease of the stomacke, and hence come a great manie of the Portugals to drinke it, and have taken it for a vice or for idlenesse, imitating the Indians to spend daies and nights about it."

The English colonists in Virginia did not hesitate to appropriate the aboriginal custom of pipe-smoking. Thomas Hariot (A Brief and True Report of the New Found Land of Virginia, 1588) dwells with enthusiasm on the virtues of the herb, "which is sowed a part by it selfe and is called by the inhabitants *uppówoc:* In the West Indies it hath divers names, according to the severall places and countries where it groweth and is used: The Spaniards generally call it *Tobacco.*" He concludes, "We our selves during the time we were there used to suck it after their maner, as also since our returne, and have found manie rare and wonderful experiments of the vertues thereof; of which the relation woulde require a volume by it selfe: the use of it by so manie of late, men and women of great calling as else, and some learned Phisitions also, is sufficient witnes." "Sucking it after their maner" means pipe-smoking which Hariot himself describes as follows: "The leaves thereof being dried and brought into powder: they use to take the fume or smoke thereof by sucking it through pipes made of claie into their stomacke and heade."

The following passages show that the English settlers soon proceeded to make their own pipes. George Waymouth, who visited Virginia in 1605, has the following notice: "They gave us the best welcome they could, spreading deere skins for us to sit on the ground by their fire, and gave us of their tobacco in our pipes, which was most excellent, and so generally commended of us all to be as good as any we ever tooke, being the simple leafe without any composition, very strong and of a pleasant sweete taste: they gave us some to carry to our captaine, whom they called our Bashabe, neither did they require any thing for it; but we would receive nothing from them without remuneration." George Percy, who visited southern Virginia in 1606, describes an entertainment given in his honor by the savages. "After we were well satisfied they gave us of their tabacco, which they tooke in a pipe made artificially of earth as ours are, but far bigger, with the bowle fashioned together with a piece of fine copper."

INTRODUCTION AND EARLY CULTIVATION OF TOBACCO IN ENGLAND

The four Atlantic states—England, France, Portugal, and Spain—received tobacco directly from America. The subject, as far as England is concerned, forms a chapter independent of the rest of Europe.

In considering the history of tobacco in England, we must distinguish between the introduction of the tobacco plant or plants and the custom of smoking tobacco, for it seems that tobacco was known or even planted in England a number of years before smoking was practised. The two earliest English botanists, John Gerard (1597) and John Parkinson (1640), are familiar with the two principal species, *Nicotiana tabacum* (in two varieties) and *Nicotiana rustica*, so that at the outset we should be justified in assuming at least two introductions. Such indeed are upheld by tradition.

Edmund Howes, in his continuation of John Stow's "Annales or Generall Chronicle of England" (1631, p. 1038), states,—

"Tobacco was first brought and made known in England by Sir Iohn Hawkins, about the yeare 1565, but not used by Englishmen in many yeeres after, though at this day commonly used by most men, and many women."

Hawkins returned from his second voyage to the West Indies on the 20th of September, 1565, and had become familiar with tobacco and smoking in Florida. John Sparke the Younger, who wrote the account of this voyage (published by Hakluyt in 1589), writes that Hawkins, ranging along the coast of Florida for fresh water in July, 1565, came upon the French settlement there under Laudonière, and continues thus: "The Floridians when they travell have a kind of herbe dryed, which with a cane, and an earthen cup in the end, with fire, and the dried herbs put together, do sucke thoro the cane the smoke thereof, which smoke satisfieth their hunger, and therewith they live foure or five days without meat or drinke, and this all the Frenchmen used for this purpose: yet do they holde opinion withall, that it causeth water and fleame to void from their stomacks." This is the earliest English notice of tobacco. It would be amazing if Hawkins and his companions should not have imitated this custom, and Hawkins may therefore have taken specimens of *Nicotiana rustica* and its seeds from Florida to England in 1565. It was from

Florida, as will be seen, that the plant was also introduced into Portugal and from Portugal into France.

It must be borne in mind, however, that Howes' statement is not coeval with the event to which he refers, but was drafted sixty-five years afterwards. In Stow's "Annales" it is entirely absent. It is therefore not consistent with the facts, as some authors have done, to attribute this and the data that follow below, contained in a book of 1631, to Stow, who died in 1606. Nor is Howes' assertion, as has been argued, corroborated by Taylor, the water-poet, who in a postscript to his versified *Life of Thomas Parr* says that tobacco was first brought into England in 1565 by Hawkins, adding, "It is a doubtful question whether the devil brought tobacco into England in a coach, for both appeared about the same time." Taylor's work was published in 1635, and his plea for Hawkins is simply copied from Howes. Nevertheless I am under the impression that Howes honestly reproduced a tradition which was current in the latter part of the sixteenth century and had come down to his own time. It is far less this tradition itself, however, than the total of the circumstantial evidence which compels us to pin our faith in Hawkins as the introducer of *Nicotiana rustica;* for this species was grown in England in the latter part of the sixteenth century, so that its presence in English soil must be accounted for in a reasonable manner. Dr. Brushfield, in 1898, formulated his opinion thus: "Tobacco was first imported into Europe about the year 1560, but not into England until a few years later. The first Englishman to notice it was Sir J. Hawkins in 1565; whether, however, he brought any to this country is unknown, most probably he did, the other alternative being its importation from Spain." In this view the botanical side of the question is disregarded, and Spain cannot be called to the witness-stand, as the Spaniards were exclusive and never took the trouble of propagating tobacco or any other American plant to any country of Europe.

On the same page of the above work, Howes makes the further statement, "Apricocks, Mellycatons, Musk-Millions and Tobacco, came into England about the 20 yeare of Queene Elizabeth" [1577], and adds in the margin, "Sir Walter Raleigh was the first that brought Tobacco into use, when all men wondered what it meant." The two different dates are not so incompatible as it would seem at first sight: in that great age of unprecedented colonial expansion and seafaring enterprise tobacco must assuredly have arrested the attention of several navigators, and the fact that different species and varieties of Nicotiana were grown in England at least in the three last decades of the century proves that several introductions at different times and presumably from different parts of America must have been effected.

In February, 1593, William Harrison completed his great work of English Chronology two months before his death (April 24, 1593). The three large folios comprising volumes II-IV of his "Great Chronologie," which he says "he had gathered and compiled with most exquisit diligence," are preserved in manuscript in the Diocesan Library at Derry, Ireland. In the fourth volume the events from A.D. 1066 up to 1593 are chronicled year by year, and in it the data referring to his own time are of particular value. Extracts covering this period are given in Furnivall's edition of Harrison's Description of England (published for the New Shakspere Society, 1876). Here we meet (p. LV) under the year 1573 the following fundamental document relating to tobacco and smoking, which has never been utilized or interpreted correctly and which is calculated to revise all former conceptions of the early history of tobacco in England.

"1573. In these daies the taking-in of the smoke of the Indian herbe called 'Tabaco,' by an instrument formed like a litle ladell, whereby it passeth from the mouth into the hed and stomach, is gretlie taken-up and used in England, against Rewmes and some other diseases ingendred in the longes and inward partes, and not without effect. This herbe as yet is not so comon, but that for want thereof divers do practize for the like purposes with the Nicetian, otherwise called in latine, 'Hyosciamus Luteus,' or the yellow henbane, albeit, not without gret error; for, althoughe that herbe be a soverene healer of old ulcers and sores reputed incurable outwardly, yet is not the smoke or vapour thereof so profitable to be receaued inwardly. The herbe [tobacco] is comonly of the height of a man, garnished with great long leaves like the paciens [Passions or Patience, *Rumex patientia* L.], bering seede, colloured, etc. of quantity like unto, or rather lesse then, the fine margeronie; the herbe it self yerely coming up also of the shaking of the seede. The collour of the floure is carnation, resembling that of the lemmon in forme: the roote yellow, with many fillettes, and therto very small in comparison, if you respect the substauns of the herbe."

This is the memorable record of a contemporary eye-witness, who in his fascinating Description of England gives ample proof of his keen power of observation of customs and manners. His notice is based on direct and personal observation, it is not copied from hearsay or books. The botanical description is even unique, almost perfect, considering the fact that the writer was not a botanist, and represents the first English description of the species *Nicotiana tabacum:* for the herb is commonly of the height of a man, garnished with great long leaves and having flowers of carnation color—characteristics of *Nicotiana tabacum* only. The herb was then planted in England, but was not yet common, and the henbane served smokers as a substitute; perhaps, however, Harrison's henbane, as suggested by the addition Nicetian (i.e. Nicotian), is *Nicotiana rustica.* Hyoscyamus, like Nicotiana, is a solanaceous plant of poisonous narcotic qualities. The first description of the tobacco plant in the botanical literature of Europe is that of the Italian botanist and physician Pierandrea Mattioli (1500-77) in his "Commentarii in Dioscoridem" (1565) under the name *Hyoscyamus niger.* The botanist Mathias de Lobel, as will be seen presently, affirms tobacco culture in England (prior to 1576) and describes pipe-smoking on the part of sailors who returned from America. Harrison therefore is in good company and upheld by the testimony of a contemporary. The tobacco plant was cultivated in England in 1573, a year before the discovery of Virginia, though not in sufficient quantity to satisfy general demand, and tobacco was smoked by Englishmen at that time from ladle-like instruments (perhaps similar to, or even identical with the subsequent pipes consisting of a half walnut, see below, p. 35). Harrison is the first English author who uses the word *tabaco*, the first who records the custom of smoking tobacco in England, and the first who describes its remedial properties and effects, and this independently of Monardes, whose work "Englished" by Frampton became known to the English public only in 1577.

Consequently the date 1577 given by Howes as that of the first introduction cannot be correct and must be discarded. The question arises, When and by whom was *Nicotiana tabacum*, ostensibly described by Harrison, introduced into England? At that time this species was widely disseminated from Mexico to the Antilles and South America; it could not have come to England from any point of North America, where *Nicotiana rustica* was the principal tobacco-furnishing plant. *Nicotiana tabacum* was introduced into Virginia from Trinidad not earlier than about 1610 (W. Strachey, Historie of Travaile into Virginia Britannia, ed. of R. H. Major, p. 31). Now it happened that on the 9th of August, 1573, Francis Drake returned to Plymouth from his expedition to the West Indies. In the same year Harrison describes *Nicotiana tabacum* which is the typical Nicotiana species of the West Indies, and records the diffusion of tobacco-smoking in England. There is no accident in history, it is governed by the law of cause and effect. In my estimation, these two events cannot be a fortuitous coincidence, but are closely interrelated. In my opinion, therefore, it is reasonable to conclude, and there is no escape from the conclusion, that tobacco was brought to England again in 1573 by Sir Francis Drake (whether by himself personally or by a sailor or member of his expedition remains immaterial), and this was *Nicotiana tabacum,* known to Harrison and subsequently to John Gerard as "the greater sort of Tabaco brought into Europe out of the provinces of America, which we call the West Indies." There are, further, two weighty testimonies to the effect that tobacco was grown in England long before 1586, the date of the return of the Virginian colonists, which in most books is erroneously taken for the year of the first introduction of tobacco and smoking. There are the two botanists, Peter Pena and Mathias de Lobel (Nova stirpium adversaria, Antwerp, 1576, p. 251), who state positively that tobacco was then cultivated in Portugal, France, Belgium, and England; and this is good confirmation of Harrison's account. And there is Richard Hakluyt, who, in his instructions written for an English factor at Constantinople in 1582, states, "The seed of tabacco hath bene brought hither out of the West Indies, it groweth heere, and with the herbe many have bene eased of the reumes," etc. Again, in this case, the West Indies hint at *Nicotiana tabacum* and at the exploits of Francis Drake. It may be noted also that H. Phillips (History of Cultivated Vegetables, 1822, Vol. II, p. 339) states that "tobacco was brought to England by Sir Francis Drake, in 1570, who that year made his first expedition against the Spaniards in South America."

On the 27th of July, 1586, the colonists settled in Virginia by Ralph Lane returned to England and disembarked at Plymouth. They offered their astounded countrymen the queer spectacle of smoking tobacco from pipes, which caused a general sensation. William Camden (1551-1623), the historiographer of Queen Elizabeth and a contemporary witness, reports this event as follows (Annales rerum anglicarum, 1615, p. 408; or History of the Most Renowned and Victorious Princess Elizabeth, 4th ed., 1688, p. 324):—

"And these men who were thus brought back were the first that I know of that brought into England that Indian plant which they call Tabacca and Nicotia, or Tobacco, which they used against crudities being taught it by the Indians. Certainly from that time forward it began to grow into great request, and to be sold at an high rate, whilst in a short time many men every-where, some for wantonness, some for health sake, with insatiable desire and greediness sucked in the stinking smoak thereof through an earthen pipe, which presently they blew out again at their nostrils: insomuch as tobacco-shops are now as ordinary in most towns as tap-houses and taverns. So that the Englishmens bodies, (as one said wittily,) which are so delighted with this plant, seem as it were to be degenerated into the nature of Barbarians, since they are delighted, and think they may be cured, with the same things which the Barbarians use."

From what has been said above it is clear that the band returning from Virginia was not instrumental in introducing tobacco cultivation into England, for this was an established fact long before that time, neither were they the first smokers on British soil. It is solely popular imagination which has vividly retained this very event and which glorified Ralph Lane, Richard Grenville, or Walter Raleigh as the first smokers.

King James, in his "Counterblaste to Tobacco" (1604), alludes to the first introduction but vaguely, "Now to the corrupted basenesse of the first use of this Tobacco, doeth very well agree the foolish and groundlesse first entry thereof into this Kingdome. It is not so long since the first entry of this abuse amongst us here, as this present age cannot yet very well remember, both the first Author, and the forme of the first introduction of it amongst us. It was neither brought in by King, great Conquerour, nor learned Doctor of Phisicke. With the report of a great discovery for a Conquest, some two or three Savage men, were brought in, together with this Savage custome. But the pitie is, the poore wilde barbarous men died, but that vile barbarous custome is yet alive, yea in fresh vigor: so as it seemes a miracle to me, how a custome springing from so vile a ground, and brought in by a father so generally hated, should be welcomed upon so slender a warrant." This "father" no doubt is Sir Walter Raleigh, but it is not necessary to concur with Edward Arber, who justly denies that Raleigh had anything to do with the introduction of the weed itself or of the habit of smoking, in the conclusion that "the king wilfully or ignorantly falsified the history of the introduction of tobacco, concocting a degrading story for his purpose." The king's remark certainly savors of malice, but he may have honestly been persuaded that Raleigh was the first introducer.

Henry Buttes (Diets Dry Dinner, 1599) states, "Our English Ulisses, renomed Syr Walter Rawleigh, a man admirably excellent in Navigation, of Natures privy counsell, and infinitely read in the wide booke of the worlde, hath both farre fetcht it, and deare bought it: the estimate of the treasure I leave to other." It may be perfectly true, of course, that Raleigh laid in a good supply of tobacco or secured it from Hariot, for his own consumption and the use of his friends. A letter of Sir John Stanhope to Sir G. Carew, dated January 26th, 1601, contains this paragraph: "I send you now no Tabacca, because Mr. Secretary, Sir Walter, and your other friends, as they say, have stored you of late; neither have I any proportion of it (that) is good, but only am rich in Aldermans Watses promises of plenty, wherewith you shall be acquainted, God willing." Raleigh may have been initiated into the art of smoking by Hariot, who had been sent out by him for the purpose of inquiring into the natural productions of Virginia. As indicated above (p. 2) after Hariot's own report, he smoked in Virginia and continued to smoke on his return to England.

E. Arber, in his valuable notes on the Introduction of Tobacco into England (1869), thinks that we have but little demonstrative proof of Raleigh's tobacco habit, but there is the testimony of John Parkinson (Theatrum botanicum, 1640, p. 711), who affirms that he knew Raleigh when he was prisoner in the Tower, and that Raleigh chose the "English Tabacco" (*Nicotiana rustica*) to make good tobacco of, "which he knew so rightly to cure that it was held almost as good as that which came from the Indies, and fully as good as any other made in England." This tobacco, however, was not thought to be so strong or sweet for the pipe, nor so efficient for diseases.

It is to Raleigh's merit that he made smoking fashionable and a gentlemanly art; his name became identified with the new national habit so thoroughly that later generations looked upon him as a kind of patron-saint of the smokers. Every one is familiar with the anecdote that Raleigh, sitting one day in a deep meditation, with a pipe between his lips, bade his man to bring him a tankard of small ale. Believing that his master's head was set on fire, he threw the liquor in his face. In fact, however, this story appears for the first time in 1611 in the Jests of Richard Tarleton, and as has been shown by G. L. Apperson (Social History of Smoking, 1914), was fastened on Raleigh as late as 1708. The tradition that Raleigh smoked a pipe or two on the morning before his execution (October 29th, 1618) appears to be well founded. The Dean of Westminster, who attended him on this morning, testifies that "he eate his breakfast hertily and tooke tobacco." Aubrey thus defends his action: "He took a pipe of tobacco a little before he went to the scaffolde, which some female (other reading: formal) persons were scandalized at; but I think 'twas well and properly donne to settle his spirits." No mention of tobacco has been discovered in any of Raleigh's printed works. His first testamentary note made shortly before his execution contains, as far as is yet known, his sole mention of tobacco and relates to that which remained on his ship after his ill-fated voyage: "Sir Lewis Stukeley sold all the tobacco at Plimouth of which, for the most part of it, I gave him a fift part of it, as also a role for my Lord Admirall and a role for himself. I desire that hee give his account for the tobacco."

Raleigh's tobacco-box was preserved at Leeds in Yorkshire, in the Museum of Ralph Thoresby, an antiquary, who died in 1725. Soon afterwards, William Oldys saw it there, and in his life of Raleigh prefixed to "The History of the World" (1736), describes it thus: "From the best of my memory, I can resemble its outward appearance to nothing more nearly than one of our modern Muff-cases; about the same height and width, cover'd with red leather, and open'd at top (but with a hinge, I think) like one of those. In the inside, there was a cavity for a receiver of glass or metal, which might hold half a pound or a pound of tobacco; and from the edge of the receiver at top, to the edge of the box, a circular stay or collar, with holes in it, to plant the tobacco about, with six or eight pipes to smoke it in." R. Thoresby himself (Ducatus Leodiensis, 1715) gives the following, slightly different description: "Sir Walter Ralegh's tobacco-box, as it is called, but is rather the case for the glass wherein it was preserved, which was surrounded with small wax candles of various colours. This is of gilded leather, like a muff-case, about half a foot broad and thirteen inches high, and hath cases for sixteen pipes within it."

John Gerard (The Herball of Generall Historie of Plantes, 1597) writes that "there be two sorts or kindes of Tabaco, one greater, the other lesser; the greater was brought into Europe out of the provinces of America, which we call the West Indies: the other from Trinidada, an Ilande neere unto the continent of the same Indies. Some have added a third sort, and others making the yellow Henbane [*Nicotiana rustica*] for a kinde thereof. Being now planted in the gardens of Europe, it prospereth very well, and commeth from seede in one yeare to beare both floures and seede. The which I take to be better for the constitution of our bodies then that which is brought from India [America]; and that growing in the Indies better for the people of the same countrey: notwithstanding it is not so thought nor received of our Tabackians; for according to the English proverbe; Far fecht and deere bought is best for Ladies."

The tobacco of Trinidad is mentioned in 1595 by Robert Dudley (Voyage to the West Indies, p. 22): "The daie followinge, beinge Sondaie, in the morninge came the salvages with two canowes aborde us, as they had promised our men, bringinge such commodities with them as their islande did afforde, saving they brought neither golde nor pearle, of the which theare are great store within the ilande, but tobacco, nutes and such kinde of fruites, the which they exchainged for knives, bugles, beades, fishinge hookes and hatchetts."

Gerard, accordingly, was of opinion that the tobacco of English growth would best suit English constitutions, as that of America would agree with Americans; but this view was not seconded by the smokers of his day.

Francis Bacon entertained no illusion as to English-grown tobacco. In his "Sylva Sylvarum: or a Natural History" (IX, 855) he writes, "Tobacco is a thing of great price, if it be in request: for an acre of it will be worth (as is affirmed) two hundred pounds by the year towards charge. The charge of making the ground and otherwise is great, but nothing to the profit. But the English tobacco hath small credit, as being too dull and earthy; nay, the Virginian tobacco, though that be in a hotter climate, can get no credit for the same cause: so that a trial to make tobacco more aromatical, and better concocted, here in England, were a thing of great profit. Some have gone about to do it by drenching the English tobacco in a decoction or infusion of Indian tobacco; but those are but sophistications and toys; for nothing that is once perfect, and hath run his race, can receive much amendment. You must ever resort to the beginnings of things for melioration."

William Barclay (Nepenthes, or the Vertues of Tabacco, Edinburgh, 1614) recommends exclusively tobacco of American growth, "Albeit this herbe disdaines not to be nourished in many gardens in Spaine, in Italie, France, Flanders, Germanie and Brittaine, yet neverthelesse only that which is fostered in India

[America] and brought home by Mariners and Traffiquers is to be used. But avarice and greedines of gaine have moved the Marchants to apparell some European plants with Indian coats, and to enstall them in shops as righteous and legittime Tabacco. . . So that the most fine, best and purest is that which is brought to Europe in leaves, and not rolled in puddings, as the English Navigators first brought home."

From the book "The Honestie of this Age, Prooving by good circumstance that the world was never honest till now, by Barnabee Rych Gentleman, Servant to the Kings most Excellent Maiestie" (1614) we receive a good idea of the increased consumption of tobacco and its sale. "There is not so base a groome, that commes into an Alehouse to call for his pot, but he must have his pipe of tobacco, for it is a commoditie that is nowe as vendible in every Taverne, Inne, and Ale house, as eyther Wine, Ale, or Beare, and for Apothicaries Shops, Grosers Shops, Chaundlers Shops, they are (almost) never without company, that from morning till night are still taking of Tobacco, what a number are there besides, that doe keepe houses, set open shoppes, that have no other trade to live by, but by the selling of Tobacco. I have heard it tolde that now very lately, there hath bin a Cathalogue taken of all those new erected houses that have set uppe that Trade of selling Tobacco, in London and neare about London: and if a man may beleeve what is confidently reported, there are found to be upward of 7000 houses, that doth live by that trade. I cannot say whether they number Apothicaries shoppes, Grosers shops, and Chaundlers shops in this computation, but let it be that these were thrust in to make uppe the number: let us now looke a little into the *Vidimus* of the matter, and let us cast uppe but a sleight account, what the expence might be that is consumed in this smoakie vapoure.

"If it be true that there be 7000 shops, in and about London, that doth vent Tobacco, as it is credibly reported that there be over and above that number: it may well bee supposed, to be but an ill customed shoppe, that taketh not five shillings a day, one day with another, throughout the whole yeare, or if one doth take lesse, two other may take more: but let us make our account, but after 2 shillings sixe pence a day, for he that taketh lesse than that, would be ill able to pay his rent, or to keepe open his Shop Windowes, neither would Tobacco houses make such a muster as they doe, and that almost in every Lane, and in every by-corner round about London. Let us then reckon thus, 7000 halfe Crowns a day, amounteth just to 319,375 poundes a yeare. *Summa totalis,* All spent in *smoake.*"

Tobacco then was an expensive pleasure. Aubrey informs us, "It was sold then for its wayte in silver, I have heard some of our old yeomen neighbours say, that when they went to Malmesbury or Chippenham Market, they culled out their biggest shillings to lay in the scales against the tobacco; now, the customes of it are the greatest his majestie hath." Compare the similar experience of the Koreans (Leaflet 18, p. 10).

C. T. published in 1615 "An Advice how to plant Tobacco in England: and how to bring it to colour and perfection, to whom it may be profitable, and to whom harmfull. The vertues of the Hearbe in generall, as well in the outward application as taken in Fume. With the danger of the Spanish Tobacco." The author's object is to instruct his countrymen in sowing, planting and perfecting this drug, as he viewed with alarm the vast sums annually spent on imported tobacco. He heard it reported by men of good judgment that there is paid out of England and Ireland near the value of 200,000 pounds every year for tobacco, and that the greatest part thereof is bought for ready money. It was sold for ten times the value of pepper, and the best of it, weight for weight, for the finest silver; it was hard to find one pound weight in five hundred that was not sophisticated. We learn that tobacco was then imported into England from the coast of Guiana, from St. Vincents, St. Lucia, Dominica, and other places, where it was directly bought of the natives. All these sorts were clean, and so was that of St. Domingo, where the Spaniards had not yet learned the art of sophistication. There was also a sort of Caraccas tobacco, which the Indians made up and sold to the Spaniards, and which was wholesome enough, but little of it came to England. This tobacco is mentioned in 1595 by Robert Dudley (Voyage to the West Indies, p. 48), who speaks of "the coast of Cracos, called the high land of Paria, one of the fruitfullest places in the worlde for excellent good tobacco, which is called for his worthiness cane tobacco."

Under Queen Elizabeth there was an import duty of 2d. a pound on tobacco, raised by James in 1604 to 6s. 10d. (equal to 25s. present value), an advance of 4000 per cent. This heavy tax nearly ruined Virginia whose economic life was based on the cultivation of the plant. In 1611 the imports of tobacco from Virginia were reduced to 142,085 pounds, one-sixth of the quantity previously exported to England. Aside from Virginia, tobacco was supplied to England from the Bermudas, where it had first been planted in St. George's Island under the first governor, Moore (1612-15), but unsuccessfully (Historye of the Bermudaes, p. 29). Under the third governor, Tucker (1616-19), some thirty thousand weight of tobacco could be despatched into the mother-country; this "proveinge good, and comeinge to a luckye markett, gave great contentment and incouragement to the undertakers to proceede lustely in their plantation." Fraudulent practices, however, were committed, and the Virginia Company of London complained bitterly to the governor, Nathaniel Butler (1619-22), anent its failure to sell a shipment of very vile conditioned tobacco, neither well cured, nor well made up. The governor, thereupon, appointed "triers of tobacco" under oath, whose duty it was to examine the crops, so that much false and bad ware was burned at the owners' doors. According to an order issued by Butler in 1621, better and poorer qualities had to be distinguished and packed separately, instead of being mixed with one another, as it had formerly been done.

In 1624 the importation of tobacco from Spain and Portugal was prohibited, and that from Virginia only allowed, so that the colony prospered again. James attempted to limit the supply at both ends by ordaining that no planter should export more than a hundred pounds a year and by creating a monopoly. Tobacco could be sold only by persons holding royal warrants of permission. These were granted for life on payment of fifteen pounds and an annual rent of the same amount.

The tobacco imported from Spanish America was called "Varinaes" up to 1639, and after that date "Spanish." It was obtained from Varina, near the foot of the range of mountains forming the west boundary of Venezuela, and watered by a branch of the Orinoco River. It was known in France as *Vérine* or *petum musqué*, and was introduced into Holland and Germany under the name *canaster* or *knaster* (from the Spanish *canastro*, "basket"), as it was rclled in cords and packed in baskets.

Coles wrote in 1657, "Tobacco prospers well about Winscomb, in Glocestershire, where I think the planting of it is now discontinued, because the store that came from thence was an hinderance to the publick revenue coming in for the custome of that which is brought from beyond seas."

By various acts passed in the reign of Charles II (1660-85), the planting of tobacco was forbidden in England in favor of the colonies, on forfeiture of forty shillings for every rod of ground thus cultivated, excepting in physic gardens, where it was allowed in quantities not exceeding half a pole of ground. Justices of peace were empowered to issue warrants to constables to search after and destroy the plants. It appears that walnut-tree leaves were used as a substitute for tobacco; for the cutting of such leaves, or any other leaves (not being tobacco leaves) or coloring them so as to resemble tobacco or selling these mixed or unmixed for tobacco was forbidden under a penalty of forfeiting five shillings a pound.

J. W. Gent (Systema Agriculturae; the Mystery of Husbandry Discovered, 2d ed., 1675, p. 156) gives the following interesting information:—

"I thought to have omitted this plant, by reason the Statute-Laws are so severe against the planters of it, but that it is a plant so much improving land, and imploying so many hands, that in time it may gain footing in the good opinion of the landlord, as well as of the tenant, which may prove a means to obtain some liberty for its growth here, and not to be totally excluded out of the husbandmans farm. The great objection is the prejudice it would bring to navigation, the fewer ships being imployed; and the lessening his Majesties revenue. To which may be answered, that there are but few ships imployed to Virginia; and if many, yet there would be but few the less; for it's not to be imagined, that we should plant enough to furnish our whole nation, and maintain a trade abroad also: And in case it should lessen the number of ships for the present, they would soon encrease again, as the trade of Virginia would alter into other commodities, as silk, wine and oyl, which would be a much better trade for them and us. And as to the lessening his Majesties revenue, the like imposition may be laid on the same commodity growing at home, as if imported from abroad, or some other of like value in lieu of it. Certain it is, that the planting of it would imploy abundance of people in tilling, planting, weeding, dressing and curing of it. And the improvement of land is very great, from ten shillings per acre, to thirty or forty pound per acre, all charges paid: before the last severe laws, many plantations were in Gloucestershire, Devonshire, Somersetshire, and Oxfordshire, to the quantity of many hundreds of acres.

"Some object, that our English-tobacco is not so good as the forreign; but if it be as well respected by the vulgar, let the more curious take the other that's dearer. Although many are of opinion that it's better than forreign, having a more *haut-gust*, which pleaseth some; if others like it not, they may in the curing of it make it milder, and by that means alter or change it as they please: It hath been often sold in London for Spanish tobacco. The best way and manner of planting and curing it, would be easily obtained by experience: many attempting it, some would be sure to discover the right way of ordering of it, and what ground or places it best affects. But that which hath been observed is, that it affects a rich, deep and warm soil well dressed in the spring before planting time: The young plants raised from seed in February or March, on a hot bed, and then planted abroad in your prepared ground, from whence you may expect a very good crop, and sometimes two crops in a year. The leaves, when gathered, are first laid together on heaps for some time, and then hang'd up (by threads run through them) in the shade, until they are through dry, and then put up and kept, the longer the better. In this, experience is the best master."

THE GREAT TOBACCO CONTROVERSY IN ENGLAND

As no other nation, the English had to fight for their tobacco, no less than for their liberty, and they put up a gallant and heroic fight for it. The struggle opened soon after the introduction of the plant and, producing a considerable literature, persisted with varying fortunes throughout the seventeenth century.

The first detailed account of tobacco was given the English public in John Frampton's "Joyfull Newes oute of the Newe Founde Worlde" (London, 1577; other editions in 1580 and 1596), which is a translation from the Spanish of Nicolas Monardes' (1493-1588) Three books on the drugs of America (Sevilla, 1574). The whole catalogue of diseases and their treatment with various preparations of tobacco thus became accessible to English practitioners, and English literature on the subject is visibly imbued by this influence. Physicians were busily engaged in analyzing the properties of the herb and discovering its use in all diseases; it was recommended as an infallible cure for nearly every ill and as a preventive of many ailments. In all these discussions the work of the doctor of Sevilla remained the fundamental source. The reader of Frampton should bear in mind that the notice entitled "A further addition of the Hearbe called Tabaco" (fols. 42-45) is not translated from Monardes, but from the French work "La Maison rustique" of Liebault (see below, p. 50) in which an account of Nicot's introduction of tobacco into France is rendered. Dr. Brushfield errs in making Monardes acknowledge the assistance he received from Nicot; not a word is said about Nicot in the Spanish original of Monardes.

The curative virtues of the tobacco plant are noted by two poets. E. Spenser, in his *Fairy Queen* (1590), makes Belphoebe include it with other medicinal herbs gathered to heal Timais (Book III, Canto VI, 32) :—

Into the woods thenceforth in haste shee went,
To seeke for hearbes that mote him remedy;
For she of hearbes had great intendiment,
Taught of the Nymphe which from her infancy
Her nourced had in trew nobility:
There, whether yt divine Tobacco were,
Or Panachæa, or Polygony,
She fownd, and brought it to her patient deare,
Who al this while lay bleding out his hart-blood neare.

This is the earliest poetical allusion to tobacco in English literature. William Lilly, the Euphuist and court-poet to Queen Elizabeth, a great smoker himself, wrote a play *The Woman in the Moone* (1597), in which Pandora wounds a lover with a spear and sends her servant for herbs to cure him :—

Gather me balme and cooling violets,
And of our holy herb nicotian,
And bring withall pure honey from the hive,
To heale the wound of my unhappy hand.

Raphael Holinshed (The First and Second Volumes of Chronicles, now newlie augmented and continued to the yeare 1586 by Iohn Hooker alias Vowell and others, 1587, fol. 209) appears to have been without enthusiasm for the weed, for he writes, "How doe men extoll the use of Tabacco in my time, whereas in truth (whether the cause be in the repugnancie of our constitution unto the operation thereof, or that the ground doeth alter hir force, I cannot tell) it is not found of so great efficacie as they write."

The praise of the healing powers of tobacco was sung in an epigram by John Davies in 1598 (Works of Marlowe, ed. of F. Cunningham, p. 268). It begins thus :—

Homer of Moly, and Nepenthe sings,
Moly the gods' most sovereign herb divine;
Nepenthe, Helen's drink, most gladness brings,
Heart's grief expels, and doth the wits refine.
But this our age another world hath found,
From whence an herb of heavenly power is brought;
Moly is not so sovereign for a wound,

Nor hath Nepenthe so great wonders wrought.
It is tobacco, whose sweet subtle fume,
The hellish torment of the teeth doth ease,
By drawing down, and drying up the rheum,
The mother and the nurse of each disease.

Both sides of the controversy are skilfully represented in Ben Jonson's *Every Man in His Humor* (Act III, Scene 2), acted on the 25th of November, 1596, and printed in 1601. Bobadilla pleads thus in favor of the case: "Signior beleeve me, (upon my relation) for what I tel you, the world shall not improve. I have been in the Indies (where this herbe growes) where neither my selfe, nor a dozen Gentlemen more (of my knowledge) have received the taste of any other nutriment, in the world, for the space of one and twentie weekes, but Tabacco onely. Therefore it cannot be but 'tis most divine. Further, take it in the nature, in the true kinde so, it makes an Antidote, that (had you taken the most deadly poysonous simple in all Florence, it should expell it, and clarifie you with as much ease, as I speak. And for your greene wound, your *Balsamum,* and your—are all meere gulleries, and trash to it, especially your *Trinidado:* your *Newcotian* is good too: I could say that I know of the vertue of it, for the exposing of rewmes, raw humors, crudities, obstructions, with a thousand of this kind; but I professe my selfe no quack-salver: only thus much: by Hercules I doe holde it, and will affirme it (before any Prince in Europe) to be the most soveraigne, and pretious herbe that ever the earth tendred to the use of man." Then Cob represents the other side as follows: "By gods deynes: I marle what pleasure or felicitie they have in taking this roguish Tabacco; it's good for nothing but to choake a man, and fill him full of smoake and imbers: there were foure died out of one house last weeke with taking of it, and two more the bell went for yester-night, one of them (they say) will ne're scape it, he voyded a bushell of soote yester-day, upward and downeward. By the stockes; and there were no wiser men then I, I'ld have it present death, man or woman that should but deale with a Tabacco pipe; why, it will stifle them all in the 'nd as many as use it; it's little better than rats bane."

It is a matter of profound regret that Shakespeare has never alluded to tobacco and smoking.

In 1602 appeared a pamphlet entitled "Work for Chimny-sweepers: or a warning for Tabacconists. Describing the pernicious use of Tabacco, no lesse pleasant than profitable for all sorts to reade. Fumus patriae, Igne alieno Luculentior. As much as to say,

Better be chokt with English hemp,
then poisoned with Indian Tabacco.

Imprinted at London by T. Este, for Thomas Bushell, and are to be sould at the great North dore of Powles 1602." The anonymous author, who calls himself Philaretes, is said to have been ordered or compelled to write this invective, presumably by James I. He alleges eight reasons against tobacco, one of which is that the first author and finder hereof was the devil, and the first practisers were the devil's priests, and therefore not to be used of us Christians. The idea is not original, for it looms up in Monardes (in Frampton's translation, fol. 38): "And as the Devil is a deceaver, and hath the knowledge of the vertue of hearbes, so he did shew the vertue of this Hearb [to the Indians], that by the meanes thereof, they might see their imaginations, and visions, that he hath represented to them, and by that meanes deceive them." Ben Jonson also (*Gipsies Metamorphosis*) calls tobacco "the Devil's own weed," and according to Joshua Sylvester, "hell hath smoke impenitent tobaccanists to choake."

Dekker, in his *The Gull's Horn-Book* (1602), thus apostrophizes tobacco: "Make me thine adopted heir, that inheriting the virtues of thy whiffes, I may distribute them amongst all nations, and make the fantastic Englishman, above the rest, more cunning in the distinction of thy roll Trinidado, leaf, and pudding, than the whitest-toothed black-a-moor in all Asia."

In 1604 appeared King James' famed "A Counterblaste to Tobacco. Imprinted at London by R. B. Anno 1604." The king's name does not appear on the title-page, nor at the end of the preface To the Reader. He simply speaks of himself as the King. The royal pamphlet has met with almost universal condemnation, and W. Bragge (Bibliotheca Nicotiana, 1880) even says that "he most Quixotically broke his lance against one of the great appetites of man." To condemn is easier than to understand. In my opinion the Counterblaste is a remarkable document of considerable culture-historical interest, which must be understood and interpreted from the spirit of the time; and there is no doubt that James was actuated by good intentions and by a solicitous care for the welfare of his subjects, even though his blind hatred of tobacco carries him too far. He condemns its use primarily out of motives of racial and national pride: "And now good Countrey men let us (I pray you) consider, what honour or policie can move us to imitate the barbarous and beastly maners of the wild, godlesse, and slavish Indians, especially in so vile and stinking a custome? Shall wee that disdaine to imitate the maners of our neighbour France (having the stile of the first Christian Kingdom) and that cannot endure the spirit of the Spaniards (their King being now comparable in largenes of Dominions, to the great Emperor of Turkie) Shall wee, I say, that have bene so long civill and wealthy in Peace, famous and invincible in Warre, fortunate in both, we that have bene ever able to aide any of our neighbours (but never deafed any of their eares with any of our supplications for assistance) shall we, I say, without blushing, abase our selves so farre, as to imitate these beastly Indians, slaves to the Spaniards, refuse to the world, and as yet aliens from the holy Covenant of God? Why doe we not as well imitate them in walking naked as they doe? in preferring glasses, feathers, and such toyes, to gold and precious stones, as they do? yea why do we not denie God and adore the Devill, as they doe?"

He goes on to refute, in the physiological terms of his time, the medicinal virtues of the drug, and after all the absurdities previously written in praise of its alleged healing powers, his arguments make rather refreshing reading. To the argument "that the whole people would not have taken so generall a good liking thereof, if they had not by experience found it verie soveraigne and good for them," he responds justly that this custom is merely based on imitation and fashion. "For such is the force of that naturall Selfe-love in every one of us, and such is the corruption of envie bred in the brest of every one, as we cannot be content unlesse we imitate every thing that our fellowes doe, and so proove our selves capable of every thing whereof they are capable, like Apes, counterfeiting the maners of others, to our owne destruction." The argument that people have been cured of diverse diseases by taking tobacco is fallacious and rests on a confusion of cause and effect; the disease takes its natural course and declines, but it is not tobacco that wrought this miracle. If a man smoke himself to death with it (and many have done), O then some other disease must beare the blame for that fault. He justly rejects the idea that tobacco could act as a panacea, a cure for all diseases in all persons and at all times. "O omnipotent power of Tobacco!" he exclaims, "And if it could by the smoke thereof chace out devils, as the smoke of Tobias fish did (which I am sure could smel no stronglier) it would serve for a precious Relicke, but for the superstitious Priests, and the insolent Puritanes, to cast out devils withall."

As to the moral evaluation of smoking, the king holds that smokers are guilty of sinful and shameful lust, that its use or rather abuse is a branch of the sin of drunkenness, which is the root of all sins, and that it disables men for military service. "In the times of the many glorious and victorious battailes fought by this Nation, there was no word of Tobacco. But now if it were time of warres, and that you were to make some sudden Cavalcado upon your enemies, if any of you should seeke leisure to stay behinde his fellows for taking of Tobacco, for my part I should never bee sorie for any evill chance that might befall him. To take a custome in any thing that cannot bee left againe, is most harmefull to the people of any land." Finally, it is a waste of national wealth: "Now how you are by this custome disabled in your goods, let the Gentry of this land beare witnesse, some of them bestowing three, some foure hundred pounds a yeere upon this precious stinke, which I am sure might be bestowed upon many farre better uses."

He condemns the prevailing custom of smoking at the dinner-table when very often men that abhor it are present. Smoking in public had increased to such a degree that men sound in judgment were at last forced to take it also without desire, "partly because they were ashamed to seeme singular, and partly, to be as one that was content to eate Garlicke (which hee did not love) that he might not be troubled with the smell of it, in the breath of his fellowes." It was accordingly an act of self-defence. A man could not heartily welcome his friend now, but straight they must be in hand with tobacco. It was a point of good fellowship, and he who would refuse to take a pipe among his fellows was accounted peevish and no good company. "Yea the Mistresse cannot in a more manerly kinde, entertaine her servant, then by giving him out of her faire hand a pipe of Tobacco." It is a great contempt of God's good gifts that the sweetness of man's breath, being a gift of God, should be willfully corrupted by this stinking smoke. "Moreover, which is a great iniquitie, and against all humanitie, the husband shall not bee ashamed, to reduce thereby his delicate, wholesome and cleane complexioned wife, to that extremitie, that either shee must also corrupt her sweete breath therewith, or else resolve to live in a perpetuall stinking torment."

He winds up his sermon as follows: "Have you not reason then to bee ashamed, and to forbeare this filthie noveltie, so basely grounded, so foolishly received and so grossely mistaken in the right use thereof? In your abuse thereof sinning against God, harming your selves both in persons and goods, and raking also thereby the markes and notes of vanitie upon you: by the custome thereof making your selves to be wondered at by all forraine civil Nations, and by all strangers that come among you, to be scorned and contemned. A custome lothsome to the eye, hatefull to the Nose, harmefull to the braine, dangerous to the Lungs, and in the blacke stinking fume thereof, neerest resembling the horrible Stigian smoke of the pit that is bottomelesse."

In 1616 the Counterblaste was reprinted in Bishop Montagu's collected edition of James' "Workes," and in 1619 the Bishop published a Latin translation of the King's works in which the Counterblaste appears as "Misocapnus ['Smoke-hater'], sive de Abusu Tobacci Lusus Regius." While the royal diatribe is sizzling, of course, with misstatements, exaggerations, and outbursts of gloomy pessimism and unrestrained animosity, it was a natural reaction against the many exorbitant claims made by the friends and defenders of the narcotic, and in his scathing denunciation of the tobacco excesses of his time the king was presumably nearly right. In our own days his phraseology has been echoed by Eliah the Prophet, and the Jameses we shall always have with us.

Nor did the king stop at purely platonic exhortations. Under the 17th day of October, 1604, he addressed at Westminster a Commissio pro Tabacco to the right Trustie and right Welbeloved Cousen and Counsellor, Thomas Earle of Dorset, high treasourer of England, who is commanded "to give order to all Customers, Comptrollers, Searchers, Surveyors, and all other Officers of our Portes, that they shall demaunde and take to our use of all Merchauntes, as well Englishe as Strangers, and of all others whoe shall bringe in anye Tabacco into this Realme, within any Porte Haven or Creek belonging to any theire severall Charges, the Somme of Six Shillinges and eighte Pence uppon everye Pound Waight thereof, over and above the Custome of Twoo Pence uppon the Pounde Waighte usuallye paide heretofore." Infractors were threatened with confiscation and blows. "If anye Merchaunte Englishe or Straunger, or other whatsoever, shall presume to bringe in anye of the saide Tabacco, before suche Payemente and Satisfactione first made, That then he shall not onelie forfeite the saide Tabacco, but alsoe shall undergoe suche furthere Penalties and corporall Punishmente as the Qualitie of suche soe highe a Contempte against our Royall and expresse Commaundemente in this mannere published shall deserve."

As stated in the introductory paragraph of this order, the object of this measure was to restrain the heavy importations of tobacco, "whereby it is likelie that a lesse Quantitie of Tabacco will hereafter be broughte into this our Realm of England, Dominion of Wales and Town of Barwick then in former tymes, and yet sufficient store to serve for their necessarie use who are of the better sort, and have and will use the same with Moderation to preserve their Healthe." The latter point is of great interest, for it does not crop up in the "Counterblaste." The king discriminates between a better and baser sort of people, and graciously concedes to the former a moderate use of the herb. By way of introduction he comments that tobacco was used and taken by the better sort both then and now only as physic to preserve health, "and is now at this Day, through evell Custome and the Toleration thereof, excessivelie taken by a nomber of ryotous and disordered Persons of meane and base Condition, whoe, contrarie to the use which Persons of good Callinge and Qualitye make thereof, doe spend most of there tyme in that idle Vanitie, to the evill example and corrupting of others, and also do consume that Wages whiche manye of them gett by theire Labour, and wherewith there Families should be releived, not caring at what Price they buye that Drugge, but rather devisinge how to add to it other Mixture, therebye to make it the more delightfull to their Taste, though so much the more costly to there Purse; by which great and imoderate takinge of Tabacco the Health of a great nomber of our People is impayred, and theire Bodies weakened and made unfit for Labor, the Estates of many mean Persons soe decayed and consumed as they are thereby dryven to unthriftie Shifts onelie to maynteyne their gluttonous exercise thereof, besides that also a great part of the Treasure of our Lande is spent and exhausted by this onely Drugge so licentiously abused by the meaner sorte, all which enormous Inconveniences ensuinge thereuppon." The king's solicitude, accordingly, centered around the misera plebs, while the nobility is dismissed with a patronizing pat on the shoulders.

Edmund Gardiner, Gentleman and Practicioner in Physicke, wrote a medical defence in 1610 under the title, "The Triall of Tabacco. Wherein, his worth is most worthily expressed; as, in the name, nature, and qualitie of the sayd herb; his speciall use in all Physicke, with the true and right use of taking it, as well for the Seasons, and times, as also the Complexions, Dispositions, and Constitutions, of such Bodies, and Persons, as are fittest: and to whom it is most profitable to take it." A new edition appeared in 1650.

Joshua Sylvester published in 1614 in folio a poem under the title "Tabacco battered; and the Pipes shattered (About their Eeares that idlely Idolize so base and barbarous a Weed; or at least-wise over-love so loathsome Vanitie): by a Volley of Holy Shot thundered from Mount Helicon. Du Bartas his Divine Weekes and Workes with a Compleate Collection of all the other most delight-full Workes Translated and written by yt famous Philomusus, Iosvah Sylvester gent: London, printed by Robert Young." The poem, like its title, is bombastic and dull: it threatens punishment with infernal rod in hell's dark furnace, with black fumes to choke, to those who on earth offended in smoke.

William Barclay's "Nepenthes, or the Vertues of Tabacco" (Edinburgh, 1614) is a vindication of tobacco, and is directed straight against the Counterblaste. He recommends tobacco either green or dry for the cure of many maladies, either as a ball made from the fresh leaves big enough to fill the patient's mouth, or as a smoke on an empty stomach ("not as the English abusers do, which make a smoke-boxe of their skull"). In his dedication to the Bishop of Murray he calls on him to defend "this sacred herb."

> A stranger plant, shipwracked on our coast,
> Is come to helpe this cold phlegmatic soyle.

He defends tobacco as having "much heavenlie vertue in store" and describes America as "the countrie which God hath honoured and blessed with this happie and holy herb."

John Deacon followed in the footsteps of James I and dedicated to him in 1616 "Tobacco tortured; or the filthie fume of Tobacco refined." This work is couched in the form of a dialogue between Capnistus and Hydrophorus. It is divided into two parts, (1) The Fume of Tobacco taken inward, is very pernicious unto the Body. (2) The Fume of Tobacco taken inward, is too too profluvious for many of our Tobacconists purses, and most pernicious to the publike State. One of the most curious attempts to prevent smoking in a family is contained in a will, dated October 20th, 1616, wherein P. Campbell leaves to his son all his household goods, "on this condition, that yf at any time hereafter, any of his brothers or sisters shall fynd him takeing of tobacco, that then he or she so fynding him, shall have the said goods."

Tobias Venner, Doctor of Physicke in Bath, published in 1621 "A Briefe and accurate treatise, concerning, The taking of the fume of Tobacco, which very many, in these dayes, doe too licentiously use. In which, the immoderate, irregular, and unseasonable use thereof is reprehended, and the true nature and best manner of using it, perspicuously demonstrated."

In this manner the struggle for or against the herb was continued, but ultimately ended in a complete triumph of tobacco, as an examination of the various manners in which it was consumed will show.

USE OF TOBACCO IN ENGLAND

It appears from Harrison's account (above, p. 7) that Englishmen took up tobacco-smoking from ladle-like pipes in 1573. From 1586 pipes were in full blast, and smoking during that early period was essentially fashionable. One of the characteristics of the gallant, the dandy of the time, was his devotion to tobacco. "To take tobacco with a grace" was one of a gentleman's accomplishments. Clusius, the botanist (Exotica, 1601, p. 310), speaks of the clay pipes made by the colonists in Virginia, and adds that from 1585 the use of tobacco increased throughout England to such a degree, particularly among the courtiers, that they had many similar tubes made after the model of those brought back from Virginia for tobacco-smoking.

John Gerard was familiar with the custom of smoking. "The drie leaves," he writes in his Herball (1597), "are used to be taken in a pipe set on fire and suckt into the stomacke, and thrust foorth again at the nosthrils against the paines of the head, rheumes, aches in any part of the bodie, whereof soever the original doth proceed, whether from Fraunce, Italy, Spaine, Indies, or from our familiar and best knowne diseases."

All the early accounts agree in stating that the smoke was expelled through the nostrils,—an imitation of Indian custom. In a play by Field (1618), a foolish nobleman is asked by some boon companions in a tavern, "Will your lordship take any tobacco?" when another sneers, "'Sheart! he cannot put it through his nose!" There were professors of the art of smoking who taught pupils the "slights," as tricks with the pipe were called. These included exhaling the smoke in globes and rings. Ben Jonson describes one Sogliardo as "an essential clown, yet so enamored of the name of a gentleman that he will have it though he buys it; he comes up every term to learn to take tobacco and see new motions." Hence Marston could make the joke, "Her love is just like a whiffe of Tabacco, no sooner in at the mouth, but out at the nose." This practice, it is said, died out after the death of

James I (1625), and from that time onward the fumes were plainly discharged from the mouth. Smoking then lost its medical aspect and developed into an honest, every-day pastime and pleasure.

In 1660 Winstanley declared, "Tobacco it self is by few taken now as medicinal, it is grown a good-fellow, and fallen from a Physician to a Complement. He's no good-fellow that's without burnt Pipes, Tobacco, and His Tinder Box."

Silver pipes are mentioned by Sir William Vaughan (Naturall and Artificiall Directions for Health, 1602, p. 22): "Cane Tabacco well dryed, and taken in a silver pipe fasting in the morning, cureth the megrim, the tooth ache, obstructions proceeding of cold, and helpeth the fits of the mother. After meales it doth much hurt, for it infecteth the braine and the liver."

In John Aubrey's Letters written by Eminent Persons we read, "They had first silver pipes. The ordinary sort made use of a walnut shell and a strawe. I have heard my grandfather Lyte say, that one pipe was handed from man to man round the table." This was done because the cost of a pipe was considerable.

Paul Hentzner (Itinerarium), a German lawyer, who visited England in 1598, has recorded the following observation: "At these spectacles [in the London theatres] and everywhere else, the English are constantly smoking Tobacco, and in this manner: they have pipes on purpose made of clay, into the farther end of which they put the herb, so dry that it may be rubbed into powder, and lighting it, they draw the smoake into their mouths, which they puff out again through their nostrils like funnels, along with it plenty of phlegm and defluxion from the head."

The clay pipe first made about 1590 soon became fashionable and the typical English pipe. It achieved fame all over Europe and was imitated in Holland and

Germany. The English became the adepts of the pipe-cult and the initiators and propagators of pipe-smoking in Europe. The first pipes had small, pear-shaped bowls and short stems, from three to six inches in length. Under the bowl was a flat heel, enabling the pipe to stand upright on a table. In 1619 the pipe-makers received their charter of incorporation from James I. The Company of Pipe-makers consisted of a master, four wardens, and twenty-four assistants. Their escutcheon bore a tobacco plant in full blossom, and their motto was "Let brotherly love continue." All pipes then were made of clay, though occasionally some were made of iron or brass. Under the reign of William III (1689-1702) the Dutch style with larger bowls and long, straight stems was adopted. Wooden pipes and briars appeared only from the latter half of the nineteenth century; briar (from the French *bruyère*, "heath") is the root of the tree heath (*Erica arborea*), a native of southern France. The English are still masters of the pipe, turning out the best pipes and the best smoking mixtures; a good English pipe makes a man feel that life is still worth living. The pipe is the emblem of strength and manliness, of peace and brotherhood, of liberty and democratic government. "The pipe," says Thackeray, "draws wisdom from the lips of the philosopher, and shuts up the mouth of the foolish; it generates a style of conversation, contemplative, thoughtful, benevolent and unaffected. May I die if I abuse that kindly weed which has given me so much pleasure." As the English had preceded all other European nations in the struggle for liberty and human rights and had set the model for constitutional and parliamentary government, history justly assigned to them the distinction of carrying this emblem all over the world.

Maple blocks were used in the old days for cutting or shredding the tobacco upon. The pipes were formerly lighted by means of live charcoal from juniper wood. King James says in his Counterblaste (1604), "In your persons having by this continuall vile custome brought your selves to this shameful imbecilitie, that you are not able to ride or walke the iourney of a Iewes Sabboth, but you must have a reekie cole brought you from the next poore house to kindle your Tobacco with?" William Barclay (Nepenthes, or the Vertues of Tabacco, Edinburgh, 1614) tells this story: "I chanced in company on a tyme with an English merchant in Normandie betweene Rowen and New-haven. This fellow was a merrie man, but at every house he must have a Cole to kindle his Tabacco: the Frenchman wondered, and I laughed at his intemperancie." Silver tongs, called ember-tongs or brand-tongs, were used in lifting the hot charcoal to light the pipe.

Many old English "clays" are provided with the maker's initials. Monograms and designs were stamped or moulded upon the bowls and stems, but more generally upon the spur or flat heel of the pipe. During the latter half of the seventeenth century English pipes were presented by colonists in America to the Indians. They subsequently became valuable as objects of barter or part purchase price in exchange for land. In 1677, one hundred and twenty pipes and one hundred Jew's harps were given for a strip of land near Timber Creek in New Jersey. When William Penn, the founder of Pennsylvania, purchased a tract of land, three hundred pipes were included in the articles given in the exchange.

It was customary for a man to carry a case of pipes about with him. In *Everie Woman in Her Humour*, a play written in 1609, there is an inventory of the contents of a gentleman's pocket, with a value given for each item. A case of tobacco-pipes is appraised at fourpence; half an ounce of tobacco, at sixpence, and three pence in coin, or, as it is quaintly worded, "in money and golde." Satirists poked fun at the smoker's pocketful of apparatus. A pamphleteer of 1609 says, "I behelde pipes in his pocket; now he draweth forth his tinder-box and his touchwood, and falleth to his tacklings; sure his throat is on fire, the smoke flyeth so fast from his mouth." In his "Epigrammata religiosa, officiosa, iocosa" (privately printed, London, 1627), John Pyne of Bearferres, of whose life no details are known, has left the following Epitaph of a certaine Tobacchonist:—

Loe heere I lye roll'd up like th' Indian Weed,
My Pipes I have pack'd up, for Breath I need.
Mans Breath's a vapour, Hee himselfe is Grasse;
My Breath but of a Weed the vapour was.
When I shall turne to Earth, Good Friends beware,
Lest it evaporate and infect the Aire.

Besides the instruments mentioned, a tobacco-box (pouches were then unknown) was indispensable to the rich young gallant. The boxes were made of silver, iron, copper, brass, ivory, mother-of-pearl, tortoise-shell, bone, or wood, curiously and artistically carved. They were usually small enough to be carried in the pocket, and contained, in addition to the weed, a pipe, the ember-tongs, flint and steel, and a priming-iron. Occasionally a looking-glass was set in the box. Tobacco-boxes were given and exchanged as tokens of friendship. In those days, when tobacco was eight or ten shillings a pound, smokers were economic and burned their tobacco to the very bottom of the bowl, pressing the ashes down by means of a stopper. The stoppers were made of wood, bone, ivory, mother-of-pearl, brass, silver, or gold, of various shapes, and adorned with figures of national heroes or heads of animals. Some smokers wore rings provided with a stud for ramming down the contents of the pipe.

During the Elizabethan period and after women smoked as well as men (cf. Howes, above, p. 4). In

Dekker's *Satiromastix* (1602) Asinius Babo, offering his pipe, observes, "'Tis at your service, gallants, and the tobacco too: 'tis right pudding, I can tell you; a lady or two took a pipe full or two at my hands, and praised it, fore the heavens." In Heywood's *Fair Maid of the Exchange* (1607), one of the characters is advised to court a girl by "asking her if she'll take a pipe of tobacco." William Prynne, the famous Puritanic inveigher against stage-plays, informs us that in his time ladies at the theatre were sometimes offered the tobacco-pipe as a refreshment instead of apples. On the title-page of Middleton's comedy, *The Roaring Girle* (1611), is a picture of the heroine in man's apparel, smoking a pipe from which a cloud of smoke is issuing. The portrait of a woman, painted about 1651, holding in her right hand a tobacco-box and gracefully wielding in her left a pipe, is reproduced in Fairholt's book "Tobacco" (p. 69).

In the *British Apollo* (Vol. I, 1708) it is stated, "Snuff, tho' the use of it has been long known to such, as were by merchandizing or other means, familiar with the Spanish customes, has been till lately, a perfect stranger to the practice of the British nation, and like our other fashions came to us from France." In the Oxford English Dictionary we are informed, "The practice of taking snuff appears to have become fashionable about 1680, but prevailed earlier in Ireland and Scotland." In general this certainly is correct, but snuff was not entirely foreign to the Elizabethan age. The following two references may serve as evidence.

Henry Buttes (Diets Dry Dinner, 1599), in his discourse of tobacco, writes, "Translated out of India in the seed or roote; Native or sative in our own fruitfullest soiles: Dried in the shade, and compiled very close: of a tawny colour, somwhat inclining to red: most perspicuous and cleare: which the Nose soonest taketh in snuffe."

Dekker, in his "The Gull's Horn-book" (1602), thus describes the approved fashion in his day: "Before the meat come smoking to the board, our gallant must draw out his tobacco-box, the ladle for the cold snuff into the nostril, the tongs, and priming-iron; all which artillery may be of gold or silver, if he can reach the price of it; it will be a reasonable useful pawn at all times, when the amount of his money falls out to run low. And here you must observe to know in what tobacco is in town, better than the merchants, and to discourse of the apothecaries where it is to be sold; then let him show his several tricks in taking it, as the whiff, the ring, etc., for these are compliments that gain gentlemen no mean respect." As Englishmen always preferred the pipe, we hear little of snuff in the first part of the sixteenth century. Irish and Scotch preferred snuff, formerly also called sneeshing, sneezing, and smutchin (from Irish *smuiteán*, "powder"; Scotch and Gaelic *smuidean*, "a mote, a particle of dust"). Howell (1650) writes in his Letters, "The Spaniards and Irish take tobacco in powder or smutchin, and it mightily refreshes the brain. I believe there is as much taken this way in Ireland as there is in pipes in England. One shall see the serving-maid upon the washing-block and the swain upon the ploughshare, when they are tired with their labour, take out their boxes of smutchin and draw into their nostrils with a quill; and it will beget new spirits in them and fresh vigour to fall to their work again."

The plague of 1665 first brought snuff into prominence in England on account of its disinfectant properties. It developed into a fashion under the reign of Queen Anne (1702-14), when French ideas and manners conquered English society and inaugurated a veritable age of snuff, which completely displaced smoking in society. To take snuff was then as essential a part of gallantry as to drink tobacco had been a century before.

A gentleman was then known by his snuff and snuff-box, and snuff-taking was universal in the fashionable world among both men and women. Alexander Pope (1688-1744), in *The Rape of the Lock,* wrote:—

> Sir Plume, of amber snuff-box justly vain,
> And the nice conduct of a clouded cane,
> With earnest eyes and round, unthinking face
> He first the snuff-box opened, then the case.

And Oliver Goldsmith (1728-74), in *Retaliation:*—

> When they talk'd of their Raphaels, Correggios, and stuff,
> He shifted his trumpet and only took snuff.

The snuff-box was the fetish of the eighteenth century, an object of luxury, a tribute of friendship and admiration, a gift to kings and ambassadors. There was an infinite number of snuffs, and there were morning, afternoon, and evening snuffs.

At first, snuff was not sold ready-made, but every one prepared it himself. It was scraped with a rasp made from the dry root of the tobacco plant; the powder was then placed on the back of the hand and thus snuffed up. Hence the name *râpé* ("rasped, grated") for a coarse kind of snuff made from the darker and ranker tobacco leaves. The rasps were carried in the waistcoat pocket, and became articles of luxury, being carved in ivory and variously enriched. The tobacconist's shop-sign, in the early days, was the figure of a Virginian or Negro or a combination of both; in the eighteenth century and until a few years ago it was replaced with the figure of a Highlander, usually with a snuff mull in his hand, credited as he was with a great fondness and capacity for snuff-taking. Walter Scott said that a Scotchman in London would walk half a mile farther to purchase his ounce of snuff where the sign of the Highlander announced a North Briton. After the suppression of the Jacobite uprising of 1745, when the wearing of the highland costume was forbidden by Parliament, the following paragraph appeared in the newspapers of the time: "We hear that the dapper wooden Highlanders, who guard so heroically the doors of snuff-shops, intend to petition the Legislature, in order that they may be excused from complying with the Act of Parliament with regard to their change of dress: alledging that they have ever been faithful subjects to his Majesty, having constantly supplied his Guards with a pinch out of their Mulls when they marched by them, and so far from engaging in any Rebellion, that they have never entertained a rebellious thought; whence they humbly hope that they shall not be put to the expense of buying new cloaths."

It has often been stated that snuff-taking is practically extinct. The latest news from London (June 12th, 1924) indicates that there is a definite increase in the consumption of snuff among women and that jewellers find a ready sale for daintily jewelled snuff-boxes.

Of the manifold forms in which tobacco is consumed the custom of chewing it is the most striking and perhaps even the most primitive. The aborigines of Australia, we now know for certain, were in the habit of chewing the leaves of *Nicotiana suaveolens,* a species native to Australia, in times prior to their contact with the whites, but they were totally ignorant of smoking the leaves. This example demonstrates well that primitive man, in testing the properties of a vegetable product, will first exercise his senses of touch, smell, and taste. The Spanish conquerors came into contact with the habit of chewing tobacco in the West Indies (account of Amerigo Vespucci) and Mexico (early accounts of B. de Sahagun and F. Hernandez). Monardes (1571) describes it as follows: "The Indians use tobacco to remove thirst which in this case they will not suffer, and likewise to stand hunger and to be able to pass days without being compelled to eat or drink. When they have to travel across a desert or unpopulous region, where neither water nor food is to be found, they avail themselves of some pills made of tobacco in this manner: they take the leaves of the plant and chew them, and while chewing, they mix them with a powder prepared from burnt river-mussels; this they mix in their mouth together till it forms a mass which they shape into pills a bit larger than peas; these are placed in the shadow to dry, are then preserved, and used in this form. Whenever they travel through territories where they believe not to find water or victuals, they take one of these pills, placing it between their under lips and teeth, and keep on chewing it continually during their journey, and thus they go along for three or four days without having to eat or drink or feeling the pinch of hunger or thirst or fatigue."

As Monardes was translated into Latin, French, Italian, and English, Europeans might easily have copied his prescription, but the fact remains that they did not. Leaves may occasionally have been chewed for medicinal purposes, but no habit of chewing for pastime or pleasure was developed. Gerard (The Herball, 1597, p. 286) observes, "The leaves likewise being chewed draw foorth flegme and water" . . . Edmund Gardiner, in his "Triall of Tobacco" (first published in 1610, new ed. 1650), says that "a sirup made of the decoction of this herbe, with sufficient sugar, and so taken in a very small quantitie, dischargeth the breast from phlegmatic matter." John Parkinson (Theatrum botanicum, 1640, p. 712) writes also that in his time the juice from the leaves of *Nicotiana rustica* was made into a syrup, or that the distilled water of the herb was taken with or without sugar, or the smoke was inhaled from a pipe, as usual. Obviously Parkinson here opposes to the common habit of pipe-smoking another less usual practice, that of taking a syrupy substance extracted from the leaf. At best, however, we are here confronted with a distant forerunner of chewing, not with chewing properly.

As far as I am able to make out, it seems that tobacco-chewing was taken up as a prophylactic against the plague which was epidemic in 1665. Samuel Pepys writes in his Diary under 7th June, 1665, "This day, much against my will, I did in Drury Lane see two or three houses marked with a red cross upon the doors, and 'Lord have mercy upon us' writ there; which was a sad sight to me, being the first of the kind that, to my remembrance, I ever saw. It put me into an ill conception of myself and my smell, so that I was forced to buy some roll-tobacco to smell to and chaw, which took away the apprehension." In the year of the plague appeared a quarto tract, entitled "A Brief Treatise of the Nature, Causes, Signs, Preservation from and Cure of the Pestilence," by W. Kemp, "Mr. of Arts," who says in regard to tobacco, "It corrects the air by Fumigation, and it avoids corrupt humours by Salivation; for when one takes it either by Chewing it in the leaf, or Smoaking it in the pipe, the humours are drawn and brought from all parts of the body, to the stomach, and from thence rising up to the mouth of the Tobacconist, as to the helme of a Sublimatory, are voided and spitten out." Derby was visited by the plague in the same year, and at the "Headless-cross the market-people, having their mouths primed with tobacco as a preservative, brought their provisions. It was observed that this cruel affliction never attempted the premises of a tobacconist, a tanner, or a shoemaker" (W. Hutton, History of Derby, 1817, p. 194).

The belief in the efficacy of tobacco as warding off the plague acted also as a new incentive to the increase of smoking. Thomas Hearne (1721), the antiquary, gives the following curious information: "I have been told that in the last great plague at London none that kept tobaconist's shops had the plague. It is certain, that smoaking it was looked upon as a most excellent preservative, in so much, that even children were obliged to smoak. And I remember, that I heard formerly Tom Rogers, who was yeoman beadle, say, that when he was that year, when the plague raged, a schoolboy at Eaton, all the boys at that school were obliged to smoak in the school every morning, and that he was never whipped so much in his life as he was one morning for not smoaking." Thomas Pope Blount (A Natural History; containing many not common observations, 1693, p. 127) writes, "Diemerbrockins, in his book De Peste, very much commends the use of tobacco in the time of plague; he says, it absolutely cured him when he had it; he also observes, that almost all those houses, where tobacco was sold, both in Spires (a city in the Palatinate) and likewise in London, were never infected, whereas the houses round about them were."

According to Penn, the chewing of tobacco was common in the reign of James, when gentlemen carried about with them small silver basins as spittoons, and Monk, the principal factor in the restoration of the monarchy, brought it into fashion; but no documentary evidence is produced by him. Apperson comments, "General Monk, to whom Charles II owed so much, is said to have indulged in the unpleasant habit of chewing tobacco, and to have been imitated by others; but the practice can never have been common."

In 1689, W. Bullock speaks of "two rowles of chawing tobacco." The London Gazette of 1725 mentions a fellow who "commonly has a chew of tobacco in his under lip"; and Smollett, in *Roderick Random* (1748), has a sailor putting a large chew of tobacco in his mouth. The *World* of 1754 pokes fun at the "pretty" young men who "take pains to appear manly; their chewing not only offends, but makes us apprehensive at the same time that the poor things will be sick." E. Baillard (Discours du tabac, 1693, p. 92) refers to chewing tobacco (*tabac machicatoire*) as relieving hunger and thirst, but does not say that it was actually used in France. In the eighteenth century a common device of tobacconists was three figures representing a Dutchman, a Scotchman, and a sailor, explained by the accompanying rhyme:

> We three are engaged in one cause,
> I snuffs, I smokes, and I chaws!

Another tobacconist had the three men on his sign, but with a different legend:

> This Indian weed is good indeed,
> Puff on, keep up the joke.
> 'Tis the best, 'twill stand the test,
> Either to chew or smoke.

The promoters of the cigar in Europe were the Spaniards, but they were exceedingly slow in making their product known to the other nations of Europe. The cigar spread in Europe only in the first part of last century. English authors of the eighteenth century, when using the word, feel obliged to explain to their readers what it means. Thus J. Cockburn, speaking in 1735 of three friars at Nicaragua, says, "These gentlemen gave us some Seegars to smoke. These are leaves of tobacco rolled up in such manner that they serve both for a pipe and tobacco itself; they know no other way here, for there is no such thing as a tobacco-pipe throughout New Spain." Victor Hugo (Les Misérables) describes a fellow "carrying in his hand a powerful cane worth two hundred francs, and as he could afford everything, carrying in his mouth a strange thing, called cigar." The first cigar-factory after Spanish model was established at Hamburg in 1788 by H. H. Schlottmann, and the cigar came into general use in Germany about 1793. Kant (Anthropologie, 1798) still uses the Spanish form *zigarro*. The Peninsular War was the occasion for both French and

English adopting the cigar from the Spaniards. The importation of cigars into England was at first prohibited; after the peace of 1815, they were admitted at the duty of 18 shillings a pound. When the duty was reduced to 9 shillings, the import reached the figure of 253,882 pounds in 1830. Cigars then were strictly an aristocratic luxury. Lord Byron (*The Island,* 1823, Canto II, 19) has sung the praise of the cigar, and has simultaneously furnished the only eulogy of tobacco that can lay claim to real poetry.

Sublime tobacco! which from east to west
Cheers the tar's labour or the Turkman's rest;
Which on the Moslem's ottoman divides
His hours, and rivals opium and his brides;
Magnificent in Stamboul, but less grand,
Though not less loved, in Wapping or the Strand;
Divine in hookas, glorious in a pipe,
When tipp'd with amber, mellow, rich, and ripe;
Like other charmers, wooing the caress,
More dazzingly when daring in full dress;
Yet thy true lovers more admire by far
Thy naked beauties—Give me a cigar!

J. W. Croker, in 1831, observed, "The taste for smoking has revived, probably from the military habits of Europe during the French wars; but instead of the sober sedentary pipe, the ambulatory cigar is chiefly used."

The cigarette was introduced into England by British officers who had served in the Crimean Campaign of 1854-56 and had taken to the cigarette smoked by their French and Turkish allies. It first became fashionable among club-men and in high social circles. Laurence Oliphant, both a man of letters and a man of fashion, is generally credited with the introduction into English society of the cigarette. At that time smokers made their own cigarettes as they needed them. About 1865 or 1866 their use had so spread that manufacturers began to cater for cigarette smokers. Even then they employed only a single man, usually a Pole or Russian, to make up cigarettes occasionally. They were perhaps in fashion by 1870, and the social history of smoking in later Victorian days is marked by the triumph of the cigarette.

TOBACCO IN FRANCE, PORTUGAL, SPAIN, AND ITALY

There were two introductions of the tobacco plant into France during the sixteenth century, due to André Thevet and Jean Nicot, respectively. Thevet was born at Angoulême in 1502, joined the Franciscan order, and studied theology without acquiring a taste for scholasticism. Though not equipped with a critical spirit and lacking solid knowledge, he was fond of travel, being stimulated by a passion for inquiring into curiosities, extraordinary or little known objects. In 1555 he accompanied N. Duardo Villegaignon as chaplain on an expedition to Brazil, which had as its object to found a French settlement on the river Ganabra or Santo Januario (the present Rio de Janeiro). He spent three months in Brazil from November 1555 to January 1556, taking part in an expedition to La Plata, where he had a narrow escape from hostile Patagonians; a Scotchman saved his life. On his return to France he published in 1557 a book on his experiences under the title "Les Singularitez de la France antarctique, autrement nommée Amerique: et de plusieurs terres et isles decouvertes de nostre temps." An English translation was printed in London, 1568, under the title "The New Found Worlde, or Ántarctike, wherin is contained wonderful and strange things." This book, somewhat bizarre and poorly organized, contains a number of interesting observations concerning the country and the life of the natives of Brazil, but two thirds of the volume deal with Africa, Peru, the Antilles, Florida, and Canada, and are compiled from oral reports or printed accounts. In this work (fol. 60) Thevet describes the use of tobacco under the name *petun* (the Tupi-Guarani word *pituma* or *pitima*) on the part of the aborigines, who rolled the leaf and wrapped it in a large palm-leaf to the length of a candle. As is well known, the natives of Brazil never availed themselves of the pipe, but only used tobacco in the form of the cigar. Thevet's description is perfectly correct; he says also that he himself tried the novel herb with some bad effects. In another passage of his work (fol. 153) he records the habit of pipe-smoking in Canada, but he does not mention that he took the plant or its seeds along to France. As late as 1575, in his "Cosmographie universelle," he advanced the claim, "I can boast of having been the first in France who brought the seed of this plant, who sowed it and named the plant in question *herbe Angoulmoisine* [after the place of his birth]. Since then, a certain individual (*un quidam*) who never made any voyage has given it his name, some ten years after my return."

This *quidam* was Jean Nicot, born at Nîmes in 1530 as the son of a notary public and educated in Paris. He was French ambassador to Portugal from 1559 to 1561. One day he went to see the prisons of the king of Portugal, and the keeper of the prisons presented him with an herb as a strange plant brought from Florida. According to another version, it was a Flemish gentleman, Damian de Goes, who in 1558 had first cultivated tobacco in the royal garden of Lisbon, the seeds having been imported from Florida. Nicot cultivated the herb in his garden in 1559, being primarily interested in its medicinal properties, and accomplished several marvelous cures. When the success of his experiments was assured, he forwarded specimens, seeds and leaves, to King François II and Catherine de Medici, the queen-mother, with proper directions as to how to apply the drug. From 1560 tobacco cultivation began to spread in France. On his return to France in 1561 Nicot offered the queen a box of powdered tobacco which she employed as a remedy for headaches.

In 1573 Nicot published in collaboration with several scholars a French-Latin Dictionary a copy of which may be seen in the Newberry Library, Chicago. Here we meet (p. 478) the word *Nicotiane* with the following definition: "This is an herb of marvelous virtue against all wounds, ulcers, Noli me tangere [lupus or other eroding ulcer of the face], herpes, and other such like things, which Master Jehan Nicot, being ambassador to the king of Portugal, sent to France, and from whom it has derived its name. See La Maison rustique, book chap. " The blank spaces after "book" and "chap." are not filled out. The book in question is a work on agriculture published in Paris, 1570, by Charles Estienne and Jean Liebault or Liebaut, who gave the first directions for the cultivation of tobacco; they also point out its medicinal virtues and refer to the Indians of Florida as smoking the leaf from tubes (*cornets*). Their information, accordingly, is based on Nicot, not on Thevet. Indeed Liebaut admits that he received oral and written accounts directly from Nicot, which are embodied in his work, and which were introduced to the English public in John Frampton's "Joyfull Newes out of the Newe Founde Worlde" (1577). He consecrates the name *nicotiane* in preference to petum, in order to honor him who first sent the herb to France.

Olivier de Serres, whose "Theatre d'agriculture" was first published in 1600, gives credit solely to Nicot, although, as will be shown below, he must have been acquainted with Thevet's work. Official France has always been prejudiced against the latter, and has heralded Nicot as the only genuine introducer of the plant. In the "Biographie universelle" it is stated under Nicot, "The Franciscan Thevet has contested to Nicot the glory of having enriched France with tobacco; but his pretention has not been favorably received, and the name *Nicotiane* first conferred upon tobacco has persisted, at least in scientific speech. It is not probable, however, that Nicot was conscious of the importance of the gift which he offered to the queen-mother, and that he foresaw that this gift would some day be thirty millions of revenue worth to the state." In Thevet's biography in the same collection, his claim is not even mentioned, while a latent animus crops up here and there: he is characterized as "known for his credulity," yet he is acquitted of ignorance and lying, and is credited at least with knowledge of languages and geography.

On the other hand, Paul Gaffarel, in his introduction to a re-edition of Thevet's "Singularitez" (1878), makes this strong plea on behalf of his hero, "The legitimate vindication of Thevet has never found a hearing. The designation *herbe angoulmoisine* which he had the right to impose on tobacco was denied acceptance, and oblivious posterity continued and continues to thank Nicot for a benefit for which it is not indebted to him. We may be permitted at least to brand this iniquitous judgment as false and to proclaim loudly that to Thevet and solely to Thevet the public treasury owes its most magnificent revenue and the majority of our readers a daily enjoyment." This panegyric is biased and overshoots the mark, for Nicot cannot be ruled out of court completely. The plain truth in the matter is that France owes her tobacco to Thevet and Nicot equally; but the division into the two camps of the Nicotophiles and Thevetophiles demonstrates sufficiently that the subject is not correctly understood.

It is perfectly clear that Thevet and Nicot introduced different plants: the species introduced by Thevet from Brazil can but have been *Nicotiana tabacum* (of some Brazilian variety), and what Nicot introduced must have been *N. rustica,* which flourished in Florida, where *N. tabacum* was at that time unknown. This condition of affairs is plainly reflected by the work of Olivier de Serres referred to above, who distinguishes two species (wrongly taken by him as the male and female plants), one with large leaves, another with small ones, the former being *N. tabacum,* the latter *N. rustica.* De Serres says, "One holds that it is the Petum of the Americans" (the term "America" at that time referred to South America), and he speaks of the "male Petum, also called tabac,"—indications that he was familiar with Thevet's work, although he avoids his name. The fact that *N. tabacum* was cultivated in France in the latter part of the sixteenth century goes to prove that Thevet's claim is correct. This settles the Nicot-Thevet controversy in favor of an equal share of honor for both. But as *N. tabacum* is the more valuable of the two species and as a commercial type is now exclusively used in France as well as elsewhere, Gaffarel is right in linking the tobacco revenue with Thevet's name.

Nicot's influence at court appears to have been overwhelming in view of the cures which the new drug accomplished in the royal family. It is curious that Thevet never made the attempt to influence the court in his favor, although he was at a time chaplain of the queen-mother and historiographer and cosmographer of the king, subsequently curator of the king's curiosities ("garde des curiositiés du roi").

The Tupi word *petun* (also spelled *petum*) introduced by Thevet from Brazil was still widely used in France during the seventeenth century, as expressly stated by Neander in his Tabacologia (1626), and still survives in Brittany and some other Départements as *betum, betun,* or *butun.* Paul Scarron (who died in 1660) even formed a verb *petuner.* In Edward Sharpham's comedy *The Fleire* (1615) appears Signior Petoune, "a traveller and a great tobacconist," a character introduced as the type of the fashionable smoker of the time. In honor of Nicot, tobacco was called "herb of Nicot, herb of the ambassador." As Catherine of Medici, queen of France, used tobacco powder for headaches and was instrumental in propagating the cultivation, such names as "herbe de la reine, herbe medicée, and catherinaire" were temporarily in vogue. The Scotch poet, George Buchanan (1506-82), fired a sarcastic epigram in Latin at the queen for her attempt "to adulterate the Nicotian plant with the name of Medici." Unfortunately, worse adulterations of tobacco than that have since been perpetrated on this world. The designation "herbe du grand Prieur" is traced to the Great Prior of France and duke of Lorraine, who made the acquaintance of the plant as guest of Nicot at Lisbon and cultivated it in his garden at home in 1560; he delighted in taking snuff to the extent of three ounces daily, and as Liebaut states, propagated it in France more than any one else because of the great reverence he entertained for the divine effects of the herb.

The first French pamphlet on tobacco is entitled "Instruction sur l'herbe petun ditte en France l'herbe de la Royne ou Medicée: et sur la racine Mechiocan. Par I. G. P. Envie, d'envie, en vie. Paris, par Galiot du Pré Libraire iuré: rue S. Iaques à l'enseigne de la Galere d'or, 1572." The author's name is J. Gohorry, and his booklet of 32 pages is entirely copied from Monardes.

Molière, in his comedy *Don Juan, ou Le Festin de Pierre,* written in 1665, places the following eulogy of tobacco in the mouth of Sganarelle (Act I, Scene 1): "Whatever Aristotle and the whole philosophy may say, there is nothing equal to tobacco; it is the passion of the gentlemen, and he who lives without tobacco is not worthy of living. Not only does it exhilarate and purify the human brain, but also it instructs the soul in virtue, and with it one learns to become a gentleman. Don't you know, as soon as one partakes of it, in what obliging manner one uses it with everybody and how delighted one is to give it away right and left wherever one may be? One does not even wait till it is requested, but one hastens to anticipate the wish of people, which shows how true it is that tobacco inspires sentiments of honor and virtue in all those who take it." The thought is similar to that expressed by Bulwer Lytton (Night and Morning, 1841), who says with reference to the pipe, "It ripens the brain, it opens the heart; and the man who smokes thinks like a sage and acts like a Samaritan."

In France tobacco first assumed the form of snuff. The king, François II, was treated with snuff against severe headaches by the queen-mother, and the courtiers hastened to imitate the practice. Snuff remained the only mode of taking tobacco on the part of gentlemen until the nineteenth century. In 1635, the free sale of tobacco was interdicted by Louis XIII. Only pharmacists were permitted to sell it for medical purposes on the prescription of a physician. In 1674 the cultivation, preparation, and sale of tobacco became a state monopoly.

The cultivation is now authorized in twenty-five Départements, but the cultivators are obliged either to sell their crops to the State or to export them. Whoever desires to cultivate tobacco must file an application to the Administration of Indirect Taxes, which furnishes the seeds and supervises the whole business. No one has the right to grow it without authorization; the prohibition is absolute, and even extends to flowerpots. The annual production amounts to 25 millions of kilograms. The preparation and manufacture are superintended by the General Direction of the Manufactures of the State (under the Ministry of Finance). The sale is directed by the Administration of the Indirect Taxes.

In Portugal, as stated (p. 49), tobacco was grown in 1558. Clusius travelled in Spain and Portugal for floristic investigations during 1560 and 1564-65, and reports (Exotica, 1601, p. 310) that he saw in Portugal the plant in blossom throughout the winter.

Nicotiana was first introduced into Spain as an ornamental garden-plant owing to its beauty, subsequently as a medicinal plant on account of its real or alleged virtues. This is clearly expressed by Doctor Monardes of Sevilla (1571) in the introduction to his brief treatise on tobacco, which has served as a model to many contemporaneous and later writers in all countries of Europe. "This herb commonly called Tabaco is a very ancient herb known among the Indians, chiefly those of New Spain. After taking possession of these countries, our Spaniards, being instructed by the Indians, availed themselves of this herb in the wounds which they received in war, healing themselves with it to the great benefit of all. A few years ago it was brought over to Spain, to adorn gardens so that with its beauty it would afford a pleasing sight, rather than that its marvelous medicinal virtues were taken into consideration. Now we use it to a greater extent for the sake of its virtues than for its beauty; and those certainly are such to evoke admiration."

The species described by Monardes is *Nicotiana tabacum*. The date of its first introduction into Spain is not exactly ascertained, various names and dates are suggested, but these accounts are not well authenticated; the exact date, moreover, is of no consequence, as Spain contributed nothing to the diffusion of the plant over Europe. Spain gave Europe only two things—the tobacco gospel of Monardes and the cigar. Monardes, it should be remembered, never was in

America, but gathered his information from the lips of voyagers and adventurers, who returned from the newly discovered land to Sevilla. The Spaniards never took to the pipe, but in accordance with the practice of the aborigines of the Antilles and Mexico adopted the cigar and cigarette. Monardes also describes the tubular pipes of Mexico, but these were used in the Spain of his time only for the purpose of obtaining relief from asthma.

Tobacco made its début in Italy under the sponsorship of two churchmen. It was first introduced into Italy in 1561 by Prospero Santa Croce from Lisbon in Portugal, where he was engaged on a diplomatic mission as nuncio of the Pope. He was made cardinal by Pius IV and died in Rome in 1589, at the age of seventy-six years. It is due to this early introduction that Mattioli in 1565 was able to describe the plant, which is *Nicotiana rustica* (above, p. 8). It then became known in Italy under the name *herba Santa Croce*. Castore Durante (Herbario novo, Rome, 1585, p. 227) writes, "At present it is found here in Rome in abundant quantity thanks to the illustrious and reverend Signor, the Cardinal Santa Croce, who brought it from Portugal to Italy." He devotes a lengthy notice to the virtues of the plant, but does not say that in his time the leaves were smoked in Italy.

Another introduction into Italy is due to Nicolò Tornabuoni, a great lover of plants. When he was papal nuncio and ambassador of Toscana at the court of France, he observed there the medicinal employment of tobacco and sent seeds to his uncle, the Bishop Alfonso Tornabuoni, at Florence. This was prior to 1574, as Cosimo I of Medici, who took a deep interest in the cultivation of the plant in Toscana, died in that year. In honor of its godfather, it was then christened *erba tornabuona*. This was *Nicotiana tabacum*. A dried specimen from this early period is preserved in the Herbarium of Ferrara (1585-98), labeled *tabacho over Herba Regina* ("tobacco or herb of the queen").

While Italy thus received the plant from Portugal and France, it took an Englishman to teach Italians how to smoke. This distinction falls on the shoulders of the Cardinal Crescenzio, who about 1610 acquired the gentle art of smoking in England or, according to another version, from an Englishman, which practically amounts to the same. In accordance with this precedent smoking and snuffing were readily adopted by the clergy and laity as well. When complaints reached the holy see from Sevilla that both ecclesiastics and seculars smoked and snuffed in the churches during service, Urban VIII issued a bull excommunicating all who would take tobacco in any form in the porches or interior of the churches. The Catholic Church, however, has always been wisely tolerant toward the use of tobacco. An Italian proverb says: Bacco, tobacco e Venere riducon l'uomo in cenere ("Bacchus, tobacco, and Venus reduce man to ashes"). As in France and Spain, the manufacture and sale of tobacco are a government monopoly in Italy. Cigars are served in the Italian army as part of the daily rations. According to Penn, Italian cigars are "incredibly vile," and bad as are the cigars sold to the public by the Régie, the military ones are worse.

TOBACCO IN CENTRAL AND NORTHERN EUROPE

The English were the most active propagators of tobacco-smoking over many parts of Europe. We noticed their influence in Italy, but it was much stronger in Scandinavia, Holland, Germany, and Russia. English sailors and soldiers, students and merchants carried the pipe victoriously wherever they went. English students at the University of Leiden appear to have been responsible for the initiation of smoking in Holland. William van der Meer, physician at Delft, who cultivated three species of Nicotiana in his garden, wrote in 1621 to Dr. J. Neander at Bremen that he did not become acquainted with pipe-smoking until the year 1590 when he studied medicine at Leiden and noticed the practice among English and French students; he tried to imitate them, but the experiment did not agree with him. At Hamburg which had commercial relations with England and Holland smoking was known at the end of the sixteenth century, and about 1650 the peasants smoked all over Germany. During the Thirty Years' War English soldiers propagated the habit as far as into Bohemia, whence it spread to Austria and Hungary. The older German form *toback* (in dialects still *tuback*) and Low German *smoken* (slang *schmockstock*, "smoking-stick," for a cigar) are witnesses of this early English influence. The plant itself was known at a much earlier date, probably through Huguenots emigrating from France, and is referred to in the correspondence of Konrad Gesner of Zürich in 1565. During the sixteenth century tobacco was cultivated in many parts of Germany, chiefly around Nuremberg, in Saxonia, Thuringia, Hessen, the Palatinate, and Mecklenburg. Tobacco was first introduced into Norway in 1616 when the country was ruled by Denmark and treated as a province of this state. Christian IV of Denmark prohibited the importation of tobacco into Norway in 1632, as he had learned that its use would do great harm to the subjects of his kingdom Norway. In 1643 he rescinded this order and levied a duty on tobacco imports. During the war period 1807-14 attempts were made to grow tobacco in various districts of the country. At present a few farmers along the west coast cultivate tobacco. In Sweden it was first planted in 1724 by Jonas Alströmer; at present it is but cultivated to a small extent in the neighborhood of Stockholm.

TOBACCO IN RUSSIA AND TURKEY

The story of the early fate of tobacco in Russia is well told by J. Crull (Ancient and Present State of Muscovy, 1698, p. 145) :—

"Formerly tobacco was so extravagantly taken, as the aqua vitae, and was the occasion of frequent mischiefs; forasmuch as not only the poorer sort, would rather lay out their money upon tobacco than bread, but also, when drunk, did set their houses on fire through their negligence. Besides (which made the Patriarch take a particular disgust at it) they used to appear before their images with their stinking and infectious breath; all which obliged the Great Duke, absolutely to forbid both the use and sale of tobacco, in the year 1634, under very rigorous punishments; to wit: For the transgressors to have their nostrils slit, or else to be severely whipt. Nevertheless, it is of late years more frequently used, than ever it was before since the time of the edict, the search being not now so strict against the takers, nor the punishment so rigorously executed. Foreigners having the liberty to use it, makes the Muscovites often venture upon it in their Company; they being so eager of tobacco, that the most ordinary sort, which formerly cost not above 9 or 10 pence per pound in England, they will buy at the rate of 14 and 15 shillings; and if they want money, they will struck their cloaths for it, to the very shirt. They take it after a most beastly manner, instead of pipes, they have an engine made of a cows-horn, in the middle of which, there is a hole, where they place the vessel that holds the tobacco. The vessel is commonly made of wood, pretty wide, and indifferently deep; which, when they have filled with tobacco, they put water into the horn to temper the smoak. They commonly light their pipe with a firebrand, sucking the smoak through the horn with so much greediness, that they empty the pipe at two or three sucks; when they whiff it out of the mouth, there rises such a cloud, that it hides both their faces and the standers by. Being debarr'd from the constant use of it, they fall down drunk, and insensible immediately after, for half a quarter of an hour, when the tobacco having had its operation, they lep up in an instant, more brisk and lively than before, when their first discourse commonly tends to the praise of tobacco, and especially of its noble quality in purging the head."

It is curious to note in this account that the Russians of the seventeenth century availed themselves of the water-pipe the history of which is given in Leaflet 18. Presumably they derived it from Turkey. The Turkish word for tobacco, *tutun* ("smoke"), is encountered in all Slavic languages, as well as in Rumanian and Neo-Greek. It seems that at the same time the water-pipe was also fashionable in Germany. At least Georg Meister (Der orientalische Kunstgärtner, Dresden, 1692, p. 59), when he observed on his travels the hooka along the Arabian coast, remarks, "As is also done in our German lands by some tobacco-fellows à la mode."

Better days came for Russian smokers under Peter the Great (1689-1725), who during his sojourn in England and on the continent became an adept of smoking. He determined to introduce tobacco into his country for the sake of the revenue it would yield. The Marquis of Carmarthen on behalf of an English company offered £28,000 for the monopoly of the sale of tobacco in Russia. For this sum the syndicate was allowed to import one million and a half pounds of tobacco a year, and the czar agreed to permit its free use among his subjects, revoking all previous edicts and laws.

In 1698 Lefort and Golovin signed in London with Sir Thomas Osborne (1631-1712) a commercial treaty by virtue of which the latter was to receive the exclusive right to import tobacco into Siberia: up to 1699 he was to import three thousand tons, the following year five thousand, and from the third year onward six thousand and more, with the obligations to pay £12,000 on the first importation and to supply the court with a thousand pounds of tobacco of first quality annually. The English Consul, Charles Goodfellow, in Moscow was Osborne's agent. In 1705 this privilege was abrogated (cf. Leaflet 18, pp. 16-17). English tobacco was then prohibited in Russia, not, however, Turkish or Russian tobacco.

It is generally asserted that tobacco was introduced into Turkey in 1605 under the reign of Sultan Akhmed I (1603-17) ; but I have found a reference in J. T. Bent's "Early Voyages in the Levant" (p. 49) from which it follows that tobacco and smoking, at least from hearsay, must have been known to the Osmans several years before that time, at the end of the sixteenth century. John Dallam, the organ-builder, when he travelled to Constantinople in 1599, tells a curious incident which happened at the time his ship met the Turkish navy not far from the Dardanelles. The Turkish captain of a galley boarded his ship and desired to receive as a present some tobacco and tobacco-pipes which were promptly granted to him. The Turk accordingly anticipated to find tobacco on an English vessel, and must have had some previous experience with the weed, which in all probability had reached Constantinople through the trade of the Levant Company of London. Indeed it was from England that tobacco was first introduced into Turkey, as we learn from George Sandys (Relation of a Journey begun A.D. 1610. Foure Bookes containing a Description of the Turkish Empire, 1615, p. 66). Sandys visited Constantinople in 1610 and writes thus: "The Turkes are also incredible takers of Opium, whereof the lesser Asia affordeth them plenty: carrying it about them both in peace and in warre; which they say expelleth all feare, and makes them couragious: but I rather thinke giddy headed, and turbulent dreamers; by them, as should seeme by what hath bene said, religiously affected. And perhaps for the selfe same cause they also delight in Tobacco; they take it through reeds that have ioyned unto them great heads of wood to containe it: I doubt not but lately taught them, as brought them by the English: and were it not sometimes lookt into (for Morat Bassa not long since commanded a pipe to be thrust through the nose of a Turke, and so to be led in derision through the Citie,) no question but it would prove a principall commodity. Neverthelesse they will take it in corners, and are so ignorant therein, that that which in England is not saleable, doth pass here amongst them for most excellent." The English, accordingly, besides introducing tobacco, taught the Turks also how to smoke it from pipes. In 1615 Pietro della Valle observed the use of tobacco at Constantinople. At first it met with violent opposition on the part of the Sultans, and the most cruel punishments were meted out to smokers.

De Thevenot (Travels into the Levant, pt. 1, 1687, p. 60) tells how at the time of his sojourn in Constantinople the Sultan used to walk through the city in disguise to see if his orders be punctually observed. "It was chiefly for tobacco that he made many heads fly. He caused two men in one day to be beheaded in the streets of Constantinople, because they were smoking tobacco. He had prohibited it some days before, because, as it was said, when he was passing along the street where Turks were smoking tobacco, the smoke had got up into his nose. But I rather think that it was in imitation of his uncle Sultan Amurat [Murad IV, 1623-40], who did all he could to hinder it so long as he lived. He caused some to be hanged with a pipe through their nose, others with tobacco hanging about their neck, and never pardoned any for that. I believe that the chief reason why Sultan Amurath prohibited tobacco, was because of the fires, that do so much mischiefe in Constantinople when they happen, which most commonly are occasioned by people that fall asleep with a pipe in their mouth, that sets fire to the bed, or any combustible matter, as I said before. He used all the arts he could to discover those who sold tobacco, and went to those places where he was informed they did, where having offered several *chequins* for a pound of tobacco, made great entreaty, and promised secrecy, if they let him have it; he drew out a cimeter under his vest, and cut off the shopkeeper's head." From about 1655 the prohibition was relaxed, and smoking both from the dry pipe and water-pipe became a general custom. In 1883 a government tobacco monopoly was introduced: the cultivation is free, but the crops must be sold to the government, which conducts the sale.

Dr. Covel, while on a journey to Adrianople, writes in his Diary under May 2d, 1675, "Here in sommer many come to take their *spasso* and recreation in the shade, sitting upon carpet with tobacco, coffee, and pure water," etc. In Bourgas (modern Lule-Bourgas) he mentions shopkeepers selling the finest tobacco-pipe heads that are to be found in Turkey (Bent, Early Voyages in the Levant, p. 173).

The following interesting account is taken from H. Phillips (History of Cultivated Vegetables, 1822, an undeservedly forgotten book) :—

"The smoking of tobacco is carried to such an excess by the Turks, that they are rarely to be seen without a pipe, and never enter into business without smoking, which often gives them an advantage over the Christians with whom they have either commercial or political transactions, as they smoke a considerable time and reflect before giving a reply to any question.

To visit them on business previously to their morning pipe, would only subject the intruder to their caprice and ill-humor. An ingenious friend, who has resided several years in Constantinople, and had opportunities of associating with the higher classes of that city, assures us that two thousand pounds is no uncommon price for a Turk to give for the amber mouth-piece of a tobacco-pipe, exclusive of the bowl or the pipe, the latter of which is made of a branch of the jasmine tree, for the summer use, while those for winter smoking are uniformly made of the branches of the cherry tree. In order to obtain them of a regular size without being tapering, the young shoots of these trees have a weight affixed at their extremities to bend them downwards, which prevents the sap from returning to the body of the tree, and causes them to swell equally in all parts. The rind or bark is carefully preserved to prevent the escape of the fumes through the pores of the wood. The wealthy Turks pride themselves on the beauty and number of their pipes; and the principal servant in their establishment has no other charge than that of attending to the pipes and tobacco, which are presented to the master or his guests by a servant of an inferior rank. These pipes are so regularly and effectually cleaned, as always to have the delicacy of a new tube, while the German pipe, on the contrary, is enhanced in value by the length of time it has been in use. We are told by the same friend that he has seen among the lower class of Armenians and Jews in Turkey, some smokers who could consume the whole tobacco of a bowl twice the size of those used in England, and draw the entire fumes into their bodies at one breath, which they discharge from their ears as well as the mouth and nostrils."

The world-wide diffusion of tobacco is one of the most interesting phenomena in the recent history of mankind and one that furnishes food for many reflections. Within the short span of three centuries tobacco has firmly established itself as a universal necessity without which mankind is unwilling to live. It has developed into one of the greatest industries of modern times, resulting in statistical figures which almost stagger imagination. Let us consider also that during the same brief period coffee, tea, and chocolate obtained a strong footing in European and American society, and likewise are now articles of international industry and commerce. None of these stimulants was known to our ancestors of three centuries ago, and now they form an integral part of world economy. The association of coffee with tobacco is very close, and their alliance has stimulated and promoted thought, scholarship, literature and art; it profoundly affected social customs, intensified sociability, and paved the way to the era of humanism.

Of all the gifts of nature, tobacco has been the most potent social factor, the most efficient peacemaker, and a great benefactor of mankind. It has made the whole world akin and united it into a common bond. Of all luxuries it is the most democratic and the most universal; it has contributed a large share toward democratizing the world. The very word has penetrated into all languages of the globe, and is understood everywhere.

B. LAUFER.

BOOKS RECOMMENDED

ARBER, E.—English Reprints. James VI of Scotland, I of England. The Essayes of a Prentise, in the Divine Art of Poesie. A Counterblaste to Tobacco. London, 1869.

This booklet contains the complete text of the Counterblaste and valuable documentary material relating to the early use of tobacco in England and France.

BRUSHFIELD, T. N.—Raleghana, Part II. The Introduction of the Potato and of Tobacco into England and Ireland. Report and Transactions of the Devonshire Association for the Advancement of Science, Literature, and Art, Vol. XXX, Plymouth, 1898 (Tobacco: pp. 178-197).

FAIRHOLT, F. W.—Tobacco: Its History and Associations. London, 1859, 1876.

PENN, W. A.—The Soverane Herbe, a History of Tobacco. London (Grant Richards) and New York (E. P. Dutton), 1902.

APPERSON, G. L.—The Social History of Smoking. London (Martin Secker), 1914.

An excellent book, both critical and entertaining.

BRODIGAN, T.—A Botanical, Historical and Practical Treatise on the Tobacco Plant. London, 1830.

Written from an Irish viewpoint and interesting for the history of tobacco in Ireland.

169

唐、宋、元时期的绘画

T'ANG, SUNG AND YÜAN PAINTINGS

BELONGING

TO VARIOUS CHINESE COLLECTORS

DESCRIBED

BY

BERTHOLD LAUFER

CURATOR OF THE FIELD MUSEUM (CHICAGO)

PARIS AND BRUSSELS

LIBRAIRIE NATIONALE D'ART ET D'HISTOIRE

G. VAN OEST AND C°, PUBLISHERS

1924

T'ANG, SUNG AND YÜAN PAINTINGS

BELONGING

TO VARIOUS CHINESE COLLECTORS

T'ANG, SUNG AND YÜAN
PAINTINGS

BELONGING

TO VARIOUS CHINESE COLLECTORS

DESCRIBED

BY

BERTHOLD LAUFER

CURATOR OF THE FIELD MUSEUM (CHICAGO)

PARIS AND BRUSSELS

LIBRAIRIE NATIONALE D'ART ET D'HISTOIRE

G. VAN OEST AND C°, PUBLISHERS

1924

PREFACE.

Several notable Chinese scholars and collectors residing at Shanghai have kindly consented to loan some of their paintings for an exhibition of Chinese pictorial art directed and organized by my friend, Mr. Loo Ching-tsai. It is hoped that this exhibit will stimulate interest and contribute toward a better understanding of Chinese painting which without doubt represents the finest manifestation of Chinese genius. Most of these paintings were selected by me at Shanghai with the assistance of Mr. Kwan Fu-ch'u to whom I am indebted for a great deal of useful information. The explanatory notes have been written by me at Mr. Loo's request. They lay no claim whatever to a scientific treatment of the subject, but are merely intended as a guide for the general public which views this exhibition and is in quest of information. The exhibit includes examples of T'ang, Wu-tai, Sung, and Yüan pictures and gives some idea of the great variety of schools, styles, and subjects. If their study will win new friends to the cause of Chinese art, I shall feel amply rewarded for having performed the task of preparing these notes.

B. LAUFER.

Peking, June 24th, 1923.

T'ANG, SUNG AND YÜAN PAINTINGS

BELONGING

TO VARIOUS CHINESE COLLECTORS.

DESCRIPTIVE NOTICES.

PLATE I. — Portrait of Buddha, attributed to Wei-ch'i I-sŏng 尉遲乙僧 from Khotan (or, according to others, from Tokhara), T'ang period, seventh century A. D. On hemp paper, 83 × 28". Figure of Buddha 25 1/2" high.

This is a unique and the most remarkable Buddhistic painting that has ever come to light. This is not the conventional, stereotyped Buddha, as repeated many thousand times after the Gandhāra and Gupta prototypes all over the East, but it is the great individual Buddha as a powerful human personality, such as we ourselves would picture in our minds when reading his beautiful sayings in the scriptures of the Buddhists. Here he has not the abstract countenance of a god, but we are here confronted with an eminently human and manly face that inspires profound sympathy, and with an expression of spiritual force and energy that in the history of art we meet again solely in Dürer's Apostles. The elongated ears are the only reminders of the hierarchical Buddha type; all other features — the high thoughtful forehead, the bushy eyebrows, the red lips, beard and moustache — are widely different. Even the nimbus is absent : it is not Buddha, the deified saint; it is Buddha, the man, the sympathizer with all creatures, the exponent of universal love. The red robe with drapery in black bordered by gold lines is also treated with great freedom, although, of course, it is ultimately traceable to the classical example of Gandhāra; there is a remote resemblance to the so-called Sandalwood Buddha; what is a unique feature is the long wooden staff the Buddha holds in his right hand. Without doubt, the artist has the intention of showing him on a walk through this beautiful grove by which he is surrounded. Who would not recollect the beginnings of the Jātaka : «Thus I have heard. At the time when Bhagavat dwelt in the Jetavana or Veluvana... »? One of these groves presented to the Buddha by one of his numerous admirers is obviously represented in this unique painting. The luxurious, tropical flora of India is beautifully depicted.

3

Two plants mainly are represented, — a large fruit-tree (perhaps the mango) styled by my Chinese friends *ki-lo-lin kwo-shu* 吉羅林果樹 («ki-lo-lin fruit-tree») and a tree with white flowers. Being deprived of any literature here at Shanghai, where I am writing, I must leave the exact identification of these trees for a later occasion. The Buddha holds one of the ki-lo-lin fruits in his left hand. There is apparently water in front of and behind the Buddha.

At the outset, this painting looks decidedly un-Chinese, and my first guess when I saw it was that it must have come from Turkestan. Needless to emphasize that a Buddha of this individual conception and highly artistic quality has never been traced in Chinese art. This picture is unique as to its technique also. At first sight it conveys the impression of embossed leather, and the pigments are glossy like varnish. There is, however, no embossing in the paper; the leaves and red fruits are represented in high relief by placing layer upon layer of thick pigments, as may be noticed in a few spots where small portions of the pigment have dropped out. It is exactly this peculiar technique which was practised by Wei-chʻi I-söng of whom it is on record that he painted «plastic or high-relief flowers» (凹凸花), and for the first time we have an opportunity of studying this method of painting in this memorable monument. This characteristic trait induced Mr. Kwan Fu-chʻu (or Kwen Vok Tsoo, as he writes himself) 管復初 to identify this picture, which, of course, is not signed or dated, with the work of Wei-chʻi I-söng, and I fully concur with him in this opinion.

The history of this remarkable painting can be traced from the numerous seals affixed to it, for the decipherment of which I am likewise indebted to Mr. Kwan. The large square seal on the top of the scroll is that of the Süan-ho period (宣和御覽 «Süan-ho imperial inspection»), which goes to prove that the painting was preserved in the gallery of the Emperor Hwi Tsung (1119-1126) of the Sung dynasty. A seal in the upper left corner indicates that it was kept in the palace called Cheng-ming 貞明內藏. A seal consisting of nine characters in the right lower corner reads 龍圖閣直學士楊時藏. This refers to the collection of a scholar of the Sung period, Yang Shi by name, who was in charge at the Lung-tʻu-ko. In the following Yüan period we find our picture in the collections of a certain Cheng Yao-an Tao-shi 鄭姚安道師 and another Tö Tsing 德靜 (with the addition 學齋 in the seal). Under the Ming the painting came into the possession of the well-known collector Hiang Tse-king 項子京. There are several other seals containing names of the Ming period : Shi Yen 時儼, Yang Yu-yün 楊又雲, and Sung Chai 松齋 («Pine-tree Studio»), name of a painter's atelier. Finally the painting landed in the temple Tʻien-siang 天祥寺 in Peking; the seal of this temple is impressed on a slip of white silk, which is pasted on the white paper border in the lower left corner. In the Kʻang-hi era the scroll was owned by the collector An Lu-tsun 安麓村 (or I-chou 儀周) and ultimately passed into the possession of Prince Kung 恭親王. Its ownership is thus established from the days of the Sung until the present time. The white paper margin deserves special mention : this paper with oblique, raised, parallel lines, called *tse-li chi* 側理紙, is said to have only been manufactured under the Sung, and probably emanates from the Süan-ho Palace.

PLATE II. — Snow Scenery by Li Chao-tao of the T'ang period 唐李昭道寒林雪景圖. On silk, in colors, 39 3/4×22 3/4 inches. From the collection of P'ang Lai-ch'en.

Two high peaks tower on the left, while on the right hand side three low mountain-ranges stretch in the distance. A temple with red walls and covered with yellow tiles, two sailing boats besides it. A wanderer guiding a donkey over a bridge carries a load on his back. A youth opens the door of a cottage thatched with straw. In another house a woman is engaged in composing a poem. In an open shed on the left-hand side a man is gazing at a waterfall. Higher up there is a temple building enclosed by a red wall and a hamlet of four thatched houses.

Li Chao-tao was the son of Li Se-sün and perpetuated his father's traditions, especially in brilliant coloring in which he was a great master.

PLATE III. — « Portrait of the White-robed Mahāsattva crossing the Sea », after Wu Tao-tse of the T'ang period 唐吳道子白衣大士渡海像; presumably copy made during the Ming period. This picture was formerly hung in the temple Siang-kwo 相國寺 in K'ai-fung, capital of Ho-nan Province, whence it was acquired for Twan Fang's collection. On bluish silk, 72×36 inches.

What is taken for the sea in the above title appears rather, or may be interpreted as well, as a cloud : the spiral designs grouped in pairs are usually symbolic of clouds. The Mahāsattva is the Bodhisattva Avalokiteçvara (Kwan-yin), as is borne out by the image on his head-dress of his spiritual father Amitābha (the Buddha of Endless Light) painted in gold on a red background. The circular, brown nimbus is somewhat unproportioned. His hair is parted in the middle, and aside from the forehead, is covered by the high head-dress. A diadem of spiral designs outlined in gold and set off from a green background encircles his hair. A white hood black and bluish along its edges is thrown over the crown and falls down his back. The face is strongly conventional and unspiritual, and hardly betrays an artist of eminence. The jewelry is mechanically copied : a gold spangle surrounds the neck, and is open in front; from the two hooks are suspended the breast ornaments which are colored green, red and blue. The fingers are slender, with nails somewhat out of proportion. He is clad in a white robe, the drapery of which is drawn in a rather thoughtless fashion. The feet are bare; they are stiffly outlined.

On the left, a descending dove holds in its beak a rosary consisting of sixty-four apparently rock-crystal beads. Below, Shan Ts'ai afloat on a lotus petal, with his hands folded for prayer. The copyist did not waste much thought on this figure, which is somewhat clumsy. Turning to the right-hand side, we notice on the top of a bowlder a red-glazed porcelain vase with long neck into which a willow branch is stuck. Avalokiteçvara is

3.

frequently accompanied by this emblem. A yellow dragon is painted on the body of the vase and winds around the latter's neck. It may justly be doubted that a vase of this type existed under the T'ang; under the Ming, however, it was a common type. A green-glazed cup is placed beside the vase. Behind it are four volumes of Buddhist books enclosed in a case (*t'ao*) which is inscribed on the exterior : «Treasure of the Law of the Great Vehicle (Mahāyāna)». A bamboo with green drooping leaves overshadows the book.

There is no doubt that the great Wu Tao-tse cannot be the author of this picture; yet it remains a valuable document for the study of the subjects which he cultivated.

The painting was presumably executed by a Buddhist monk on the basis of a paper rubbing taken from a stone engraving. As is well known, many of Wu Tao-tse's works are preserved, in their outlines at least, as engravings in stone tablets.

PLATE IV. — Autumn Landscape in the style of Wang Wei. 20 3/4×9 1/2 inches.

A lake with rushes growing along its banks and framed by distant mountains. In the foreground there are three cottages, a fourth higher up; a railed stairway leads up to a high cliff surmounted by a high pavilion in which two men are seated. The leaves of the maple-trees bear the autumn colors red and yellow. A traveller on donkey's back is crossing a bridge.

Fan picture taken from an album which contained also, among other fan pictures, plates XIII and XVII.

PLATE V. — Picture of Two Horses, ascribed to Han Kan of the T'ang dynasty 唐韓幹雙馬圖. On silk, 18 3/4×11 inches.

A groom with black cap, clad in a loose jacket with short sleeves and green belt and wearing high black boots, is astride an unsaddled, trotting white horse, and guides a spirited black horse which is a little distance ahead of him. He therefore leans a bit forward in a strenuous effort to keep up with his charge. Either horse is equipped with a headstall; in the white horse, the mane is erect, and the end of the tail tied up; in the black horse, both the mane and the tail are drooping. The bits are distinctly shown in the black courser. There is no background, and in its powerful spirit and motion the picture is worthy of the great name of Han Kan, who, together with Ts'ao Pa, ranks as the foremost horse-painter of the T'ang period. The two square seals in the upper left corner are presumably of imperial origin. Formed part of a large album in the collection of the late Viceroy Twan Fang, who had acquired it in 1907. Plate XVI belonged to the same album.

PLATE VI. — «Portrait of the Jovial Monk (Pu-tai Ho-shang)» attributed to Sin Ch'eng of the T'ang dynasty 唐辛澄布袋和尚像. On silk, with jade rollers.

Despite the Chinese title conferred upon this painting, I am under the impression that the subject is an Arhat (Lo-han) confined to the narrow space of an oval. A similar Arhat of the same style of composition is in the collection of Field Museum, Chicago. Both have particularly one feature in common that the neck is not at all represented and that the head turtle-like is fitted to the trunk. The face is characterized by a broad nose with prominent nostrils, beard and moustache. A ring is stuck in the right ear, which, as frequent in Buddhistic portraits, is exaggerated in size. Hands and feet are not delineated. The robe is merely suggested by a few lightning-like dashes of the brush. The Arhat is seated on a brown round mat apparently made of coir fibre, the weaves being carefully brought out. There is a large square seal almost faded away on the top, and there are several small seals in the lower left corner and another in the upper left one.

The artist, Sin Ch'eng, travelled in Se-ch'wan during the period Kien-chung 建中 (A. D. 780-784) of the T'ang dynasty, and Buddhist subjects were his specialty.

PLATES VII AND VIII. — The collection contains two very remarkable productions ascribed to the celebrated monk and painter of the T'ang dynasty, Kwan Hiu (A. D. 832-912) 貫休. They were kept in celebrated private collections of China, and can be traced from generation to generation.

1. Portrait of the Bodhisattva Mañjuçrī (Wen-shu) 文殊像. Ink-sketch on paper, 41 1/4 × 20 3/4 inches.

Under rocks and spreading pine-tree branches, the Bodhisattva of wisdom and learning, adorned with a nimbus and looking toward the left, is seated on the back of a powerful lion. He clasps a Ju-i sceptre (a magic wand fulfilling one's every wish) in his right hand. A large hat plaited of bamboo strips and surmounted by a jewel hangs over his back. This is not the usual, conventional type of Bodhisattva decorated with the jewels of an Indian prince, but it is the monk, the Arhat, who poses as Mañjuçrī, or Mañjuçrī in the disguise of an Arhat. The tonsure of the Buddhist monk is marked on his head, and his hair covers the ears and part of the forehead. His long bushy eyebrows lend him strength of character, and he is well equipped with whiskers, moustache, and beard. His right foot is shod with a sandal, while his left foot, as well as his left hand, are hidden. The whole figure is full of life and vigor, and seems well capable of subduing the king of the beasts who tamely crouches under his feet, and appears to be imbued with the spirit of Buddhist sympathy. The scroll contains the seals of Mi Yüan-chang 米元章 (that is, the painter Mi Fei) and Kia Se-tao 賈似道 of the Sung, and of Hiang Mo-lin 項墨林 of the Ming.

2. Portrait of the Bodhisattva Samantabhadra (P'u-hien) 普賢像. Ink-sketch on paper, 41 1/4 × 20 3/4 inches.

Exact counterpart to the preceding work : an Arhat who poses as Samantabhadra, or Samantabhadra in the disguise of an Arhat. It may not be amiss to recall the fact that Buddhist monks were sometimes regarded as incarnations of Bodhisattvas : thus, in the beginning of the seventh century, the two priests Han-shan 寒山 and Shi-tö 拾得 were looked upon by their contemporaries as incarnations of Mañjuçrī and Samantabhadra. Such popular conceptions may have inspired in Kwan Hiu the subject for this pair of scrolls. A portrait of Samantabhadra as the genial monk is known as having been painted by Yen Li-pen (end of seventh century, cf. the writer's «Jade», pp. 341-350). Samantabhadra («the Perfectly Good One»), who is the incarnation of goodness and happiness, looks toward the right with a sharp, penetrating eye, holding a rosary in his right hand and a gnarled stick in his left. Like Mañjuçrī, he is characterized by a large nimbus and a tonsure around which the hair is roughly arranged. His heavy eyebrows fall down in bunches of spirals, and his face is as hairy as that of Mañjuçrī. He wears ear-rings, and is equipped with a hat of the same style as his companion. A long, loose garment envelops his entire body. The palm of his left foot is visible, the toes being slightly turned upward. The elephant, who symbolizes care, caution, gentleness, and dignity, is of the genial kind, and one of the marvelous beasts with six tusks (Sanskrit *shad-danta*), as conceived by the Buddhist mythology of India.

The biography of Kwan Hiu 貫休 is contained in the *Sung kao seng chwan* (ch. 30), and has been translated into French by Sylvain Lévi and Édouard Chavannes (*Journal asiatique*, 1916, sept.-oct., pp. 298-304). The following data are derived from this work. Kwan Hiu was born in A. D. 832 at Teng-kao 登高, a village in the district Lan-k'i 蘭溪, prefecture of Kin-hwa 金華 in Che-kiang Province. His family name was Kiang 姜, while Kwan Hiu was the monastic name conferred upon him after his ordination. His appellation (字) was Tö-yin 德隱 or Tö-yüan 德遠, his title (*hao*) was Ch'an-yüe 禪月 («Moon of Contemplation», Sanskrit Dhyānacandra). At the age of seven years, he entered as a novice the monastery Ho-ngan-se 和安寺 in his prefecture, and received instructions from the master of the dhyāna Yüan Cheng. Every day he recited a thousand words from the Sūtra of the Lotus of the Good Law (Saddharma-puṇḍarīka), and never forgot what he heard but once. He had a friend in a near-by temple, with whom he discussed poetry and exchanged thoughts. The monks were astounded at their extraordinary qualities. After taking the monastic vows, his reputation as a poet was in the ascendancy. Thereupon he betook himself to Yü-chang in Kiang-si, where he resided in the monastery Yün-t'ang yüan 雲堂院, in order to propagate there the Sūtra of the Lotus of the Good Law and the Çraddhotpāda çāstra. He understood perfectly the sense of these two books and explained them in a temperamental manner. He was highly esteemed by the governor of that place, Wang Ts'ao; and the latter's successor erected an altar for the purification of the body by bathing and the purification of the soul by confession, and appointed Kwan Hiu as guardian of this altar.

In A. D. 894, at the age of 63 years, he wended his way to Hang-chou on a political mission to Ts'ien Liu, king of Wu and Yüe, whom he offered five pieces of poetry. He re-established good relations between the king and the last emperor of the T'ang dynasty. The king erected an inscription tablet in honor of the military leaders who had cooperated with him in pacifying the country of Yüe, and honored Kwan Hiu by having his poems engraved on the reverse of this tablet.

Kwan Hiu was a skilful painter of miniatures, and had mastered the Six Principles of Art laid down by Sie Ho. He excelled particularly in ink-sketches. At the request of a pharmacist he erected a hall of the Arhat. Before he painted one of these venerable personages, he prayed every time and obtained the figure of the Arhat in a dream. He retained it in his mind, and his portraits did not conform with the adopted standard.

In 896 he was in the province of Hu-pei in the service of another local despot, but was slandered and deported. He took refuge in the country of Shu, and was received in the capital Ch'eng-tu with great honors by Wang Kien, who in 907 founded the small dynasty of the first Shu. He received from him the title «Great Master of the Moon of Contemplation». He died at Ch'eng-tu in 912, at the age of eighty-one years. The ruler of Shu was deeply grieved, and had him interred with the ceremonial due to a functionary. He erected in his memory a stūpa called the Stūpa of the White Lotus (Pai-lien-t'a). Kwan Hiu equalled Chang Hü 張旭 in talent. In 913 his literary works were collected by T'an Yü 曇域, one of his disciples, who added a postscript to this edition. After a portrait which was extant he is described by his biographer as a fat man of small stature.

Reference has been made above to his ability to receive the inspiration for his Arhat figures from dreams or visionary contemplation. To the west of Hwi-chou fu in An-hwi Province is a temple containing a Hall of the Arhat Corresponding to a Dream (Ying Mong Lo-han Yüan 應夢羅漢院). The monk of this temple, Ts'ing Lan, was the companion of Kwan Hiu, who made for him the images of the Sixteen Arhat. The story goes that the emperor seized them to be placed in his palace, whereupon, at the remonstration of a monk who appeared to him in a dream, he returned them to the monastery. This anecdote was presumably concocted to explain the curious name of the Arhat Hall of Hwi-chou, which was originally suggested by Kwan Hiu's dream-pictures. This is the more probable, as there is on record another Arhat Hall Corresponding to the Dream in Fu-chou 撫州, Kiang-si, where Eighteen Arhat painted by Kwan Hiu were shown.

Kwan Hiu was the Arhat painter par excellence and deeply impressed his contemporaries and the succeeding generations. In the Catalogue of the Gallery of the Sung Emperors (*Süan ho hwa p'u*) are mentioned twenty-six pictures of the Arhat from his hand; further, portraits of Vimalakīrti, Subhūti, a famous priest, and a renowned Indian priest. During the tenth and eleventh centuries, a set of his Arhat paintings was preserved in the temple Yün-t'ang west of Yü-chang, and received regular offerings. Whenever the prefect addressed these Arhat with prayers for rain, the request was granted. The temple Sheng-yin 聖因 on the West Lake near Hang-chou preserves paintings of the Sixteen Arhat by the artist, which K'ien-lung examined in 1757; he wrote a eulogy for each of these.

4.

The work of Kwan Hiu is also connected with the transformation of the group of Sixteen into Eighteen Arhat, the latter being the usual number found in modern temples. In times prior to our artist, the regular number was sixteen. In the records concerning him, both sixteen and eighteen are mentioned. Thus it seems that, on the one hand, he painted the traditional series of sixteen, and, on the other hand, the two additional Arhat, one mastering the dragon, the other subjugating the tiger. The fame of Kwan Hiu's works has doubtless contributed to render the series of the Eighteen Arhat popular, which was definitely recognized as the standard group about a half century after his death.

In the collection of P'ang Lai-ch'en of Shanghai (in his Catalogue, p. 20) is described a colored painting on silk representing an elephant which carries a flower-basket on its back and is attended by a foreigner 番人. This picture is ascribed to Kwan Hiu; on what grounds, is not stated. It is not on record that this artist painted secular subjects, although, of course, the elephant is a sacred animal and may be regarded as a Buddhistic motive. The subject, however, savors rather of Yen Li-pen.

PLATE IX. — Portrait of Yen Tse-ling, by Wei Hien of the Wu-tai period (tenth century) 五代衞賢嚴子陵像. On dark brown, heavy silk, 71 × 31 inches long; old ivory handles.

The portrait is a masterpiece and represents one of those dear genial old men at whose sight we regret that we were not privileged to know him. He is a model of good breeding and refinement; his high wrinkled forehead indicates a high degree of intellectual development, but with all spiritualization he remains eminently human and simple-hearted. His hands are gently folded in front of his chest, the palms and finger-tips being turned downward. His hair is tied up with a light blue ribbon. A long red garment lined with white cloth falls down to his feet, the folds being indicated by heavy black lines; it is held together by a double cord which terminates in two tassels. His feet are shod with straw sandals; the left foot is shown sideways, the right foot in a three-quarter view. Tse-ling is the title of Yen Kwang 嚴光, a friend of the emperor Kwang Wu (A. D. 25) of the Han dynasty. Yen Kwang declined to accept an invitation to court, but preferred living in the country, tilling the ground and devoting his hours of leisure to fishing. On one occasion when the old friends met, the emperor insisted on their sleeping together. During the night Yen Kwang put his foot on the imperial abdomen. The following morning the grand astrologer announced that a strange star had been seen occupying the imperial place. At this the emperor laughed and said : «It is only my old friend Yen Tse-ling, with whom I was sleeping last night.»

Wei Hien was a native of Ch'ang-an (Si-an fu), and twenty-five of his works were preserved in the gallery of the Sung emperors (*Süan ho hwa p'u*, ch. 8, p. 3), among these a number of portraits, Arhat pictures, and snow scenes.

Plate X. — «Lion-cat» by Li Ai-chi of the Wu-tai period (tenth century) 五代李靄之獅猫圖. On paper, 21 3/4 × 18 inches.

The artist was one of the famous cat specialists of China, and this picture, as attested by the seal of the Süan-ho Palace on the top, was deemed worthy of being included in the Gallery of the Emperor Hwi Tsung. The cat with long, black hair, the tips of the hair being somewhat greenish, with white moustache and yellow eyes, the slit of the eye in shape of a crescent, is apparently represented in the act of pouncing upon a rat. It is a thoroughly naturalistic portrayal full of life and motion.

Plate XI. — Landscape by Li Ying-k'iu of the Sung period (tenth century) 宋李營邱溪山深秀圖, better known as Li Ch'eng 李成 (title Hien-hi 咸熙). Ink-sketch on silk, 43 1/2 × 19 3/4 inches.

Two pointed peaks rise in the distance and three powerful rocks of rectangular form stand as firm as the Pyramids. Below there is a peaceful farmer's house thatched, but tiled along the ends of the roof; two men standing in the door are engaged in conversation, and a donkey is turning a mill. Two laborers carry firewood suspended from the ends of a pole. Two men return home to their village in a row-boat; two others, astride donkeys, are crossing a bridge, followed by two coolies carrying their baggage. Beyond the bridge there is a hamlet, and a villager drives two donkeys before him. Three men are climbing up a rocky platform. The artist's seal is contained in the right lower corner. Li Ch'eng, as may be read in Giles' Introduction, was a genius and perhaps the most forward landscapist of the Sung era, who died a premature death at the age of forty-nine in consequence of his fondness for wine. Though he was very productive, his pictures are very rare.

Plate XII. — Picture of a Boy playing with a Cat, by Su Han-ch'en 蘇漢臣 of the Sung period. Dated A. D. 1072. Brown silk.

The figure of the boy is drawn in sharp outlines, pigments being sparingly used, his garment being in yellow brown. His face is refined and intelligent. He his bare-footed, his shoes being placed on the floor. His large eyes gaze at some distant point. The front part of his head is adorned with a round tuft of hair, two bunches of hair falling down in front.

The occiput is bold and perhaps artificially rounded. Seated on a low, hide-covered drum, he holds a white and yellow cat with black tail.

Signed Han-ch'en, with date 熙寧壬子; that is 1072.

Plate XIII. — Landscape by Kwo Hi 郭熙 of the Sung period. On silk, 10×9 1/2 inches.

Rectangular mountains form the background; stately bare trees rise in the foreground. A sage is inside of an open pavilion erected on four pillars and with a cross on the top of the roof. On the left there is a torrent with a grove of distant trees. Kwo Hi is regarded as one of the greatest Sung painters. Fan picture which has formed part of an album (cf. Plate IV).

Plate XIV. — «Bamboo in the Moonlight» by Su Tung-p'o or Su Shi of the Sung period 宋蘇東坡(軾)月竹圖. Ink-sketch on paper, 18 1/4×12 1/5 inches.

Su Shi is equally renowned as a poet, philosopher, and statesman, and his bamboo studies are worthy of high admiration. This picture was taken from an album and subsequently mounted on a scroll. The bamboo leaves occupy only the upper left portion of the paper and fall sideways and downward; some cover the full moon. The picture is signed and dated (1075) by the artist.

Plate XV. — «Boat returning in Wind and Rain» by Chao Ta-nien of the Sung period 宋趙大年風雨歸舟圖. On silk, 39×14 inches.

A heavy shower sweeps over the mountains, bending the rushes which grow on the two banks of the river. A ferry boat is hurrying for shelter. An elderly gentleman bare-headed and bearded protects himself with an umbrella as best as he can, but apparently does not feel very comfortable in the storm. The owner of the boat, bare-legged, equipped with a conical hat, a loin-cloth and a straw rain-coat, makes an heroic effort to brave the elements. Color is used as little as possible. The painting is signed by the artist in the lower right corner, «Made by Ta-nien Chao Ling-jang». The large square seal on the top reads 益王之章 («Seal of Yi Wang»). Chao Ta-nien counts among the best landscapists of the Sung epoch.

Plate XVI. — A poem (過洞庭詩 *Kwo tung t'ing shi*) written in bold, heavy, thick strokes by Mi Fei 米芾 (1051-1107) of the Sung. On green silk, 18×10 1/4 inches; eight seals.

The famous painter's style of writing is quite in harmony with his style of painting. Formerly in Twan Fang's collection (cf. Plate V).

PLATE XVII. — Two Sages by Li T'ang 李唐 of the Sung period, who flourished under the Emperor Hwi Tsung (1100-1125). 10 1/3 × 9 inches.

Two sages are seated on the top of bowlders along the bank of a pond and enjoy themselves by playing the lute. They arrived at this spot in a house-boat. The scenery is made up by willows, pines, and mountains enveloped by morning mists. Fan picture formerly belonging to the same album as Plates IV and XIII.

PLATE XVIII. — «Picture of a Lonesome Pine-tree and a Scholar» by Ma Shi-jung of the Sung period 宋馬世榮孤松高士圖. In colors on silk, 62 1/2 × 41 1/2 inches.

A Chinese anticipation of Heine's poem of the Solitary Pine. Massive mountains with bold outlines sharply cut out like silhouettes form the background. The ragged old pine sends its roots partially above ground, and its trunk is bent from old age. The upper branches are bare, and the top is cut off. A gentleman scholar in whom the whole scenery is focused stands beneath the tree, resting his left hand on the trunk. A youth is bringing his lute (*k'in* 琴) and his master is apparently ready to compose a song inspired by the scenery. On the left-hand side a farmer's cottage is visible. It is a very impressive painting of vigorous style and deep sentiments.

Ma Shi-jung was the son of Ma Hing-tsu 馬興祖, but — and this is the most noteworthy fact about him — he was the father of the celebrated Ma Yüan 馬遠. Of Ma Yüan's wonderful work we may form a correct conception by studying the long landscape roll in the Freer Gallery of which Mr. Freer, during his lifetime, had a photographic reproduction made and distributed, accompanied by an essay of Mr. Laurence Binyon. In comparing the present scroll with this masterpiece, we must confess that the son was a by far greater genius than his father, but that the son inherited from him, not only his genius, but also very concrete methods of drawing trees and mountains. Ma Yüan's bold sweeping and inimitable lines are indeed forecast in this production, although it does not equal his best work.

PLATE XIX. — «Gazing at the Plum-blossoms», ink-sketch by Hia Kwei of the Sung period 宋夏珪觀梅圖. 32 × 56 3/4 inches.

The nature sentiment expressed by this landscape is focused on the sage or scholar with high cap who stands in the open door of his cottage viewing the fresh plum-blossoms. The cottage rises above a stone or brick foundation enclosed by a balustrade, and is overshadowed by stately trees and massive rocks of square outlines with a few pines crowning the top. A temple building is concealed between the rocks. Hia Kwei belongs to the best landscapists of the Sung and flourished under the Emperor Ning Tsung (1194-1224) when he was appointed academician in the Han-lin College.

PLATE XX. — «Picture of Cloud Dragon» by Ch'en So-weng of the Southern Sung 南宋陳所翁雲龍圖. Ink-sketch on silk, 78 1/4×40 5/8 inches. From the collection of P'ang Lai-ch'en of Shanghai.

A pair of dragons with sealy bodies and sword-like claws rising from the waves of the sea and soaring in the clouds. The dragon is conceived of, not as an animal, but as a powerful rain-sending deity of majestic character, as a live personification of one of the great forces of nature, full of life and spirit. The artist was famous for painting such dragon spirits or spiritual dragons. The young or female dragon on the lower left side is looking up toward the male. Masses of clouds are spread all over the surface, and the dark rainclouds are most skilfully and efficiently shaded. There is no signature, but the artist has affixed his seal in the lower right corner.

Ch'en's name (名) was Yung 容; his title (字) was Kung-ch'u 公儲; his *hao,* as given above, So-weng. He was a native of Ch'ang-lo 長樂, and graduated as *tsin-shi* in the period Twan-p'ing (1234-1237) during the reign of the Emperor Li Tsung.

PLATE XXI. — «Releasing an Ox for the Vernal Sacrifice», painting by Chu Hi of the Sung period 宋朱羲春郊放牧圖. On silk, 69×38 1/2 inches.

A youth is astride a black water-buffalo with yellow eyes, which slowly steps down a gentle slope covered with rushes and bowlders in the left lower corner. The boy guides the animal by means of a rope attached to its cartilage and passing through its left nostril. He wears only a short jacket with wide, loose sleeves; his legs and feet are bare. His hair falls loosely down from the tonsure on the crown of his head. He claps his hands while looking in the direction of a flock of eight orioles (*hwang-li* 黃鸝) on the wings, evidently harbingers of the spring.

The scene is overshadowed by stately poplars and willows outlined in bold, black strokes. The leaves are green, but pigments are sparingly used, and ink black is the predominant color. No signature, but three seals in the lower left corner. The attribution of the picture to Chu Hi, who was a native of Kiang-nan, appears well founded, for this artist owed his reputation to his extraordinary ability to paint oxen, and is compared with Tai Sung 戴嵩 of the T'ang dynasty, who was likewise a specialist in bovines, and of whom Giles records a very interesting story. A brief notice of Chu Hi is contained in the *Li tai hwa shi hwi chwan* (ch. 8, p. 1).

PLATE XXII. — «Fisherman's Return in a Boat during a Winter Night» by Hü Tao-ning of the Sung period 宋許道寧雪夜歸舟圖. On silk, 39 3/4×19 1/2 inches.

An evening snow landscape in combination with a genre scene. The background is formed by tall, pointed peaks. In the upper left section there is a waterfall partially screened by the evening mists. A fisherman returns home in his boat across a lake or river. A bridge is laden with snow. Behind bare trees appears a cottage, probably the fisherman's dwelling. Nine collectors' seals are attached to the scroll.

PLATE XXIII. — «Picture of Fading Lotus and two Herons» by Ai Süan of the Sung period 宋艾宣殘荷雙鷺圖. Painting in colors on dark silk, 63 7/8×40 1/4 inches. Light blue porcelain handles. From the collection of P'ang Lai-ch'en of Shanghai.

One of the white herons is on the wing, the other stands on the bank, its beak slightly open. Rushes grow on the right-hand side, but the lotus is predominant, and is treated with grace and elegance. The stalks are drawn in bold outlines; and a favorite motive, a leaf withered along its edges, is introduced with great skill. It is a picture of the autumn, of great naturalistic beauty, and excels in «life's motion» (生動).

Ai Süan was a native of Chung-ling 鍾陵 and made a specialty of flowers, trees, birds, and mammals.

PLATE XXIV. — «Picture of an Egg-plant (*Solanum melongena*)» by Ts'ien Shun-kü of the Southern Sung dynasty 南宋錢舜舉紫茄卷. Roll on brownish paper, 31 1/4×12 1/2 inches. From the collection of P'ang Lai-ch'en of Shanghai.

Naturalistic representation of an egg-plant with leaves, two flowers, and three fruits, with excellent coloration and shading, particularly in the leaves, perfectly true to life. Simultaneously it is a wonderfully clever and spiritual composition of great vigor and beauty; it is a gem. The accompanying poem is written by the artist himself and signed Ts'ien Süan 錢選 Shun-kü of Wu-hing 吳興 (in Che-kiang), his place of birth, with the addition of three seals. The high esteem in which this roll was held is attested by four testimonials written by reputed scholars of the Sung period and another by Weng Fang-kang 翁方綱 in the period K'ien-lung (1736-1795). During the K'ang-hi period (1662-1722), the painting was in the collection of Liang Tsiao-lin 梁蕉林, who has attached two seals to it; it also bears three seals of Hiang Mo-lin 項墨林, a well-known collector and connoisseur of the Ming period.

Ts'ien Süan (*hao* Yü-t'an 玉潭 and Ts'ing-kü lao-jen 淸癯老人) lived in the latter part of the thirteenth century at the close of the Sung dynasty to which he remained loyal. Under the reign of the Mongols he led an itinerant life, engaged in poetry and painting until his end. He was equally reputed as a landscapist and a painter of genre scenes, birds and flowers. The masters whom he took as models were Chao Ch'ang 趙昌 (eleventh century) and Chao Ts'ien-li 趙千里, the great landscapist. The lines of his brush are so fine that they are compared with silk floss. His masterpiece is said to be the picture «Shi Lo inquiring of the monk Fu-t'u-teng for the path to salvation» in the collection of P'ang Lai-ch'en.

PLATE XXV. — «Autumn Landscape with Bamboo» by Lady Kwan (Kwan Tao-sheng) of the Yüan period 元管夫人(道昇)水竹幽居圖. Ink-sketch on silk.

Powerful peaks emerge from the background, and are enveloped by autumn mists. Waterfall and pavilion framed by tall bamboos. Below, a pile-dwelling with rounded roof rises from the water. No human figures. The picture is a masterpiece and the work of a genius (神品); it was doubtless painted in close contact with nature, and it is interesting to note that, as stated in the accompanying legend, it was done in the ninth month of the autumn. The date (second year of the period Chi-chi 至治) corresponds to our year 1322, which makes the picture just six hundred years old. The writing is that of the artist herself who signs Kwan Tao-sheng of Ts'ien-shwi 天水, adding the name of her studio Ch'eng-pi-hüan 澄碧軒. A line in praise of the picture is added by her husband Tse-ang; that is, Chao Mong-fu. She specialized and excelled in bamboo, plum-tree, and epidendrum. She is said to have originated a certain style of bamboo, and is the authoress of a treatise on bamboo drawing; which still enjoys authority.

PLATE XXVI. — Bamboo Study by Kwan Tao-sheng 管道昇畫竹卷. Roll on paper, 44×6 3/8 inches.

An indescribable little gem, a painted poem. Bamboo groves are spread over bowlders which are screened by roaming morning mists. Distant mountains are in the background. Small rushes grow on the banks of a river. There is excellent perspective in that the distance between the bamboos is marked by heavy lines in front and fainter lines farther back. The same peculiar feature, the absence of human figures, is characteristic for this painting also.

Signed in the lower left corner : Kwan shi Tao-sheng. Four testimonials or poems at the end : by Ku Ying 顧瑛, Sheng Chao 盛著, Ye Tö-hwi 葉德輝, who was the last owner of the roll, and Shön I-pin 沈儀彬, a lady whose husband's name is Sü 徐, and who wrote in the ninth year of the Republic.

PLATE XXVII. — « Picture of the T'öng-wang Pavilion » by Wang Chen-p'öng of the Yüan dynasty 元王振鵬滕王閣圖. Ink-sketch on silk, 28 1/2 × 14 1/4 inches, in form of a roll.

This picture is made in illustration of the celebrated poem of Wang P'o 王勃 on the T'öng-wang Pavilion erected under the T'ang in Kan-chou south of Po-yang Lake in Kiang-si. When the buildings were completed, the General Yen 閻 invited the scholars to compose a poem on the pavilion. Wang P'o chanced to pass through the place, seized his brush, and rapidly jotted down his verses.

This monumental work, which formerly was in the possession of Li Hung-chang, is so finely drawn that it is difficult to realize that it was executed by means of a brush. The whole composition is compressed and concentrated into a given narrow space, the technical term for this scheme being « boundary painting » (*kie-hwa* 界畫), an artistic principle inaugurated by Ma Yüan and Hia Kwei. Wang Chen-p'öng (*tse* P'öng-mei 朋梅, *hao* Ku-yün 孤雲, « Solitary Cloud ») made a specialty of this sort of pictures, but did also landscapes, figure and genre painting of the usual kind. The lengthy legend added by the master's own brush and signed with his name harmonizes in style with the composition; the characters are on so small a scale that they are difficult to recognize even by the aid of a magnifying lens. The theme was written in the period Hwang-k'ing (1312-1314) under the reign of a Mongol Emperor who was the protector of the artist and showed him many favors.

The following passage taken by Giles from a native art-critic renders Wang full justice, and the characteristics of his work are fully reflected in this wonderful picture. « He excelled in drawing boundary pictures, and his brush-work and coloring were rhythmical and harmonious. He painted every hair and every thread, and was closely accurate in his delineation of left and right, high and low, looking down and looking up, crooked and bent, square and round, level and straight; but spirit is volatile, and cannot be fettered by method. »

It would be futile to attempt to describe the subject of this painting. Solely an architect could do it justice. It undoubtedly represents a fundamental document for the history of Chinese architecture; and in a country, where so few ancient buildings have survived, the monuments of pictorial art are of paramount importance for the reconstruction of the past and the study of houses and palaces. For the Han period we have the bas-reliefs in stone and, above all, numerous models in pottery of structures of all kinds; for the T'ang, Sung, and Yüan periods we must chiefly utilize the examples which have survived from the hands of the great masters of architectural painting, — a recognized standard subject of academic art. Two main groups of buildings are represented on our roll, each provided with double roofs. The roofs are adorned with turned-up or flying eaves, and are decorated with ornaments at the ends of the ridges. A tent of cloth resting on two poles is erected on the terrace of the upper story in the building on the right-hand side. A few human figures are barely outlined, and quite proportionately are so reduced that the eyes are not indicated.

The Chinese describe this well by the phrase 內中寸人豆馬 : «inside of the building figures of men appear like an inch and horses like a bean». The artist, of course, intends to show the building from a distant point. The structure on the left-hand side rises from a solid stone terrace. A straight stairway leads up to the central edifice that stands free on all sides. We may presume that there is a similar stairway on the three other sides. The extensive hall posed on columns occupies the entire space of the building. To the left there is an open square pavilion on four columns and surmounted by a double roof; a house-boat lies at anchor in the water.

The roll is justly entitled «a rare treasure of the world» 稀世之寶. It is a marvel of drawing, skill and endurance. There is a testimonial at the end by Hu Yen of Nan-chang 南昌胡儼.

Wang is familiar to us as the author of a series of fine-spirited pictures describing incidents in the life of Confucius and years ago published by me in *The Open Court*. I also acquired a long roll painted by him in colors and depicting in a wonderfully spiritual and refined manner the life of happy Taoist sages in the solitude of mountains. Assuredly he was a versatile and resourceful genius.

Plate XXVIII. — «Picture of a Lonely Pavilion with Old Trees» by Ni Tsan (1301-1374) of the Yüan dynasty 元倪瓚幽亭古木圖. On paper styled *lo-wen* 螺紋 («spiral pattern») celebrated at that time, 41×19 1/2 inches.

An open pavilion erected on four poles, the roof being thatched with reeds and surmounted by a knob in the centre. It is framed by dead trees and dim mountains in shadowy outlines. The melancholy character of the landscape is intensified by the absence of any human or animal figures, — a feature for which the painter was noted. As Giles records after a Chinese source, «he painted forests, bamboos, and rocks. There was none of the dust of markets or courts about his compositions. In his landscapes he put in no human figures. He was extremely terse and refined, appearing to be tender, but in reality virile. Painters of the Sung dynasty are easy to copy; painters of the Yüan dynasty are difficult to copy. The style, however, of painters of the Yüan dynasty can be caught, with the single exception of that of Ni Tsan.»

The name Yün-lin 雲林 («Cloud Forest») which he conferred on himself is quite characteristic of his visionary style. His title (字) was Yüan-chen 元鎮. His manner of pale coloring is known as *tan hwa fa* 淡畫法.

The scroll is endorsed by his contemporary Lu Mong-tö 陸孟德 and passed through the hands of many famous collectors of paintings as Ye Yün-ku 葉雲谷 and the Fang Mong-yüan 方夢園 collection of Yang-chou.

PLATE XXIX. — Two Magpies by Ch'en Chung-mei 陳仲美 of the Yüan period.

The picture is confined in space to the lower portion.

Mountains in faint outline. A magpie perches on the stump of a tree. Below, another bird perches on a twig and looks downward.

PLATE XXX. — Mongol Horseman returning from the Hunt, by unknown artist (no signature or seal). Presumably Ming copy after a painting of the Yüan period.

He is clad in a brown fur coat with white fur cuffs, and wears a fur cap adorned with a red tassel at the end. The hunter's left hand is clasped around his bow, while with his right hand he governs the green bridles. The bow-case is of gilded leather decorated with fish scales; the quiver contains seventeen arrows. He has shot and captured a white-spotted deer which he carries in front of him. The black horse turns its head around, being represented en face, wherein the painter has not been quite successful. The man is shaggy and erect; the croup is of red cloth covered with metal strips and clasps in relief.

LIST OF PLATES.

PRINTED

MAY TENTH OF THE YEAR THOUSAND NINE HUNDRED AND TWENTY FOUR BY THE IMPRIMERIE NATIONALE, PARIS, FOR MESSRS G. VAN OEST AND C°, PUBLISHERS AT PARIS AND BRUSSELS, ON « LAFUMA PUR CHIFFON » PAPER. COLLOTYPE PLATES PRINTED BY M. LÉON MAROTTE, PARIS.

Plate I. Portrait of Buddha, attributed to Wei-ch'i I-söng.

Plate II. Snow Scenery, by Li Chao-tao.

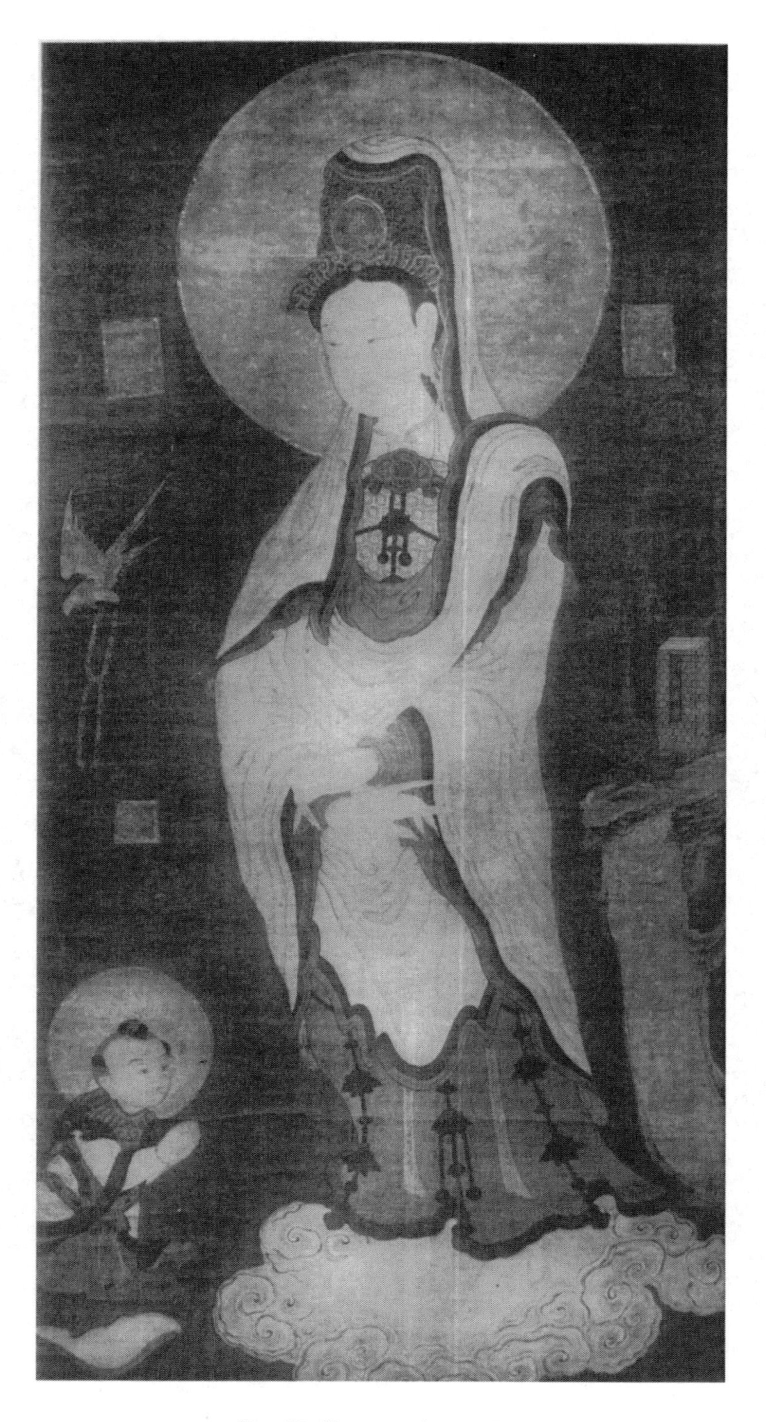

Plate III. Kwan-yin, after Wu Tao-tse.

Plate IV. Autumn Landscape, in the style of Wang Wei.

Plate V. Two Horses, ascribed to Han Kan.

Plate VI. Pu-t'ai Ho-shang (or Arhat?), attributed to Sin Ch'eng.

Plate VII. Mañjuçrí, attributed to Kwan Hiu.

Plate VIII. Samantabhadra, attributed to Kwan Hiu.

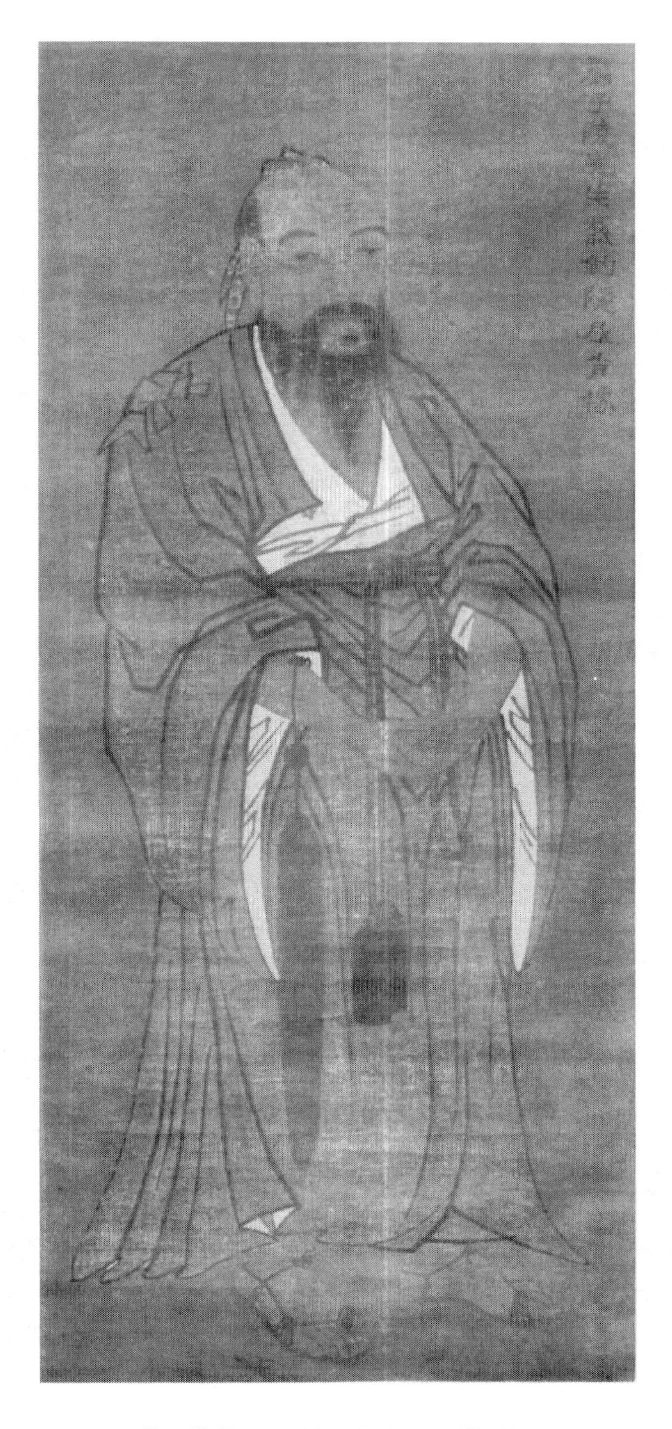

Plate IX. Portrait of Yen Tse-ling, by Wei Hien.

Plate X. « Lion-cat », by Li Ai-chi.

Plate XI. Landscape, by Li Ch'eng.

Plate XII. Boy playing with a cat, by Su Han-ch'en.

Plate XIII. Landscape, by Kwo Hi.

Plate XIV. Bamboo in the Moonlight, by Su Shi (Su Tung-p'o).

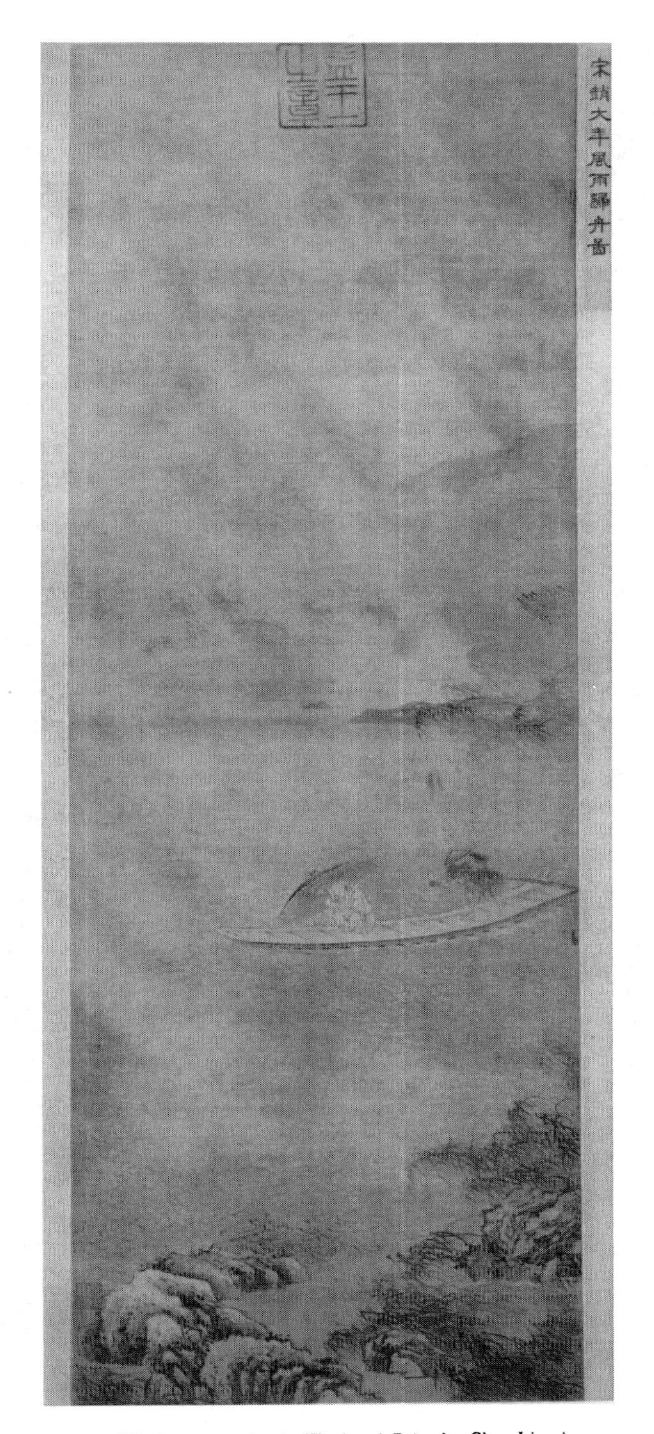

Plate XV. Boat returning in Wind and Rain, by Chao Ling-jang (Chao Ta-nien).

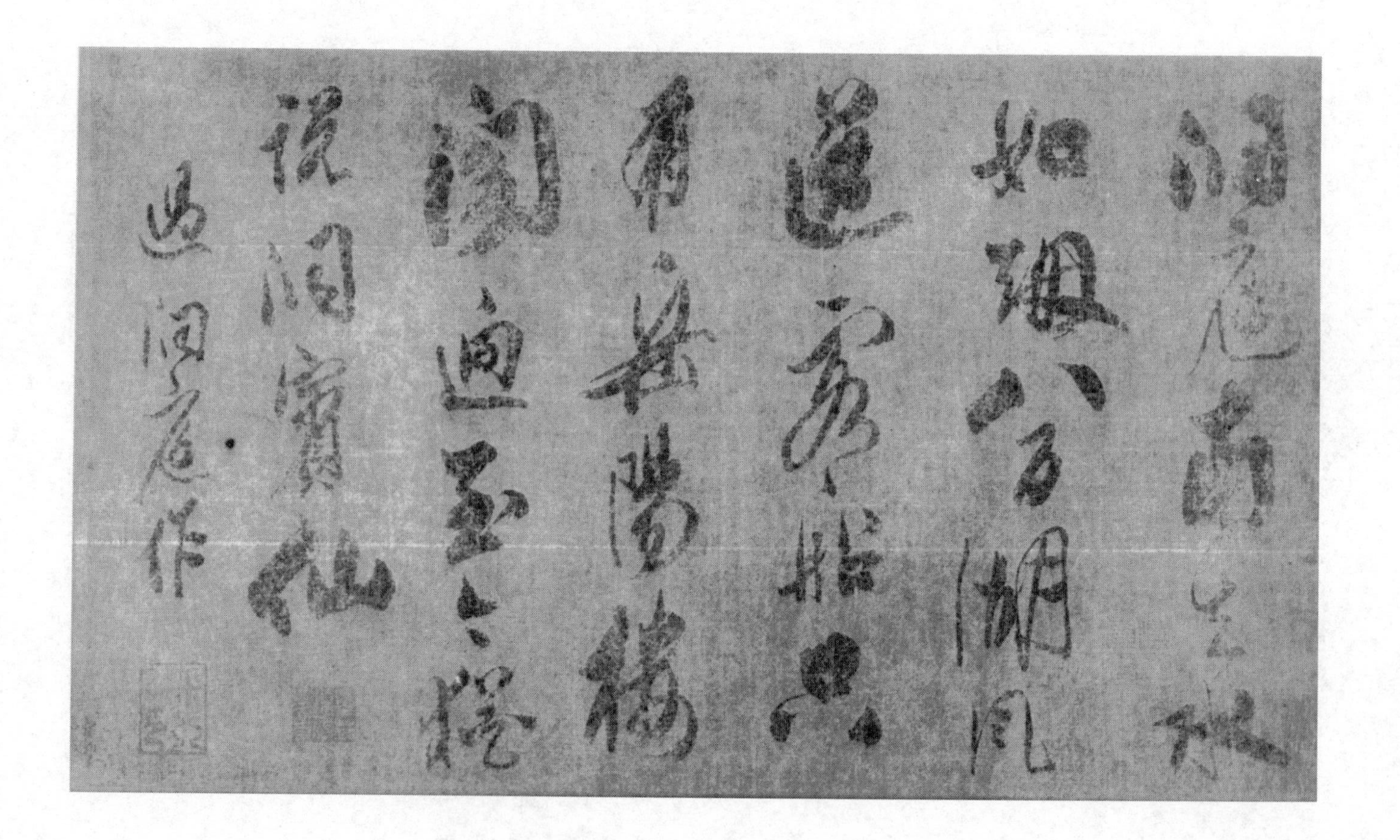

Plate XVI. Autograph of a poem, by Mi Fei.

Hélio Léon Marotte Paris

Plate XVII. Two Sages, by Li T'ang.

Plate XVIII. A lonesome Pine-tree and a Scholar, by Ma Shi-jung.

Héliog. Léon Marotte Paris

Plate XIX. Gazing at the Plum-blossoms, by Hia Kwei.

Hélio Léon Marotte Paris

Plate XX. Dragon in the clouds, by Ch'en Jung.

Plate XXI. Releasing an Ox for the Vernal Sacrifice, by Chu Hi.

Plate XXII. Fisherman's Return during a Winter Night, by Hü Tao-ning.

Hélio Leon Marotte Paris

Plate XXIII. Fading Lotus and two Herons, by Ai Süan.

Plate XXIV. Egg-plant, by Ts'ien Süan.

Plate XXV. Autumn Landscape with Bamboo, by Kwan Tao-sheng.

Plate XXVI. B.

o-sheng.

Plate XXVII. The T'öng-wang Pavilion, by Wang Chen-p'öng.

Plate XXVIII. Lonely Pavilion with old trees, by Ni Tsan.

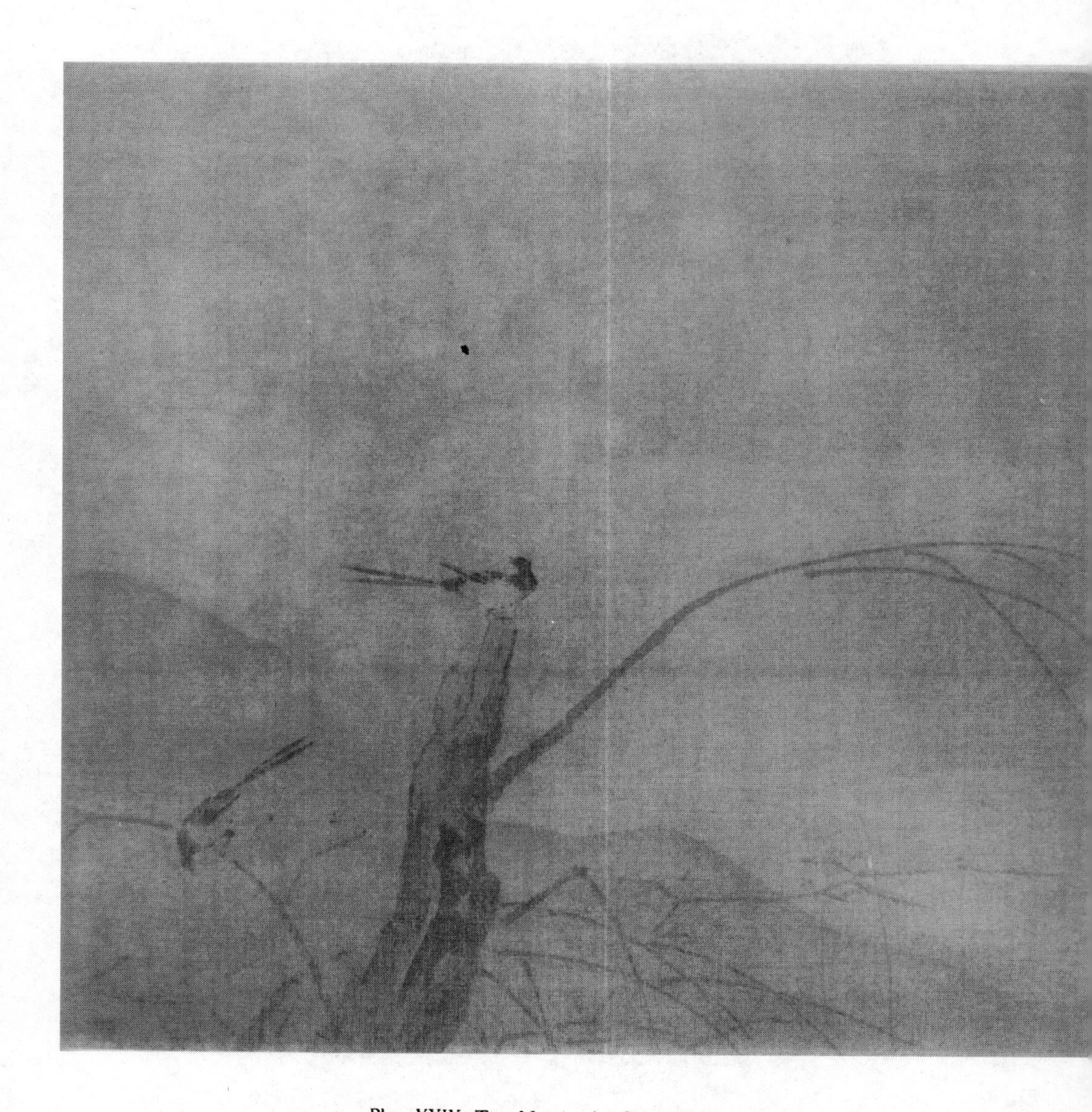

Plate XXIX. Two Magpies, by Ch'en Chung-mei.

Helio Léon Marotte Paris

Plate XXX. Mongol Horseman returning from the hunt.
(Ming copy of a Yüan original?).

170

中国篮子

CHINESE BASKETS

by

BERTHOLD LAUFER

Curator of Anthropology

38 Plates

ANTHROPOLOGY DESIGN SERIES NO. 3

FIELD MUSEUM OF NATURAL HISTORY

CHICAGO

1925

CHINESE BASKETS

PREFACE

The baskets of the North American Indians have attracted a great deal of attention, resulting in an enormous literature both technical and popular in character, which is concerned with their technique and designs. The basketry of the natives of the Philippines, Hawaii, and New Zealand has elicited a few monographs, but, as far as I know, the baskets of China have never been studied by any one. The mere fact of their existence is not even mentioned in the current books on China and the Chinese. Yet, since earliest times, baskets have occupied a prominent position in Chinese civilization, although naturally, owing to climatic and soil conditions, no ancient baskets have survived. In the ancient Book of Songs (Shi king) and in the ancient Rituals occur many terms for various kinds of baskets, but contemporaneous descriptions of these are lacking, and we hardly receive more than a very general idea of their appearance. We read of round baskets of bamboo, of square shallow baskets of bamboo or straw, or of square baskets in which eatables or clothes were preserved. The young bride offered to her father-in-law fruit in a basket. It was a famous maxim in the good old times that men and women should not touch each others' hands and that when a man wanted to make a gift to a woman, she should receive the object in a basket. We also hear of baskets used for specific industrial purposes; thus, for instance, as fish-traps set at the openings of dams, and, above all, in the silk industry, which was the main occupation of women. The tender leaves of the mulberry on which the silkworm feeds were gathered in deep baskets, and a square type of basket served for depositing the cocoons. Even at present basket trays play an important part in the rearing of silkworms (Plate I). Baskets also were utilized in funeral ceremonies and filled with cereals as food for the departed soul, being placed near the coffin. This custom is still perpetuated by the farmers living in the environment of Peking, who bury with their dead an oval basket plaited of willow-twigs. In ordinary life this vessel serves the purpose of a grain-measure, holding one pint; when used as a mortuary vessel, the wooden hoop around the rim is removed, and food is stored up in the basket (for illustration see Laufer, Chinese Pottery of the Han Dynasty, Plate 75, No. 5).

There is a certain degree of interrelation of basketry, pottery and metal vessels. In the case just cited the mortuary basket is a substitute for a pottery urn which is usually deposited in the grave in the provinces of Chi-li, Shan-tung, and Kiang-su. There are two ancient types of rectangular baskets known as *fu* and *kuei* and used for boiled grain at sacrifices and ceremonial feasts; they are preserved only in pottery and bronze (for illustrations, see *op. cit.*, Plate 25, and Art in America, October, 1925, Fig. 4), which give us at least a vague idea of what they may have looked like in basketry. On the other hand, there are ancient sacrificial vessels of tazza shape in pottery and bronze, which at present have survived in basket form in the worship of Confucius and his disciples (Plate 38).

There is a fundamental difference between the baskets of northern China and those of the central and southern portions of the country. In the north they are part and parcel of the rural population, plain, practical, strong, durable, chiefly for agricultural purposes, as collecting and carrying earth and manure, winnowing, storing grain, or used as means of transportation (hamper and dossers). These being exclusively of ethnological interest have not been included in this publication, which is devoted to the artistic baskets whose home is in the Yangtse Valley and the country stretching southward. Here we meet in full development the flower basket with a great variety of shapes and graceful handles, the picnic basket with padlock, the neat travelling basket in which women carry their articles of toilet, and the "examination basket" in which candidates visiting the provincial capital for the civil service examinations enclosed their books and writing-materials, also the cozy for tea-pots, more practical and efficient than our thermos-bottles, and the curious pillow of basketry weave. The basket boxes with raised and gilded relief ornaments are also characteristic of the south.

In accordance with the general interest in Chinese art and art industries in this country, there is now also a considerable interest in and demand for Chinese baskets. They make a ready appeal to our esthetic sense on account of their elegance of shape, variety of design, and other artistic features, not to speak of usefulness of purpose. Chinese genius has developed baskets along lines unknown in other countries; it was not merely satisfied with creating pleasing forms and attractive decorations, but also endeavored to combine basketry with other materials like wood, metal, and lacquer, and to enliven and embellish its appearance through the application of processes originally foreign to the industry. The covers of many baskets display a finely polished, black lacquer surface on which landscapes or genre pictures are painted in gold or red. Others are decorated with metal fittings (of brass or white metal) finely chased or treated in open work. Delicate basketry weaving is applied to the exterior of wooden boxes and chests, even to silver bowls and cups, as may be seen in the exhibits (at present at the south end of the west gallery). In this association with other modes of technique Chinese basketry has taken a unique development which should be seriously studied and considered by our own industrial art-workers.

In accordance with the object of this design series, this publication does not aim at a scientific study of the subject, but is primarily intended for the designer, craftsman, and art student. Nevertheless it is hoped that the technical student of basketry also will find it useful and instructive, as full information as to weave and design is given in the plates for each object. The locality where each object was made is noted, and as the collection is fairly comprehensive, it gives an adequate view of what types of baskets are made in middle and southern China.

The technique of some Chinese baskets is described and illustrated in an article by L. Parker, "Some Common Baskets of the Philippines" in The Philippine Craftsman (Vol. III, No. 1, Manila, 1914), and in another study by E. M. Ayres and L. Duka, "Basketry Weaves in Use in the Philippines," in the same journal (Vol. V, No. 5, 1916), as many types of baskets used in the Philippines are either made by Chinese or by the native tribes after Chinese models. Illustrations of Japanese baskets may be consulted in the monograph of J. Conder, "Theory of Japanese Flower Arrangements" (Transactions of the Asiatic Society of Japan, Tokyo, Vol. XVII, 1889, Plates 11, 44, 60, 61, 64) and in an article by C. Holme, "The Uses of Bamboo in Japan" (Transactions of the Japan Society of London, Vol. I, 1893; see also Vol. V, 1902, p. 50 and Plate V).

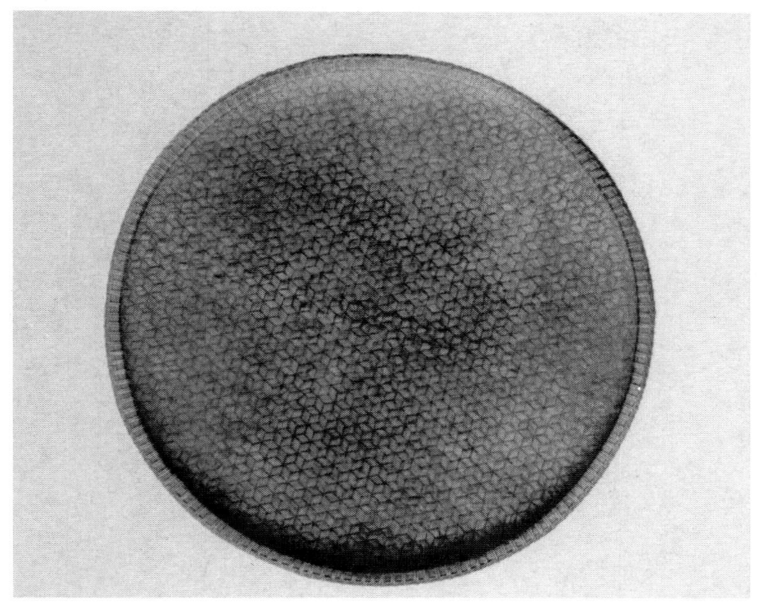

FLAT BASKETRY TRAY.

Used in the rearing of silkworms. Made of bamboo in twilled, checker, twined, and coiled weaves, on strong foundation of bamboo strips.

AN-HUI PROVINCE.

Height 2½ inches. Diameter 15⅛ inches.

Cat. No. 126216.

OLD RECTANGULAR FLOWER BASKET.

Made in checker, twined, and coiled weaves of brown bamboo strips with designs in black. Handle of three bamboo strips painted red. Base of wood. Date inscribed on base in black ink, "24th year of the period Kwang-sü," corresponding to our year 1898.

KI-AN FU, KIANG-SI.

Height (without handle) 5¼ inches. Length 12 inches. Width 8 inches.

Cat. No. 126186.

OLD RECTANGULAR FLOWER BASKET.

Made of brown bamboo, in checker, twined, and coiled weaves.

KI-AN FU, KIANG-SI.

Height (without handle) 5¼ inches. Length 21 inches. Width 13 inches.

Cat. No. 126185.

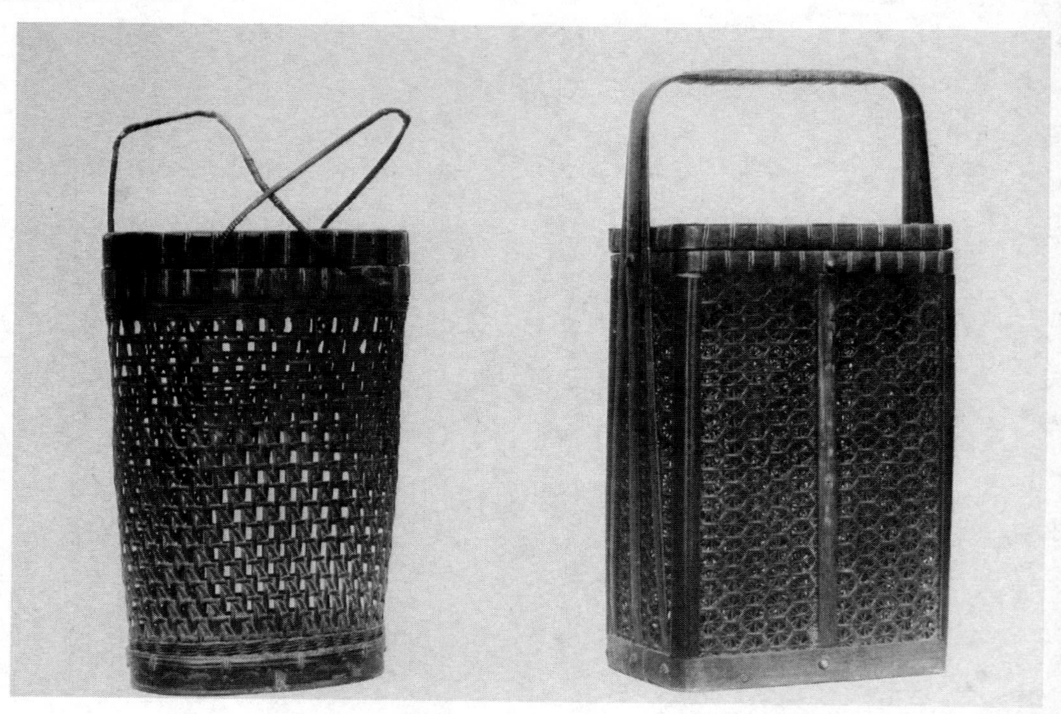

OLD CYLINDRICAL BASKET.

With cover and wooden base, carved with geometric design. Made of bamboo split, in open, twined, and twilled weaves. Two coiled handles.

KIA-HING, FU-KIEN.

Height 12 inches. Diameter 13½ inches.

Cat. No. 126196.

OLD RECTANGULAR BASKET.

Made of bamboo split in open weaves, forming hexagons. With cover, coiled handle, and wooden base.

FU-KIEN PROVINCE.

Height 12 inches. Width 7 inches.

Cat. No. 126199.

OLD CYLINDRICAL BASKET.

Made in three compartments, of checker and twined weaves, edges and cover coated with red lacquer. The bamboo handle is fitted with ornamental metal work which is adorned with bosses and raised, punched designs. The top is surmounted by a metal handle in form of a branch with a peach.

HING-HWA, FU-KIEN.

Height 8 inches. Diameter 7 inches.

Cat. No. 126198.

OLD OVAL BASKET.

Made in two compartments, of two layers of weaving, twilled and fine strip in screen-like weave. Fitted with bamboo bands carved with key patterns. Inscribed in black ink on handle (continued on opposite side), "Made with a pure heart by Madame Ku of Li-men in 1888."

HING-HWA, FU-KIEN.

Height 11 inches. Length 11 inches.

Cat. No. 126197.

OLD CIRCULAR FLOWER BASKETS.

Made of bamboo, varnished, in open and twined weaves. Base, rim, and handle lacquered red.

WEN-CHOU, CHE-KIANG.

Height (without handle) 3½ inches. Diameter 13½ inches.

Cat. No. 126180.

Made of brown bamboo, in checker and twined weaves. With twined handles.

KI-AN FU, KIANG-SI.

Height 3½ inches. Diameter 12¼ inches.

Cat. No. 126187.

CIRCULAR BASKETS.

With twisted handles. Made of brown bamboo in checker and twined weaves.

KI-AN FU, KIANG-SI.

Height (without handle) 3, 2¾, and 3½ inches, respectively. Diameter 9¼, 8¼, and 8 inches, respectively.

Cat. No. 126192. Cat. No. 126189. Cat. No. 126188.

CIRCULAR BASKETS.

With twisted handles. Made of brown bamboo in checker and twined weaves.

KI-AN FU, KIANG-SI.

Height 3½ and 3 inches, respectively. Diameter 8 and 6½ inches, respectively.

Cat. No. 126188. Cat. No. 126191.

OLD OVAL COVERED BASKET.

Made in checker, open, twilled, twined, and coiled weaves. On cover, the two characters *fu* and *lu* ("happiness" and "prosperity") in dark brown. Partially painted in red.

Height 5¼ inches. Length 9 inches.

Cat. No. 126201.

OLD FLOWER BASKETS.

Circular in shape, with hexagonal base. Made of bamboo in open, coiled, and twined weaves. Strips of black woven in coils on the rim. Handle of three strips carved and painted red, as is the base.

KIA-NING FU, FU-KIEN.

Height (without handle) 3 inches. Diameter 12¾ inches.

Cat. No. 126420.

Oval in shape. Made of cane strips, in open, twined, and coiled weaves. Base painted red.

AN-HUI PROVINCE.

Height 5½ inches. Length 13 inches.

Cat. No. 126184.

OLD FLOWER BASKET.

In shape of a boat, with double handle. Made of dark bamboo in open, twined, and coiled weaves. Traces of gold paint on base and rim.

WEN-CHOU, CHE-KIANG.

Height 8½ inches. Length 14 inches. Width 10½ inches.

Cat. No. 126183.

OLD FLOWER BASKETS WITH DOUBLE HANDLES.

Oval in shape. Made of dark bamboo in open, twined, and twilled weaves.

WEN-CHOU, CHE-KIANG.

Height (without handle) 6¾ inches. Length 14½ inches.

Cat. No. 126182.

Made of brown and black bamboo split and varnished, in checker, twined, and open weaves. The base is formed by a bamboo strip lacquered red.

NING-PO, CHE-KIANG.

Height 5¼ inches. Length 13¼ inches.

Cat. No. 126178.

OLD CIRCULAR FLOWER BASKETS.

Made of bamboo, in open, checker, twined, and coiled weaves. With handle consisting of three bamboo strips strengthened by cane.

WEN-CHOU, CHE-KIANG.

Height (without handle) 3⅝ inches. Diameter 12½ inches.

Cat. No. 126181.

Made of bamboo, in checker, twined, and open hexagonal weaves. Strips and handle are painted red.

WEN-CHOU, CHE-KIANG.

Height (without handle) 3½ inches. Diameter 10 inches.

Cat. No. 126179.

OLD ROUND BASKET (TOTAL VIEW).

Made of fine bamboo split of two shades of brown, varnished, in checker, twined, twilled, and interlaced weaves. For details, see following Plate.

NING-PO, CHE-KIANG.

Height 8½ inches. Diameter 11¾ inches.

Cat. No. 126177.

COVER AND BOTTOM OF BASKET ILLUSTRATED IN PRECEDING PLATE.

The cover (on the left) is decorated with a phœnix soaring over peonies on the inner side, with a mat design. The bottom of the basket (on the right) is decorated with a design of lotus-flowers and seed-receptacles.

WICKERWARE BASKET.

Of coil weave, fitted with a porcelain tea-pot. The spout projects through a perforation in the lock, making it possible to pour without removing the pot from the cozy. The hinges are in shape of double coins, and the hook for fastening is in form of a fish.

CANTON, KWANG-TUNG.

Height 7½ inches. Diameter 9 inches.

Cat. No. 126174.

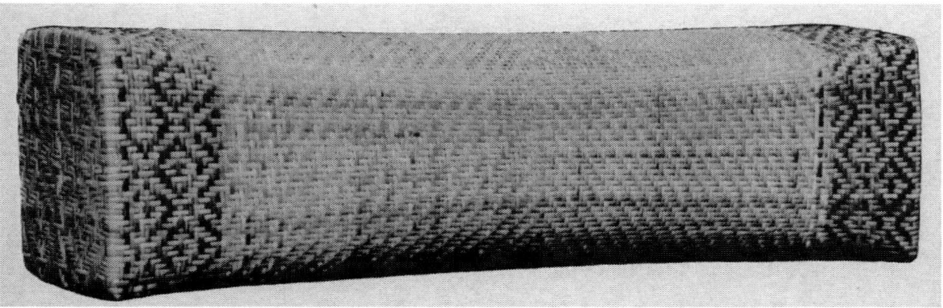

PILLOW OF RATTAN STRIPS.

Hollow. The ends are ornamented with strips of black in twilled weave of square pattern.

CANTON, KWANG-TUNG.

Height 5 inches. Length 15¼ inches. Width 5¼ inches.

Cat. No. 126176.

CIRCULAR BAMBOO CARRYING BASKET.

On the cover, in twilled weave of brown and tan, an ornamental form of the character *ki* ("joy") doubled, expressing the wish that all joy may be doubled. The edge of the cover is of checker and twined weave. The basket proper is made of an hexagonal open weave of three layers; this design is styled "grape-design." Base, trimmings, and handle of bamboo.

NING-PO, CHE-KIANG.

Height (without handle) 6 inches. Diameter 9½ inches.

Cat. No. 126170.

OLD CARRYING BASKET.

Made in two compartments with cover and handle of lacquered wood. The exterior is covered with basketry in hexagonal weaving of split bamboo of two shades. The design in the panel represents the magic wand in a bamboo tube of Chang Kwo-lao, the second of the Eight Immortals. The black-lacquered cover (on the left side) is decorated with a painting in gold and red representing a battle-scene as seen on the stage.

FUCHOW, FU-KIEN PROVINCE.

Height 13¾ inches. Diameter 17½ inches.

Cat. No. 126209A.

OLD BASKETS WITH COVERS AND HANDLES.

Octagonal in shape, with wooden bottom. Made in twined, coiled, and interlaced weaves of extreme fineness. The handle is lacquered red.

Height (without handle) 4¼ inches. Diameter 9½ inches.

Cat. No. 126202.

Made of bamboo split, in open twined, twilled, and coiled weaves. The cover is surmounted by a wooden knob.

FU-KIEN PROVINCE.

Height 8¼ inches. Diameter 11 inches.

Cat. No. 126195.

CIRCULAR CARRYING BASKET.

Made in two compartments, with black-lacquered handle, of fine bamboo split, in checker and twined weaves. The cover is decorated with a painting in gold outlines, representing two women, one mending a garment, the other painting silk fans.

CH'ENG HIEN, SHAO-HING FU, CHE-KIANG.

Height 9⅛ inches. Diameter 8⅛ inches.

Cat. No. 126166.

CIRCULAR BASKETS.

Made of brown bamboo, in checker weave of two layers. No handles.

KI-AN FU, KIANG-SI.

Height 3⅛ and 3 inches, respectively. Diameter 7 and 4¾ inches, respectively.

Cat. Nos. 126193, 126194.

OLD OVAL BOX.

For keeping cakes. Covered with basketry in checker weave and trimmed with brass mounts and double handle. Hinged lid with brass lock. Cover, base, and edges lacquered black.

Height 3¾ inches. Length 10 inches.

Cat. No. 126215.

CIRCULAR LACQUER BOX.

The brown lacquer is ornamented with designs in gold and red. The centre of the cover is occupied by an ornamental form of the character *shou* ("longevity"), surrounded by five bats (*wu fu*) symbolizing five kinds of blessing (*wu fu*): old age, wealth, health, love of virtue, and natural death. On the sides are the eight Buddhistic emblems of luck, alternating with the character "double joy" (*shwang hi*).

FUCHOW, FU-KIEN.
Height 3½ inches. Diameter 8½ inches.
Cat. No. 126220.

CIRCULAR BASKET WITH COVER.

Made in coiled, twined, and interlaced weaves of extreme fineness. Partially painted gold and red. The handle is surmounted by a ring.

K'IEN-LUNG PERIOD (1736-95).
Height 4¾ inches. Diameter 8½ inches.
Cat. No. 126203.

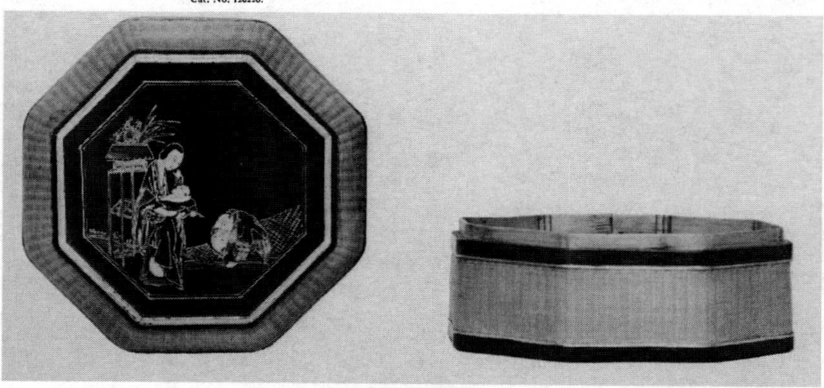

OCTAGONAL BASKET.

Base, trimmings, and cover of wood. Basketry in double layers: on the exterior, split bamboo in checker and twined weaves; in the interior, twilled weaving in yellow and brown. The painting on the black-lacquered cover represents a mother holding a child; two boys are engaged in wrestling.

CH'ENG HIEN, SHAO-HING FU, CHE-KIANG.
Height 6⅛ inches. Diameter 10½ inches.
Cat. No. 126165B.

OCTAGONAL BASKET.

Made, on the outside, of fine bamboo split woven in checker and twined weaves, lined with twilled weaving of brown shade. Base and trimmings of wood. The painting in the black-lacquered panel on the cover represents a mother seated on a bench, with two boys, in a courtyard. One of the boys is leaning toward a painted screen.

CH'ENG HIEN, SHAO-HING FU, CHE-KIANG.

Height 6⅛ inches. Diameter 10½ inches.

Cat. No. 126165C.

CIRCULAR CARRYING BASKET.

Made of fine bamboo split in checker and twined weaves. Interior lined with twilled weaving. Base, trimmings, and handle of wood. The cover (on the left) is lacquered black and painted with a landscape in gold. Along the base a raised key-pattern gilded.

CH'ENG HIEN, SHAO-HING FU, CHE-KIANG.

Height 5¾ inches. Diameter 11½ inches.

Cat. No. 126164.

OCTAGONAL BASKET.

Of same technique. The painting on the black-lacquered cover represents two mainah birds perching on a branch. On the black-lacquered base raised designs in gold of flowers and bats alternating.

CH'ENG HIEN, SHAO-HING FU, CHE-KIANG.

Height 6 inches. Diameter 9¾ inches.

Cat. No. 126162.

CARRYING BASKET FOR SENDING PRESENTS OF FOOD.

Octagonal in shape. Made in three compartments, the upper one in open work. Outside of fine bamboo split woven in checker and twined weave. Wooden handle and trimmings of compartments are coated with black lacquer. Around base, band of floral designs in gold. The black-lacquered cover is decorated with a painting in gold, representing two sages on a paved road in front of a terraced building.

CH'ENG HIEN, SHAO-HING FU, CHE-KIANG.
Height 14½ inches. Diameter 13⅜ inches.
Cat. No. 126160A.

CARRYING BASKET FOR SENDING PRESENTS OF FOOD.

Octagonal in shape. Made in three compartments, the upper one in open work. Outside of fine bamboo split woven in checker and twined weave. Wooden handle and trimmings of compartments are coated with black lacquer. Around base, band of floral designs in gold. The black-lacquered cover is decorated with a painting in gold, representing two sages on a paved road leading to a terraced building.

CH'ENG HIEN, SHAO-HING FU, CHE-KIANG.

Height 14¾ inches. Diameter 13⅞ inches.

Cat. No. 126160B.

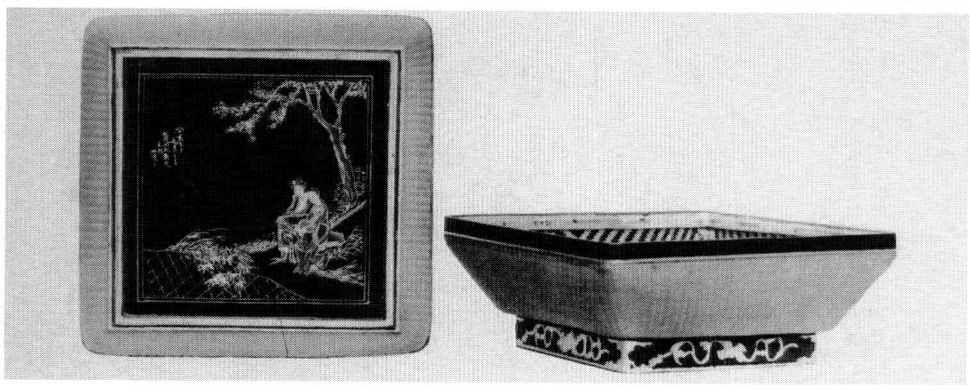

SQUARE BASKET.

Of same technique as those in Plates XXIII and XXIV. The painting on the black-lacquered cover represents two women seated on the bank of a brook overshadowed by a tree.

CH'ENG HIEN, SHAO-HING FU, CHE-KIANG.

Height 6¾ inches. Each side 10 inches long.

Cat. No. 126161.

CIRCULAR BASKET.

Of same technique. On the base, plum-blossoms and leaves in gold. The black-lacquered cover is decorated with a painting in gold and red outlines: a woman seated on a stone bench under a tree and engaged in sewing.

CH'ENG HIEN, SHAO-HING FU, CHE-KIANG.

Height 4½ inches. Diameter 8¼ inches. Cat. No. 126163.

OCTAGONAL BASKET.

Of same technique. The painting on the black-lacquered cover represents mother and son walking under a willow.

CH'ENG HIEN, SHAO-HING FU, CHE-KIANG.

Height 6¼ inches. Diameter 10⅝ inches. Cat. No. 126165A.

OLD RECTANGULAR BASKET.

Fitted with tray inside. Used for books and papers by candidates going to the provincial capital to compete in the civil service examinations, hence called *k'ao lan* ("examination basket" The bamboo foundation is covered on the outside with twilled weaving in brown, the designs being brought out in black. In the centre of both long sides the character *fu* ("good luck"), implyir a wish for the success of the candidate. Hinged handle wrapped with cane, hinged cover, lock and bosses of metal.

Height 9 inches. Length 11¼ inches.

Cat. No. 126200.

RECTANGULAR BASKET.

Made in two compartments, with wooden base and trimmings of bamboo. Basketry of twined, checker, twilled, and coiled weaves on top, in tan and brown. Key pattern and swastika along base. On each of the small sides two ornamental forms of the character *shou* ("longevity"). In the lower panel of the long side the two characters *man t'ang*, on the opposite side (not shown in the illustration) the two characters *fu lu*. They form the sentence *fu lu man t'ang*, which means, "May your hall be filled with happiness and prosperity!" The handle is carved with floral designs.

HANGCHOW, CHE-KIANG.

Height 13 inches.

Cat. No. 126168.

SQUARE PICNIC BASKET.

Made in two compartments with shallow, wooden tray. The exterior is covered with twilled weaving and trimmed with bamboo strips painted black and carved with key patterns. Bas and top of wood. The bamboo handle is carved with floral designs and the emblems of the Eight Immortals in relief and ornamented with metal fittings. Closed by metal bar with padlock attached. Four characters are painted in black on the cover: "Outdoors are the green mountains."

NING-PO, CHE-KIANG.

Height 11 inches. Each side 9½ inches long.

Cat. No. 126205.

OCTAGONAL PICNIC BASKET.

Made in two compartments with shallow tray and fitted with metal trimmings and lock. The exterior is covered with a twilled weaving and trimmed with bamboo strips dyed black and carved with key-pattern. The bamboo handle is finely carved with raised floral designs and the emblems of the Eight Immortals.

On the cover, in black lacquer, the character *fu* ("happiness").

NING-PO, CHE-KIANG.

Height 9¾ inches. Diameter 8½ inches.

Cat. No. 126204.

On the cover, in black lacquer, four characters reading, "The fifteenth day of the month (i. e. the day of the full moon) is an opportune time for a long outing."

NING-PO, CHE-KIANG.

Height 8¼ inches. Each side 8 inches long.

Cat. No. 129206.

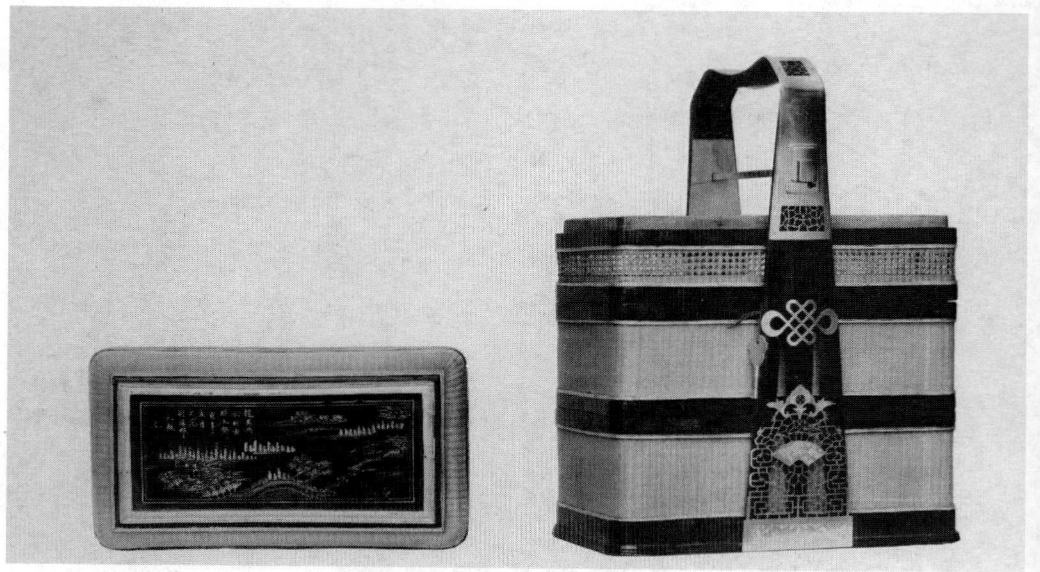

TRAVELLING BASKET.

Of rectangular shape, made in three compartments. The upper compartment is woven in open work; the remainder, in checker and twined weaves of fine bamboo split. The bamboo handle is lacquered black and ornamented with fine metal fittings with designs in open work and engraved. It is fitted with lock and sliding bar. The cover (on the left) is painted with landscape in gold set off from a background of black lacquer.

CH'ENG HIEN, SHAO-HING FU, CHE-KIANG.

Height 10½ inches. Length 10 inches.

Cat. No. 126167.

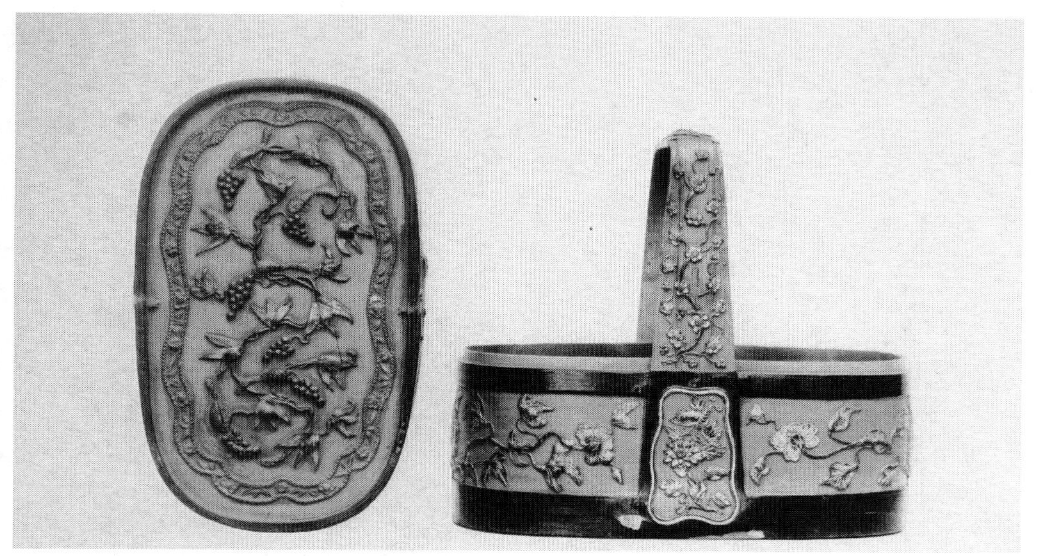

OLD TRAVELLING BOX OF BASKET SHAPE.

Used by women for carrying toilet articles. Made of wood, in imitation of a basket form, in two compartments, and lacquered red and black. Decorated all over with ornaments in gilded relief. On the cover, squirrels gnawing at bunches of grapes (a frequent and favorite motive in Chinese decorative art). On the handle, a spray of plum-blossoms. On the sides, leaf and floral designs. The medallion in the centre shows a butterfly on a flower.

HANGCHOW, CHE-KIANG.

Height 7½ inches. Length 13½ inches.

Cat. No. 126210.

OLD TRAVELLING BOX OF BASKET SHAPE.

Used by women for carrying toilet articles. Made of wood in two compartments and lacquered red and black. Decorated with ornaments in gilded relief on cover, sides, and handl representing plants, Ju-i sceptre (on handle), Taoist genius in medallion (on base), double coins and Swastikas enveloped by fluttering bands.

HANGCHOW, CHE-KIANG.

Height 10 inches. Length 12 inches.

Cat. No. 126213.

OLD TRAVELLING BOX OF BASKET SHAPE.

With handle and shallow tray. Used by women for carrying toilet articles. Made of wood stained brown and lacquered in black and red. Decorated with designs in gilded relief on cover, sides, and handle. On either side of the handle, in gilded relief within red panel, a vase with branch of plum-blossoms and ornament of natural stone formation as used on the desk. On the cover, gourd vine with leaves and fruits and the two genii of Harmony and Union, framed by a band of key-pattern.

HANGCHOW, CHE-KIANG.

Height 7⅜ inches. Length 12¼ inches.

Cat. No. 126212.

OLD TRAVELLING BOX OF BASKET SHAPE.

Used by women for carrying toilet articles. Made of wood in two compartments and lacquered red and black. Decorated with ornaments in gilded relief on covers, sides, and handles. On the long sides an all-over pattern of plum design. In the beaded medallion a cat leering at a butterfly (a famous motive of pictorial art); below, two dragon-like creatures; above, on the handle, spray of plum-blossoms. On the cover, two figures in a garden, enclosed in a medallion surrounded by Swastika pattern; border of gourd with leaves and fruits.

HANGCHOW, CHE-KIANG.

Height 7¼ inches. Diameter 12¼ inches.

Cat. No. 126211.

OLD TRAVELLING BOX OF BASKET SHAPE.

Used by women for carrying toilet articles. Made of wood in two compartments and lacquered red and black. Decorated with ornaments in gilded relief on cover, sides, and handles. On the side, flowers, birds, and butterflies; a Taoist genius, enclosed in a medallion. On the cover, in a countersunk medallion, a procession of five figures: in the centre the Chwang-yüan on horseback; that is, the successful candidate who won the first place at the triennial Palace Examination. He is preceded by two gong-beaters and followed by two standard-bearers, his title "Chwang-yüan" being inscribed on the tablets.

HANGCHOW, CHE-KIANG.

Height 10½ inches. Length 11¾ inches

Cat. No. 126214.

OLD CIRCULAR LACQUERED BOX.

Combination of box with basketry. Made of wood lacquered black and covered with fine bamboo basketry on the outside. The painting on the cover in gold represents a landscape with an open pavilion and a sage with attendant on the bank of a river. Inscription, in vermilion on bottom, yields the date 1726. There is a seal inside of the cover with the name Li Fang.

YUNG-CHENG PERIOD (1722-35).

Height 4 inches. Diameter 11 inches.

Cat. No. 126219.

OLD CIRCULAR LACQUERED BOX.

Combination of box with basketry. Made of wood lacquered red on the exterior and black in the interior. The outside is equipped with fine bamboo basketry. The painting on the cover in colors and gold outlines represents a painter with attendant, a lady leaning over a table and admiring one of his pictures. Inside of the cover there is a seal in vermilion with the name Hu-chi.

Height 4 inches. Diameter 9¾ inches.

Cat. No. 126218.

CEREMONIAL TAZZA.

Used for offerings on the altar in the Confucian temples. Made of red-lacquered wood covered with basketry of checker weave. Cover surmounted by gilded knob. Edges gilded.

SUCHOW, KIANG-SU.

Height 10¼ inches. Diameter 5⅝ inches.

Cat. No. 126207.

OCTAGONAL JARDINIÈRE.

Of basket form. Made of wood and coated with black lacquer in the interior, with red carved lacquer on the exterior. There are eight countersunk panels with reliefs of various flowers in gold and brown. The eight panels on the base are decorated alike with a cloud pattern encircling an ornamental form of the character *shou* ("longevity") in the centre. Handle and edge are finely decorated with meanders and star designs enclosed in hexagons. The lower portions of the handle are in the form of a Ju-i sceptre.

K'ANG-HI PERIOD (1662-1722).

Height 14 inches. Diameter 11 inches.

Cat. No. 126208.

171

爬树鱼

THE CHINA JOURNAL OF SCIENCE & ARTS

[Registered at the Chinese Post Office as a Newspaper]

Vol. III JANUARY, 1925 No. 1

CONTENTS

Editors: Arthur de C. Sowerby, F.Z.S. (Science). John C. Ferguson, Ph. D. (Literature & Arts).

Assistant Editor and Manager: Clarice S. Moise, B.A.

Secretary: Nan L. Horan

Contributions of a suitable nature are invited and all MSS. not accepted for publication will be returned.

Books for review should be sent to the Editor as early as possible.

The subscription for the year (six issues) is $10.00, *Shanghai currency, or its equivalent. In the U.S.A. and Canada, Gold* $6.00; *in Great Britain and Europe,* 25/-. *Postage free.*

Crossed cheques (Shanghai currency) or P. O. O. should be sent in payment of the annual subscription from Outports, Europe and America direct to the Manager.

Office: 102, *The Ben Building,* 25 *Avenue Edward VII, Shanghai, China.*

Printed by the North-China Daily News and Herald, Ltd., for the Proprietors

Photo by B. T. Prideaux

A fine specimen of the Chinese Giant Salamander, recently purchased in Shanghai, and said to have been imported from Canton Province. This is probably the Na Yü of Dr. B. Laufer's article.

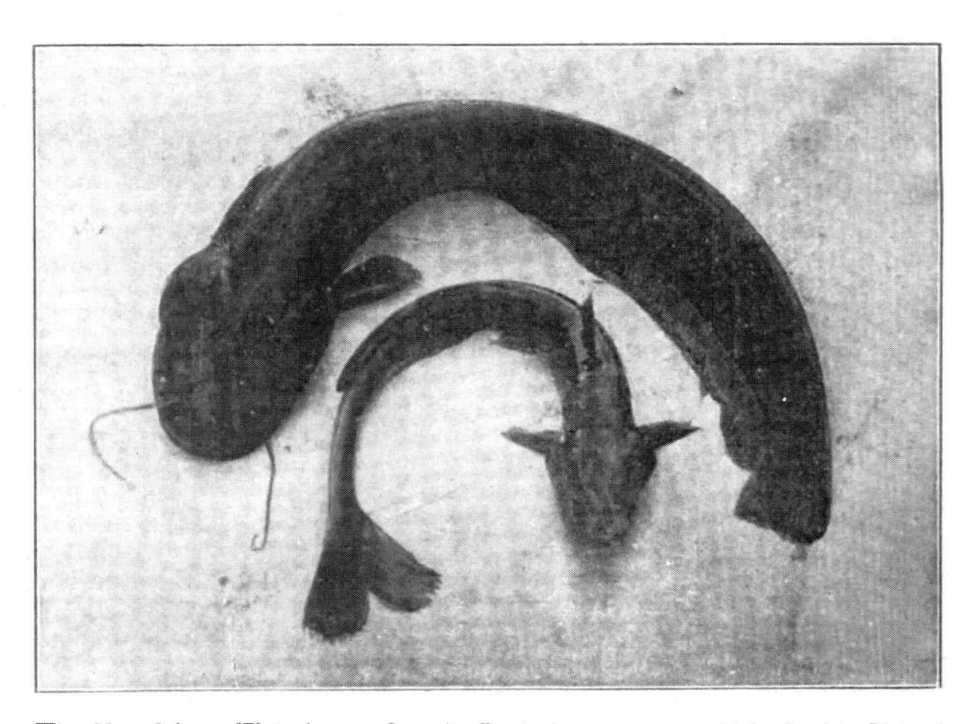

The Sheatfish, or Wels (upper figure), *Parasilurus asotus*, which the Na Yü of Dr. B. Laufer's article is said to resemble. The lower figure is of another Catfish, *Pseudobagrus emarginatus*, described by Sower by from the Yalu River.

(See Pages 37 and 42)

THE TREE-CLIMBING FISH

BY

BERTHOLD LAUFER

(Field Museum, Chicago)

In the spring of this year the Chicago public was agitated by newspaper stories of tree-climbing fishes, and ex-Mayor William H. Thompson laid a wager of $25,000 that he would locate the tree-climbing fish and have a film of motion-pictures taken of it; for which purpose he sent out a yacht into the South Sea Islands. On August 3, the Chicago *Sunday Tribune* published the reproduction of a woodcut representing *Anabas scandens* in the act of scaling a tree and derived from Frank Leslie's *Illustrierte Zeitung* of 1867. It happened that on the same day I was reading in the *Tung chai ki shi* 東齋記事, written by Fan Chen 范鎮 of the Sung period (reprinted in *Shou shan ko ts'ung shu*), and quite accidentally lighted upon the following passage (ch. 5, p. 3b) :—

"In Shu (Se-ch'wan) there is the *na* fish 魶魚 which is clever in climbing trees 善緣木, and which has a voice like a wailing baby. Mong-tse's saying 'climbing a tree to seek for fish' alludes to something that is impossible to obtain. This is an unheard-of matter."

The passage in Mong-tse will be found in Legg's edition, Chinese Classics, Vol. II, p. 145.

My curiosity being thus aroused, I pushed my inquiry a little further and found the following data. The *I yü t'u tsan pu* 異魚圖贊補, a fish-book by Hu Shi-an 胡世安 in the appendix styled *Jun tsi* 閏集 (p. 7b),* contains this notice: "The *na* fish 魶魚, identified by some with the *hia* 蝦, is provided with feet like the *i* or *ni* 鯢. It has a large head and a long tail. It cries like an infant. It climbs trees without falling. According to the *Fang wu lio* 方物畧 (that is, the *I tu* 益部 *fang wu lio*, a book on the products of Se-ch'wan by Sung K'i 宋祁, A.D. 998-1061,

---

*Wylie (Notes on Chinese Literature, p. 154) states that this work was written after 1630, but the edition before me, in two volumes, bears a brief prefatory note signed by Hu Shi-an and dated 1618.

cited also in the 餠源), the *na* lives in the gorges and valleys of western Se-ch'wan, particularly in the Ya River. In shape it resembles the *ni*, and has feet which enable it to clamber trees. Its voice is like the wailing of a child. The people of Shu (Se-ch'wan) keep it and feed it." Then follows the quotation from Fan Chen given above; his book is cited here as *Tung chai pi lu* 筆錄.

The earliest notice of the climbing fish is preserved in the *Shu chi* 蜀志 (Records of Se-ch'wan,) the exact date of which is not known, but which must be an ancient work, as it is frequently quoted in the *Ts'i min yao shu*, a book on agriculture of the fifth or sixth century. The *Shu chi*, as cited in the *Pen ts'ao kang mu*, says, "The fish *na* has its habitat in the mountains, passes, and valleys west of Ya-chou; it resembles the sheat fish (*nien* 鮎, *Silurus asotus*). Being provided with feet, it is able to climb trees. It has a voice like an infant, and is edible."

Hung Mai 洪邁 (1124-1203), in his *I kien chi* 夷堅志, has this story: "Outside of the temple on Mount Ta-kwei 大龜山, in the district of Ying-shan 應山縣 (in Te-an fu, Hupeh), there is a pond in which a fish called *ya erh yü* 牙兒魚 is produced. It is equipped with four feet, and is able to ascend the bank of the pond and climb up a tree. It emits the sound *i-ying* like a baby. The largest have a weight of over a catty." The poet Su Shi 蘇軾 (1036-1101), when he lived at Wu-ch'ang, is said to have kept a fish resembling a sheat-fish and provided with feet; it was able to walk over the ground.

The *Kin ch'wan so ki* 金川瑣記 (Records of the Kin-ch'wan territory in north-western Se-ch'wan), written by Li Sin-heng 李心衡 from Shanghai, refers to the same species, without giving its name, by saying that one or two kinds of fish have the same shape as the *i* or *ni yü* 鯢魚, and are provided with four feet, while they are but four or five inches in length; according to the natives, they are capable of curing pain of the heart. The author obtained several specimens in the market, but did not have the courage to taste them (p. 30b of the edition of *Siao fang hu chai yü ti ts'ung ch'ao*, section 8).

The Provincial Gazetteer of Se-ch'wan (四川通志, ch. 74, p. 27b) mentions the species among the products of Ya-chou-fu, giving also the term *t'i* 鮷 for it and adding the colloquial name *wa-wa yü* 娃娃魚 ("baby fish"). The information given is identical with that recorded previously. The Geography of the Ming Dynasty 大明一統志 is cited as locating the species in Jung-king 榮經, a district in the prefecture of Ya-chou, and in the mountains west of Ya-chou. Li Shi-chen's *Pen ts'ao kang mu* (ch. 44, p. 12b) gives the name *hai erh yü* 孩兒魚 ("baby-fish") as a synonym of *t'i* 鮷 and distinguishes under this term two different species: one living in lakes and rivers in shape and colour like the sheat, with abdominal fins resembling feet and uttering a sound like *ya-ya* 軋軋; this is the *t'i*; the other, living in mountain-torrents, in shape and voice like the *t'i*, but solely endowed with the faculty of climbing; this is the *ni*. Hence the term *na* 魶 is dealt with by Li Shi-chen under the heading *ni* 鯢, and is characterized by him as a word peculiar to Se-ch'wan. As the word *ni* anciently referred to a marine mammal (sometimes explained as a female whale), stories relating to this sea animal were transferred to the climbing fish. Kwo P'u 郭璞 (A.D. 276-324) defines the fish *ni* as resembling the sheat, with four feet, the front feet like those of a monkey, the hind feet like those of a dog, wailing like a child, the biggest from eight to nine feet long. In this case fish and sea mammal are confounded; the fish in question is hardly a foot in length. The biggest yarn in regard to it is told by Ch'en Ts'ang-k'i 陳藏器, author of the *Pen ts'ao shi i* 本草拾遺, written in the K'ai-yüan period (A.D. 713-741) :—

"The *ni* lives in mountain-creeks. It resembes the sheat-fish (*nien*), and has four feet and a long tail. It is able to ascend trees. At the time of a great drought it fills its mouth with water, ascends a mountain, covers its body with grass and foliage, and opens its mouth. When birds arrive to drink this water, it swallows them. Its voice is like the crying of a baby." This story is repeated in the *Ta Ming i t'ung chi* with reference to the climbing fish of Se-ch'wan and by Chang Chu 張澍 in his *Sü K'ien shu* 續黔書 of 1805 (ch. 4, p. 16b, ed. of *Yüe ya t'ang ts'ung shu*). A grain of truth may underlie this figment inasmuch as *Anabas scandens* differs from all other fishes in that its gullet bones are divided into a large number of small leaflets between which water catches, so that the fish can live a long time out of water.

Sulaymān (*ca.* A.D. 851) speaks of a fish which jumps out of the water, climbs up a coconut palm and drinks the sap of the plant; thereupon it returns to the sea. Qazwīnī mentions the island Salāhat in the Indian Ocean, where lives a fish which comes out of the sea, climbs the trees and sucks the juice of fruits; thereupon it falls down as though it were intoxicated, and is caught by the inhabitants. Bākuwī, with reference to the same island, writes that the fishes climb up the camphor-trees and lick them; a kind of sugar flows from this spot and is collected (G. Ferrand, Textes relatifs à l'Extrême-Orient, pp. 41, 305, 464).

The first reliable information in European literature on the climbing-fish (*Anabas scandens* Cuvier) was given by J. E. Tennent (Ceylon, 4th ed., Vol. I, 1860, p. 215). "It grows to about six inches in length, the head round and covered with scales, and the edges of the gill-covers strongly denticulated. Aided by the apparatus in its head, this little creature issues boldly from its native pools and addresses itself to its toilsome march generally at night or in the early morning, while the grass is still damp with the dew."

Good information on the Anabantidae or climbing-perches is given by David Starr Jordan (Fishes, p. 580, New York, 1907; cf. also p. 677). I do not know, of course, what species is represented by the climbing-fish of the Chinese records, and should gratefully appreciate any information on the subject that residents of Se-ch'wan, especially of Ya-chou, woulld be good enough to send me or the editor of this Journal. A photograph of the species would be very welcome.

Note.—In the above interesting contribution from Dr. Laufer, who is well known for his valuable researches in the field of Chinese Archeolgy, he appears to assume that the Chinese writers referred to are alluding to a fish. In our opinion they are all referring to the giant salamander

(*Megalobatrachus*). This creature answers to the descriptions of the *na yü* in every way except in regard to its " wailing like a child." The Chinese writers mostly say that the *na yü* is like the sheat fish, or *nien yü* (*Parasilurus asotus*, also known as wels, and a member of the catfish family, *Siluridæ*), but that it has legs, can climb trees and wails like a child. One of the colloquial names is given as *wa wa yü* (or little child fish). Except for the last characteristic, for which we cannot vouch, this description suits the giant salamander. It is extraordinarily like the sheat fish in general shape, having the same flat and broad head, the same wide mouth, the same shaped body and tail. It is also covered with slime or mucous, which gives the sheat fish its Chinese name, *nien* (=fish-slime), and, of course, has legs. It is able to climb up fairly steep surfaces, providing they are rough enough, *e.g.*, the bank of a tree. The colloquial name of the giant salamander is *wa wa yü* ; but we have been unable to secure any classical name for it. The Cantonese name for the giant salamander is *hai-ko yü* (海狗魚), which may be interpreted either as " sea dog fish," or " seal fish," one of the Chinese names for the seal being *hai-ko*, or sea-dog. The habitats and distribution of the *na yü* as given by the Chinese writers agrees with those of the giant salamander, which is an inhabitant of mountain streams in Szechwan, Hupeh, South Shensi, Kweichow, Hunan, Canton and Kwangsi, possibly also of mountainous areas of the Lower Yangtze Valley and South-east China. Finally as far as is known, there is no member of the *Anabantidæ*, or climbing-perches, in the regions given as the habitat of the *na yü*, nor do we know of any climbing fish in these parts. The *Anabantidæ* are confined to India, the Malay Archipelago, the Philippine Islands and South Africa. Round the coast and in the tidal creeks of China, from the mouth of the Yangtze southward, the peculiar little fish known as the mud-slipper (*Periophthalmus*) occurs. It is able to climb the sloping roots, trunks and lower branches of trees that reach into the water, by the use of its pectoral fins which are distinctly leg-like, and by a vigorous use of its tail ; but we are not aware that any member of this genus occurs in the areas cited for the *na yü*. It is conceivable, however, that some confusion has arisen in the minds of the Chinese writers between this littoral species of fish and the fresh-water, highland-inhabiting giant salamander.—Ed.

Since writing the above we are able to report from personal observation that the giant salamander does actually make a noise like the first cry of a new-born babe when handled at all roughly or held out of the water any length of time. This seems to place the identity of the *na* as the giant salamander beyond question. There is no similarity between the sheat fish and the climbing perch. The accompanying illustrations of a giant salamander and a sheat fish show the marked similarity of the two creatures very well.—Ed.

Owing to lack of space in this issue, the second half of Mr. Fu-liang Chang's article entitled " Some Chinese Trees and Tree Products," which began in our November (1924) issue, has had to be held over till the next issue.

172

中国古代青铜器

$1.00 A COPY $6.00 A YEAR

ART IN AMERICA
AND ELSEWHERE

AN ILLUSTRATED BI-MONTHLY MAGAZINE

VOLUME XIII · NUMBER VI
OCTOBER, 1925

EDITED BY
DR. W. R. VALENTINER
AND
FREDERIC FAIRCHILD SHERMAN

PUBLISHED AT
105 MIDDLE STREET, PORTLAND, MAINE, AND
28, EAST EIGHTY-FIFTH STREET, NEW YORK CITY

LONDON: MESSRS. BROMHEAD, CUTTS & CO., LTD.
18 CORK STREET, BURLINGTON GARDENS

PRINTED IN THE UNITED STATES OF AMERICA

ENTERED AS SECOND-CLASS MATTER OCTOBER 4, 1922, AT THE POST OFFICE AT PORTLAND MAINE, UNDER THE ACT OF MARCH 3, 1879

ART IN AMERICA *AND ELSEWHERE*
AN ILLUSTRATED BI-MONTHLY MAGAZINE
VOLUME XIII · NUMBER VI . OCTOBER 1925

ARCHAIC BRONZES OF CHINA

CASTING in bronze may justly be regarded as the oldest of national arts of ancient China. It is that province of art in which the national soul is most typically and felicitously crystalized. Art — I say advisedly, not artcraft: the archaic bronzes virtually belong to the realm of art, and their makers were full-fledged artists, not artisans. Only the epigones of the T'ang, Sung, and more recent periods, degraded the art of bronze into the level of an industrial process; theirs was the technique, not the spirit. It is the spirit which makes art and imbues it with life, and it is religious fervor which spurred the early artists to supreme efforts and which created the admirable casts of the metal founders of the Shang dynasty (1783-1123 B. C.), almost at the threshold of civilization. This was a spontaneously creative epoch of forms, types, designs, symbols, and expressions of religious sentiments. True it is these humble metal founders were not conscious of being artists, nor did they stamp their names on their products.

Like the nations of western Asia and the prehistoric peoples of Europe the Chinese of the third and second millenniums B. C. passed through a bronze age of long duration, while iron but gradually came into use from about 500 B. C. Implements were cast in copper or clay moulds, but the process of casting as far as the large vessels are concerned was that *à cire perdue,* moulding the surface in wax. It is amazing that vessels, and many of great dimensions and complexity, were anciently produced in a single cast, inclusive of the bottom and handle or handles. The bronze experts of China are inclined to look upon this point as a characteristic feature of an archaic bronze and in their examination first inspect the bottom of a vessel; if it turns out that the latter is cast separately and soldered in the piece in question forfeits its claim to ranking in the San Tai (the three dynasties Hia, Shang, and Chou, as the archaic period is styled). In most of the Sung and later bronze vases and jars, bottom and even handles are moulded separately. A strikingly large variety of metal alloys was utilized, different alloys being employed for different classes of objects. Bells and mirrors, e.g., had specific formulas. We have several books of ancient Rituals which determine exactly the shape, alloys, measurements, capacity, weight, and ornaments for each type of bronze vessel, and their forms were defined according to the nature of the offerings, which were wine, water, meat, grain, or fruit, and according to the character of the deity to whom the vessel was dedicated.

In opposition to the spontaneous productions of the Shang period the art of the Chou (1122-247 B. C.) is ritualistic, impersonal, sacrosanct, and hierarchic in character, to some extent it is even lofty, sublime and transcendental. There is no trace of realism, but this subconscious art is formed of strictly national elements untouched by outside currents, and is refreshing in its groping for naive expression of ideas. The human figure, with a few exceptions, is almost absent. Plant designs do not appear in decorative art. All principal designs are of geometric style and receive a symbolic interpretation evolved from the minds of agriculturists. The ancient Chinese were a nation of farmers, and farmers have always formed the bulwark of Chinese society. Being keenly interested in weather and wind and all natural phenomena exerting an influence on fields and crops, their attention turned toward the observation of the sky and the stars, and this occupation resulted at an early date in a notable advance in the knowledge of astronomy. Hence we encounter interpretations of ornamental forms such as thunder and lightning, clouds, winds, and mountains. Animals are always strongly conventionalized and among them we meet the tiger, the elephant, the rhinoceros, the tapir, the domesticated sheep and ox, fantastic birds, and a variety of reptiles. Of insects we find represented with predilection the cicada, whose wonderful life-history excited admiration, and who developed into an emblem of resurrection. Above all, numbers play a prominent role in the cosmogony of the Chou period; everything in the old rituals was reduced to a fixed pattern or standard of numbers and categories reflected in celestial phenomena. Geometrical calculation resulted in the construction of images of the principal cosmogonic deities and emblems of rank. By studying carefully the form and designs of a Chou bronze and counting its characteristic features or the number of designs it is possible in some cases to solve its mystery as though it were a cross-word puzzle.

The majority of ancient bronze vessels were not found in graves, but were accidentally discovered embedded in the ground and even in rivers. Other bronzes were handed down as heirlooms in families from father to son, or were preserved in temples, libraries, and private museums. Many bronzes are covered with lengthy inscriptions of archaic style made in the cast. These inscriptions frequently give us a clew as to the purpose for which the vessels served, or the events which prompted their production. It was a common occurrence that the emperor bestowed valuable bronzes on his vassal kings and princes or on deserving ministers of state. Many men had bronzes cast to mark or commemorate an important event in their career, and in this case dedicated them to the memory of their parents, as the Chinese invariably attribute to their ancestors whatever good luck may fall to their lot. Thus, e.g., we read in a lengthy inscription: "On a certain day the emperor Mu of the Chou dynasty dwelt in the ancestral temple, and accompanied by his chief minister, ordered the annalist to issue a diploma in favor of Mr. Sung who was to be promoted to a high office. A black silken robe, a girdle with a buckle, jade ornaments, a standard and bridles adorned with tiny bells were conferred upon him by his majesty. Mr. Sung prostrated himself before the Son of Heaven, expressing his profound gratitude and extolling the imperial benevolence and glory. In order to celebrate this occasion he ordered this precious bronze vessel to be cast in memory of his venerable deceased father and his venerable deceased mother, animated as he was by the desire to cultivate filial piety and to solicit their constant and powerful protection." As demonstrated by this inscription, a bronze vessel of this class served no practical purpose, but remained a family treasure. The characteristic point is that Mr. Sung, on the memorable day of his promotion, turns to his dead parents and ascribes his success to their good influence; even in this case the casting of a bronze was a religious act inspired by deep religious sentiments.

A three-footed bronze goblet of the Shang period used in pouring out libations of wine in the worship of Heaven, the supreme deity (Fig 1), is now in the Freer Art Gallery, Washington. This type has been explained as being derived from an inverted helmet to which three feet are added. With a stretch of imagination we might be disposed to argue that the hero of ancient days, when celebrating a victory, doffed his helmet on the battlefield, offering in it a potation to the gods, and that subsequently the helmet was chosen as the model for a libation-cup. On second thought, however, this explanation is hardly convincing; the Chinese never were so warlike that a military headgear would have commended itself as an emblem worthy of being introduced into the ritualistic cult, nor is the alleged coincidence perfect. Another interpretation seems more plausible. This type of vessel is styled *tsio,* and this word is a general term for small birds. I am inclined to think that the form of this vessel has grown out of the figure of a bird resting on its nest. This theory is confirmed by the fact that there are specimens provided with a cover terminating in a bird's or animal's head. In all probability they were all provided with covers, but most of these are lost. Animalized forms in vessels are typical of ancient Chinese art. The bird, I imagine, was a messenger who carried man's prayers to the god of Heaven. The three feet indicate plainly that the vessel was put over a fire and it is obvious that the wine made of millet or rice was heated in the vessel itself. As is wellknown wine is always taken hot in China. The part forming the bird's head is chamfered into a spout. The two spikes surmounted by knobs (explained as "posts, supports") and set on the edges were probably made for the purpose of lifting the hot cup from the charcoal fire. There is also a symbolic interpretation of these spikes; they are compared with the stalks of cereals — evidently in allusion to millet or other grain from which the sacrificial wine was prepared.

During the Chou period, when the Son of Heaven performed in the spring the ceremony of ploughing the fields, he was assisted by all the great ministers of state, all princes present at court, and the grand prefects. The Son of Heaven himself ploughed three furrows; the great ministers, five; the other ministers and the princes, nine. At their return to the palace the Son of Heaven assembled his companions in his

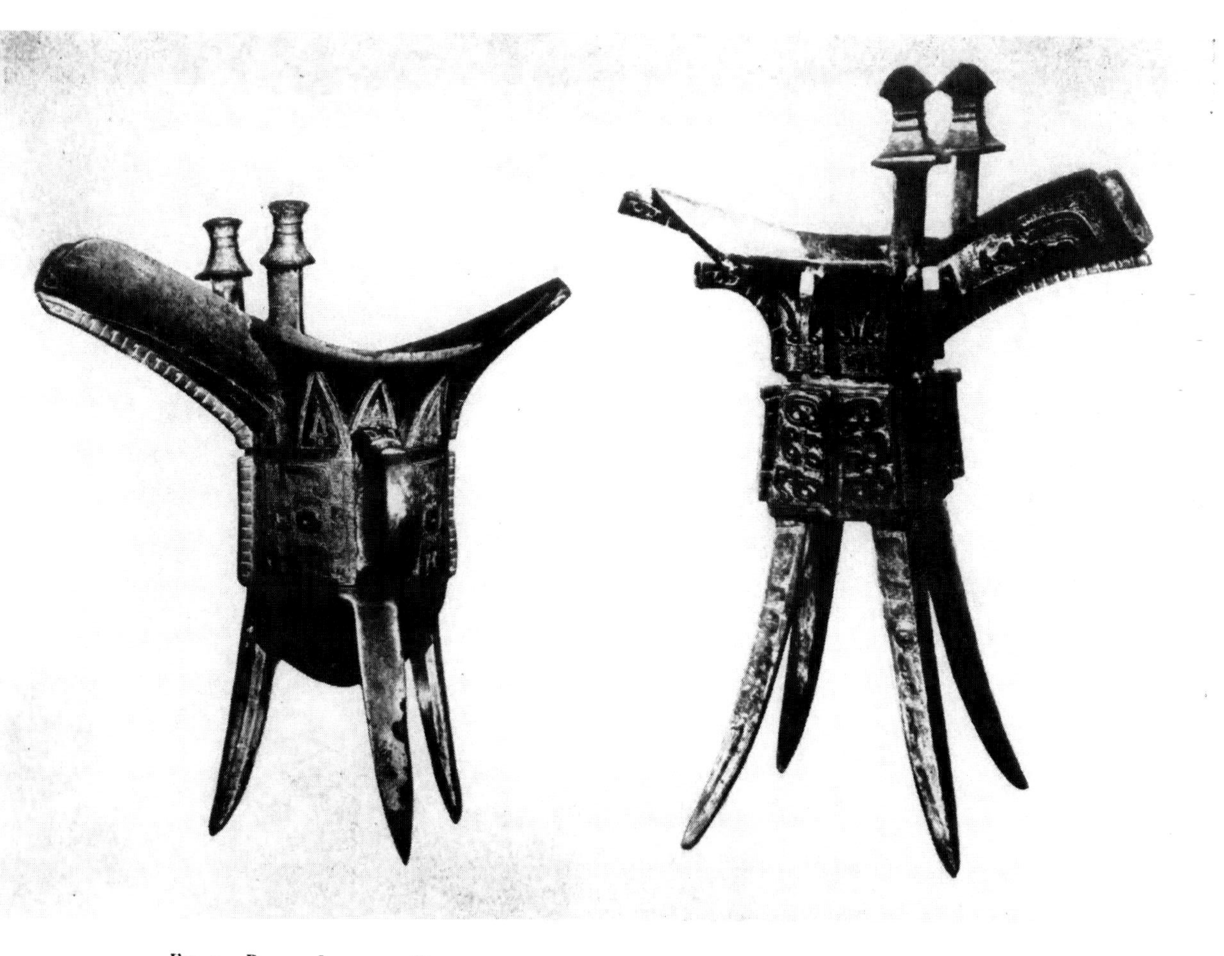

Fig. 1. Bronze Libation-Cup
Shang Period (1783-1123 B. C.)
The Freer Gallery, Washington, D. C.

Fig. 2. Square Bronze Goblet
Shang Period (1783-1123 B. C.)
Collection of Mr. Edsel Ford, Detroit, Mich.

Fig. 3. Bronze Beaker
Shang Period (1783-1123 B. C.)
Isabella Stewart Gardner Museum, Boston

chief apartment and raising this goblet addressed them thus: "I offer you this wine in compensation for your trouble." The service of this cup was also required for the ceremonies held in the ancestral temple when the master of the house offered wine from it to the representative of the dead ancestor. It contained but one pint *(sheng)*, but was regarded as more honorable and dignified than larger vessels holding three or even four and five pints. Such goblets were also carved from jade.

Under the Chou they were regarded as valuable presents exchanged by the vassal kings. Under the T'ang they were still used by the emperors in the solemn ceremonies addressed to the dieties Heaven and Earth on the summit of the sacred Mount T'ai in Shan-tung. Under the Ming it was a favorite type, but degraded into profane purposes; during the marital ceremony bride and groom alternately drank wine from a cup of this shape for the following reason: only the emperor had the privilege of worshipping Heaven; all others were allowed to invoke Heaven but once in their lifetime, at their wedding ceremony, and solely on this occasion could use this type of goblet which otherwise was an exclusive imperial prerogative. In the age of the Manchu it was frequently imitated in plain and decorated porcelain, also in silver, either for ornamental purposes or for the nuptial ceremony.

The example shown herewith is a superb specimen which excels in all essential characteristics associated with the *tsio* of the Shang dynasty. It is well balanced in its proportions, and in its bold outlines it stands with the convincing force of a masterpiece. The three feet rise in elegant curves. The body is divided into four sections formed by three projecting and denticulated ridges and the single loop-handle that springs from a conventionalized zoömorphic head. Both symmetry and a studied asymmetry, simultaneously applied, has always been one of the great principles underlying Chinese art. The loop-handle unexpectedly breaks the symmetry of arrangement, adding a pleasing effect to the whole work. The designs are chased with wonderful clarity, being compositions of plain and convolute spirals, the projecting eyes in the centre hinting at a watchful or all-seeing deity. To the artist of that archaic period the production of a sacred vase was a religious duty, and his creation was a reality imbued with the power of life and vision. The triangular patterns in the upper panel are interpreted as mountains.

The vessel illustrated in Fig. 2 is unique. It is known as "phoenix goblet." It is now in the collection of Mr. Edsel B. Ford of Detroit. In its structure, it is a *tsio* "made square," the squaring-up process affecting not only the form of the vessel but even extending to the spiral designs. In conformity with the square shape we are confronted here with four spear-shaped legs. Likewise each leg has four sides while in the preceding type it is trilateral. The two outer sides of each leg are ornamented with a conventionalized human or animal head dissolved into geometrical combinations of angular spirals: the eyes are plainly indicated by small strokes in quadrangular enclosures, and the nose is forcibly brought out. Eight tooth-shaped ridges dissect the four surfaces of the vessel into eight panels; each side is divided by a groove into an upper and a lower section. The decorative elements, according to Chinese conception, represent reclining or sleeping silkworm cocoons; and these designs, again, are so combined on each side as to form a face which may be interpreted as that of the Silk Goddess. These designs, in undercut relief, are set off from a background of spirals delicately traced and symbolic of thunder and lightning. On the exterior of the spout we encounter the motive of the "phoenix *(fung)* dancing in the clouds," the clouds being expressed by spirals. The term "phoenix," it should be understood, is merely a convenient word used by us, but, as a matter of fact, bears no relation to the phoenix of the occident. This square bronze goblet was dedicated to the deity Earth and served for libations of wine in honor of this deity. Earth was conceived by the ancient Chinese to be square and female, and four was its sacred number, while one and three were the sacred numbers of Heaven. In fact, the lower square section of the bowl rising above the four legs bears a most striking resemblance to the jade image under which Earth was worshipped. Silk was looked upon as one of the precious gifts of Mother Earth, the first discovery of silk, the rearing of silkworms, as well as the spinning of silk, are ascribed by tradition to a woman's initiative. The empress took a profound interest in the welfare and promotion of the silk industry. In the imperial worship performed by the Manchu dynasty silk was offered in the sacrifice to Earth and was buried in the ground. Finally, the "phoenix dancing in the clouds" is an emblem of love and veneration and symbolizes the empress. It occurs on the ancient jade girdle-ornaments worn by women and buried with them in the grave as an emblem of resurrection. All these facts combined prompt us to the conviction that this vessel had an extraordinary place assigned to it and enjoyed a specific function in the rituals performed by the empress in her homage to Earth and Silk. This phoenix goblet, as it is called, is unique and, as far as I know, the only one of its class in existence. Even the emperor K'ien-lung in all his glory did not possess a piece like this in the museum of his palace. The beauty of this bronze is enhanced by a rich patina of the brown of autumn leaves, interspersed with specks of malachite blue-green.

The bronze beaker (type *ku),* illustrated in Fig. 3, is of the Shang period, being ten and a half inches in height. It is at present in the collection of Mrs. Jack Gardner, Boston. It is recorded to have been discovered in an ancient well at Wu-ch'ang on the Yang-tse, capital of Hupeh Province. It is equally beautiful for its well-balanced proportions, its noble simplicity, purity of form and design, and the exquisite quality of the patina. This type was first produced under the Shang, and was subsequently adopted by their successors, the Chou. Judging from a famous passage in the Confucian Analects *(Lun yü,* VI, 23) it appears that this vessel underwent some changes in the age of the great sage, but, nevertheless, retained its old name. Confucius denounced the government of his time, which indulged in high-sounding phrases without applying the wise principles of the ancients, and illustrates the folly of using words that do not express the reality underlying them by an allusion to the vessel *ku,* which literally means a "corner." Confucius maintained that the term *ku* referred essentially to a vessel with corners, while the vessel thus named and made in his time had none. By these "corners" we have to understand the four slightly projecting, dentated ribs around the stem and foot, as they appear in this example and as they were regularly made under the Shang. At the time of Confucius the form of the vessel had apparently undergone a change, while its ancient name was retained.

The spiral composition is chased with unequalled vigor and firmness, and the asymmetry in the arrangement of the designs is a noteworthy feature. The two raised knobs in the middle portion and on the foot are intended for eyes and hint at the fact that the artist meant to bring out the head of some mythical creature in the seemingly arbitrary combination of these scroll designs. As the spirals symbolize clouds, and the peculiar lanceolate designs, in combinations of four or six, are explained as representing the winds, we shall not err in regarding this head as that of a Storm God moving over the clouded sky.

Some of these *ku* are entirely bare of ornamentation, others are decorated from top to bottom; others, again, like the specimen here illustrated, are ornamented in the middle and lower portions, a few, also, in the middle portion only. All, however, are built in three sections, plainly set off by grooved zones, and have the same slender, graceful body and flaring trumpet-shaped opening. Our example embodies all characteristics of the Shang period as evidenced by comparison with other known specimens in the collection of the late Viceroy Tuan Fang. The entire vase which served as a wine-vessel is coated with an exceedingly beautiful, lustrous, deep olive-green patina. No other nation can boast of having conceived a vase that could rival this type in grace and beauty of form and sense of pleasing proportions.

The square bronze vase (frontispiece) is now in the Freer Art Gallery of Washington. It is called a phoenix sacrificial vessel *(fung tsun),* being thirteen and three-quarters inches high, and presents a relic of the Chou dynasty (1122-247 B. C.). This majestic piece is constructed in three sections clearly set off from one another, although the whole piece is cast in one mould. As in the case of the *ku* (Fig. 2) the corners are provided with projecting ribs, and each of the four sides is divided into two panels by a similar rib running through the centre. The composition of each zone, however, presents a unit, the same subject being repeated on each of the four sides. The upper panel is occupied by eight triangular fillets which are intended to symbolize mountains; for this reason they always have their place on the neck of a vase, the point or summit reaching its edge. Being suggestive of a towering mountain scenery they lend the vase a feeling of loftiness and sublimity and readily appeal to our imagination. As the triangular bands are filled in by cloud and thunder patterns we have a symbolic representation of mountains overcast with clouds ready to pour down fertilizing rain on the fields. Such was the wish of the farmer, and in this simple, impressionistic manner he conveyed his thoughts. In the lower segment of the upper zone we note in strong relief a pair of conventionalized animals facing each other, their bodies being formed of spiral designs, their eyes being indicated by ovals. In the rectangles forming the base is brought out a pair of similar or identical creatures. The two birds confronting each other in the middle zone exhibit a certain tendency to realism, especially in the bold outlines of their tail-feathers, while circles, half-circles, spirals, and curves are resorted to in order to make up the composition. This bird is possibly intended for the fabulous *fung* (so-called phoenix) for which this vase is named.

It is finely incrusted with a deep greenish-brown olive-like patina on three sides, the fourth exhibiting a light green tinge.

Bells occupy a prominent place in Chinese antiquity, and belong to the most admirable achievements which the Chou artists have created in bronze. Elaborate rules for the making of bells are formulated in the *Chou li,* the old State Handbook of the Chou dynasty, which with minute detail sets forth the court ceremonial, the function of the officers and regulations for their guidance, as well as the productions of the imperial workshops. Bells were invented in China independently of the occident; the ancient Chinese bell is a type of its own, and also differs considerably from the globular and spherical bell subsequently introduced with Buddhism from India. The independence of the Chinese type is demonstrated by its peculiar flat form and the absence of a clapper, the instrument being struck outside by means of a wooden mallet. It was chiefly used in the ancestral hall to summon the spirits of the departed in order to partake of offerings of meat and wine. A bell was also suspended in front of the banqueting hall and was sounded as a call to the guests. It likewise had an orchestral function in accompaniment with other musical instruments; and music, as in Plato's republic, formed an integral part of Chinese education and ceremonial. Music, archery, knowledge of rites and good manners were the essential points of good breeding. Most of the early bells have the two coats set with bosses, arranged, according to a fixed scheme, in groups of three, distributed over three rows, three times three being enclosed in a rectangle, so that eighteen appear on each face, making a total of thirty-six; there are many bells, however, without any bosses, and a few have twenty-four of them. Much speculation has been rife among Chinese and other archaeologists as to the function of these bosses. Wang Fu, author of a catalogue of bronzes in the museum of the Sung emperors, has compared them with nipples, which he takes as an emblem of nutrition, arguing that nipples are represented on bells because "the sound of music means nutrition to the ear." The simile with nipples, however, does not occur in any ancient text, above all, is absent in the *Chou li,* which speaks merely of knobs. It can hardly be imagined that these bosses — of which, by the way, there is a large variety of different shapes, many of these showing no resemblance whatever to nipples — should have served a purely ornamental or esthetic purpose. They were doubtless made with a practical end in view, and, as supposed by some Chinese authors on music, for regulating and harmonizing the sounds of bells, while later generations forgot this practical purpose and merely applied the bosses ornamentally. The bell now in Miss Buckingham's collection in the Art Institute of Chicago is remarkable for its imposing simplicity and grandeur of conception. It belongs to the Chou period, being sixteen and one-quarter inches in height, and is a truly classical example of Chou art that inspires a feeling of reverence such as we may receive from the lofty arches of an old Gothic cathedral. The *Po ku t'u lu* (chapter 23, p. 14) illustrates a Chou bell very similar to our example except that it is adorned with eight dragons (or perhaps lizards) instead of four, two being added on the right and left sides. The thirty-six nipple-shaped bosses (eighteen on each face) are perfectly modeled, and the five vertical lines of the central zone, as well as the raised meander bands, are delineated with unsurpassed precision and firmness. The entire bell is coated with a beautiful blue-green patina speckled with gold and brown, which was produced by chemical action underground.

A rectangular bronze vessel of the Chou period (Fig. 4), now in Miss Buckingham's collection in the Art Institute, Chicago, is perfectly unique, and none like it is traceable in any Chinese catalogue of bronzes. It is composed of two equal parts, completely symmetric, each in the shape of a rectangle, posed on a hollow base with sides slanting outward. Each single part could form a vessel in itself, and such a single vessel was frequently used in ancient times for holding millet in sacrifices, being known under the name *fu*. In the origin this vessel was a basket, defined by the ancient dictionaries as "square outside and round inside, used to hold boiled millet in State worship." To every student of basketry, baskets which consist of two equal halves perfectly fitting one over the other (for instance, globular baskets composed of two hemispherical pieces) are wellknown, and such baskets are still made in China. The supposition seems to be well justified that the caster of this bronze derived his inspiration from such a double basket; hence the name *shuang fu* has been proposed for this novel type. The *fu* were also carved from wood or moulded from clay. A few specimens of this type in Han pottery have survived; but the favorite material for it was bronze. In the collection of the emperor K'ien-lung there were sixteen bronze *fu*, figured in the *Si ts'ing ku kien* (chapter 29), but he had no double *fu* like this one. Conventionalized animal-heads are cast in prominent relief on the narrow sides of the upper and lower portions, and small zoömorphic faces (two on each long side, and one on each narrow side) are so fitted to hold the two parts closely together. The slanting sides of the upper and lower bases have gracefully cut-out arched openings, making four feet in the corners. The long, massive bands of meander patterns laid around the body in an elaborate composition are very delicately traced. The patina which covers the entire object on the exterior and interior is very extraordinary in its delightful shades of light blue and green.

Miss Buckingham recently acquired an exquisite bronze vase of the early Han period (206-22 B. C.), exhumed from a grave in the perfecture of Chang-te in Ho-nan Province and sixteen and a half inches in height. It has a large globular body adorned with a pair of heavy movable rings, corresponding in type to the wellknown Han pottery vases which served for burial purposes. Traceable to the culture of the Chou dynasty this type was subsequently adopted by the Han and developed into one of the most popular vases of that period. The Field Museum of Chicago has also several such vases of cast iron. The present example is distinguished by two remarkable features: it is invested with a heavy coating of gold foil, being the only gilt vase of this class has ever come under my notice; it is, further, adorned with an inscription which reads, "Eastern Palace, number seven." This demonstrates that this vase made for imperial use, for the decoration of a palace chamber, and formed one of a series. The surface is partially covered with thick green patina which in combination with the lustre of the gold produces an extraordinary effect.

B. Laufer

CHICAGO, ILL.

TWO PORTRAITS BY BARTHEL BEHAM IN NEW YORK

I

EVERY historian of the Fine Arts coming from Europe is struck by the fact that the best period of German painting (that is to say the fifteenth and sixteenth centuries) is but poorly represented in American public collections. The Museum of Fine Arts in Boston, which boasts of a beautiful large Triptych by the Master of St. Severin, is an exception. In the Metropolitan Museum in New York we look in

Fig. 4. Rectangular Bronze Vessel
Chou Period (1122-247 B. C.)
The Chicago Art Institute, Chicago, Ill.

173

中国的象牙

IVORY IN CHINA

BY

BERTHOLD LAUFER

CURATOR OF ANTHROPOLOGY

ANTHROPOLOGY

LEAFLET 21

FIELD MUSEUM OF NATURAL HISTORY

CHICAGO

1925

The Anthropological Leaflets of Field Museum are designed to give brief, non-technical accounts of some of the more interesting beliefs, habits and customs of the races whose life is illustrated in the Museum's exhibits.

LIST OF ANTHROPOLOGY LEAFLETS ISSUED TO DATE

1.	The Chinese Gateway	$ .10
2.	The Philippine Forge Group	.10
3.	The Japanese Collections	.25
4.	New Guinea Masks	.25
5.	The Thunder Ceremony of the Pawnee	.25
6.	The Sacrifice to the Morning Star by the Skidi Pawnee	.10
7.	Purification of the Sacred Bundles, a Ceremony of the Pawnee	.10
8.	Annual Ceremony of the Pawnee Medicine Men .	.10
9.	The Use of Sago in New Guinea	.10
10.	Use of Human Skulls and Bones in Tibet . . .	.10
11.	The Japanese New Year's Festival, Games and Pastimes	.25
12.	Japanese Costume	.25
13.	Gods and Heroes of Japan	.25
14.	Japanese Temples and Houses	.25
15.	Use of Tobacco among North American Indians .	.25
16.	Use of Tobacco in Mexico and South America . .	.25
17.	Use of Tobacco in New Guinea	.10
18.	Tobacco and Its Use in Asia	.25
19.	Introduction of Tobacco into Europe	.25
20.	The Japanese Sword and Its Decoration	.25
21.	Ivory in China	.75

D. C. DAVIES
DIRECTOR

FIELD MUSEUM OF NATURAL HISTORY
CHICAGO, U. S. A.

FIELD MUSEUM OF NATURAL HISTORY
DEPARTMENT OF ANTHROPOLOGY
CHICAGO, 1925

LEAFLET NUMBER 21

Ivory in China

Ivory occupies a very prominent place in the art of the Far East, and Chinese carvers in ivory have always stood in the front rank of their craft. But those who have hitherto written on the subject have merely treated it as an art industry, extolling Chinese mastery of technique, skill in execution, and grace of design. Correct as this judgment may be, it is based on more or less modern productions which are distinguished for technical ingenuity rather than for artistic merits. The archæology of ivory and the older real works of art created in this substance have almost wholly been neglected. The object of the present study is to fill this gap, to set forth the importance of ivory in the early antiquity of China, to trace the sources of supply and the development of the ivory-trade, and to interpret the art of ivory in its relation to Chinese life and culture. This essay is divided into five chapters dealing with the elephant in China and the trade in elephant ivory, folk-lore of the mammoth and trade in mammoth ivory, trade in walrus and narwhal ivory, ivory substitutes, and objects made of ivory. It is occasioned by a collection of ivory carvings made by me in China in 1923 (Captain Marshall Field Expedition) and recently placed on exhibition, and may serve as a guide to this collection.

THE ELEPHANT IN CHINA AND TRADE IN ELEPHANT IVORY

The fact that the elephant was known to the ancient Chinese may come as a surprise to many readers. The former existence of the animal on Chinese soil is well authenticated by linguistic, pictographic, historical, and archæological evidence. Not only have the Chinese an old, indigenous word for the pachyderm, but they also possess this word in common with the eastern branch of the family of peoples to which they belong and the languages of which are closely related. The ancient Chinese designation of the elephant was *dziang* or *ziang;* in the modern dialects of the north it is *siang,* in Shanghai *ziang,* in Canton *tsöng,* in Hakka *siong,* in Fu-kien *ch'iong.* In Burmese we correspondingly have *ch'ang,* in Siamese *chang,* in Shan *tsan* or *sang,* in Ahom *tyang,* in Mo-so *tso* or *tson,* in Angami Naga (Assam) *tsu.* This fact of language warrants the conclusion that all these tribes must have been acquainted with the animal from ancient times and even in a prehistoric period when they still formed a homogeneous stock. The Tibetans, akin to the Chinese in language, are outside of the pale of this development and designate the elephant as the "great bull" or the "bull of Nepal" (in the same manner as the Romans when they first saw elephants in the war with Pyrrhus spoke of "Lucanian oxen"), thus indicating that they made its acquaintance only in late historical times on coming in contact with India and Nepal (seventh and eighth centuries A.D.).

The written symbol for the elephant was conceived in ancient China in that early epoch when writing was still in the purely pictographic stage. The primeval pictogram denoting the elephant unmistakably represents it with its principal characteristics,— the trunk, a large head with two protruding tusks, and body with four feet and tail (Figs. 1, 3-6). In Fig. 1

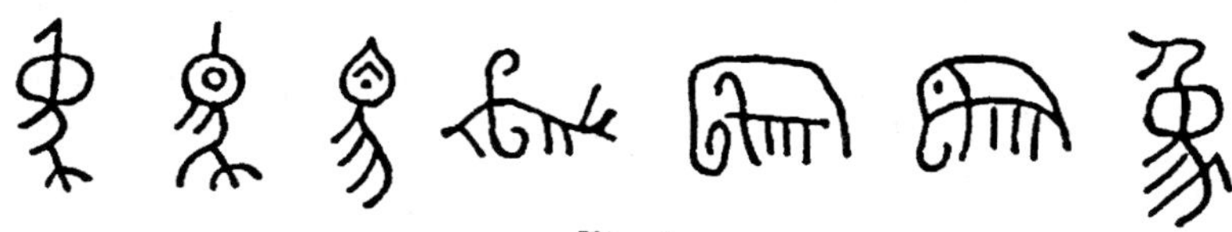

Fig. 1
Archaic Forms of the Written Symbol for the Elephant.

Fig. 2
Elephant from a Bell of the Shang Period, about 1500 B.C.

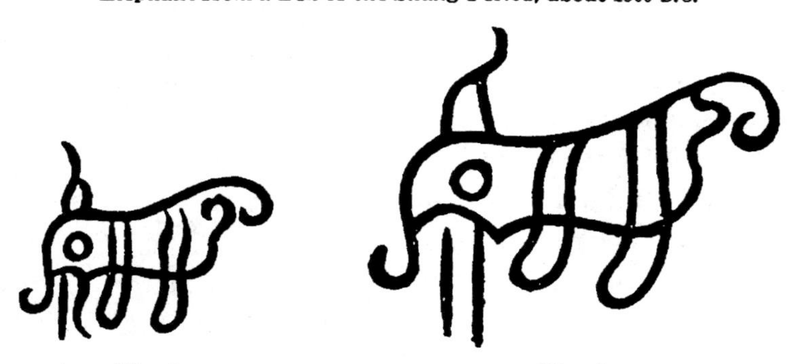

Fig. 3 Fig. 4
Symbols of the Elephant from Inscriptions on Bronzes of the Chou Period.

Fig. 5
Elephant from a Bronze Beaker of the Chou Period.

Fig. 6
Elephant from a Bronze of the Shang Period, applied to a Seal of Later Date.

Fig. 7
Elephant Head in the Pictographic Writing of the Mo-so.

seven different old forms of the character are reproduced; these finally led to the stage in No. 7 in about A.D. 100, which approaches very closely the modern form (Fig. 13 on p. 21).

In the inscriptions cast on the archaic bronze vessels of the Shang (1783-1123 B.C.) and Chou dynasties (1122-247 B.C.) the symbol of the elephant is not infrequently represented. Fig. 3 is reproduced from the *Po ku t'u lu* (chap. 2, p. 24), the catalogue of bronzes in the possession of the Sung emperors, published by Wang Fu in A.D. 1107; it occurs on a bronze tripod vessel ascribed to the Chou period. Fig. 4, of the same type, is from a vessel of the same period in a Japanese collection. Fig. 5 is taken from a bronze beaker in the Imperial Museum of Peking. Fig. 6 represents an elephant figure applied to a seal and said to go back to the Shang period. Fig. 7 is the sign for the elephant in the pictographic writing of the Mo-so, an aboriginal tribe in Yün-nan Province.

The most remarkable representation of the elephant in the Shang period (1783-1123 B.C.) occurs in a bronze bell discovered in Shan-tung Province and inscribed with the name of an emperor who reigned 1506-1491 B.C. The rim of this bell is decorated with a row of elephants of naturalistic style (Fig. 2; cf. L. C. Hopkins, Development of Chinese Writing, 1909, p. 15). Under the Chou we usually meet the hieratic, strongly conventional forms, but also very artistic applications of elephant designs to the decoration of bronze vessels (Figs. 8-9).

In the ancient Rituals (*Li ki* and *I li*) are mentioned two types of ceremonial vessels designated as "elephant goblets." The Chinese commentators have exerted their ingenuity in explaining what these vessels are. One says that they were adorned with ivory; another holds that the entire vessel was made in the shape of an elephant; another interprets that it was decorated with the picture of an elephant; others again take the word *siang* in the sense of "form, image, picture" and conclude that the goblet was adorned with the design of a phœnix. Considering the archæological facts, i.e., the bronze vessels which have come

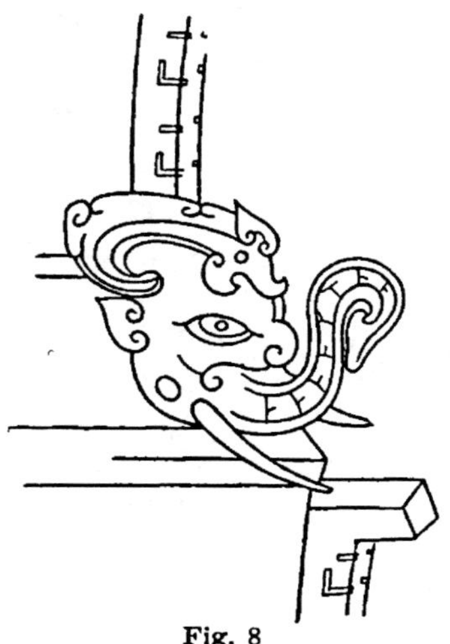

Fig. 8
Elephant Head Projecting from the Side of a Bronze Beaker.

down to us from the archaic period of the Shang and Chou dynasties (1783-247 B.C.), we find a goodly number of these provided with feet shaped into elephant's heads terminating in a trunk, the latter forming the foot of the vessel. This motive is particularly con-

Fig. 9
Elephant Heads as Decorations on a Bronze Vessel.

spicuous in the tripod colanders (called *hien*) which represent the combination of a stove with a cooking-vessel used for steaming grain and herbs in ancestral worship; a charcoal fire was built in the hollow tripod base which is separated by a hinged grate from the upper receptacle holding the articles to be steamed. A good example of a vessel of this type is on exhibition in the centre of Case 1, Blackstone Chinese Collection. The artistic motive of the elephant-foot in bronzes has persisted in Chinese art throughout the centuries down to the K'ien-lung period (1736-95); it is likewise visible in the Han mortuary pottery (206 B.C.—A.D. 220).

In the monuments of the Han period there are highly naturalistic representations of the elephant in scenes carved on tomb sculptures. One of these illustrated in Fig. 10 is depicted on one of the eight stone

Fig. 10
Elephant on Bas-relief of the Han Period, First Century A.D.

slabs forming the remnants of a mortuary chamber and found on the hill Hiao-t'ang-shan northwest of the city Fei-ch'eng in western Shan-tung. The elephant mounted by three mahouts equipped with iron hooks is shown in a long procession of figures forming the retinue of a "barbarian" prince. It is certain that the elephant did not exist in Shan-tung at that time, but it is equally certain that the Han artist must have drawn the animal from life.

In the beginnings of history the Chinese were restricted to what is now northern China in the valley of the Yellow River, and physical and climatic conditions of the country then were to some extent different from what they are at present: the mountain-ranges were still crowned by dense forests haunted by great numbers of wild beasts among which were elephants. As the farmers (and the Chinese were a nation of farmers) gradually advanced and cleared the jungle, the elephant gradually retreated farther south, or was exterminated. This may have been accomplished by the beginning of the first millennium B.C., but the recollection of the animal survived in the minds of the people for many centuries later. By the middle of the first millennium B.C. the habitat of the elephant became restricted to the Yang-tse Valley, extending from far-west Se-ch'wan to the sea, and the regions still farther south and west, as will be demonstrated hereafter in detail.

An ancient saying, presumably of proverbial character (recorded in the *Tso chwan* under the year 548 B.C.) was to the effect that the elephant has tusks which lead to the destruction of its body, because of their use as gifts.

In the old Book of Songs, the earliest extant collection of Chinese poetry, an allusion is made to elephant-tusks brought as tribute by the wild tribes bordering the river Hwai, which flows through the provinces of An-hui and Ho-nan and empties its waters into the Hung-tse Lake.

Elephant-teeth and rhinoceros-hides were among the products sent as taxes by the two provinces Yang-chou and King-chou,—the former covering the territory south and north of the Yang-tse delta; the latter, the present area of the provinces of Hu-nan and Hu-pei.

In early antiquity elephant ivory was perfectly known and wrought into articles of every-day use like spikes or pins for scratching the head and tips for the ends of bows. Ivory ranked next to jade and gold.

The emperors of the Chou dynasty (1122-247 B.C.) had five kinds of chariots of state, three of which were covered with leather. In the first the ends of the principal parts were decorated with jade; in the second, with gold; in the third, with ivory; while the fourth was of plain leather; and the fifth, of wood. The ceremonial leather cap worn by the emperors was adorned with jade ornaments of various colors, and in the place where it fitted over the nape of the neck, had a foundation of ivory. Confucius is said to have possessed an ivory ring five inches wide.

Memoranda or writing-tablets (*hu*) used by the feudal princes and great prefects were made of ivory, while the emperor had the prerogative of using a polished jade slab for the same purpose. This example shows again that ivory ranked next to jade in value. The ivory tablet of the feudal princes was rounded at the top and straight at the bottom to symbolize that they should obey the Son of Heaven. The tablet of the great prefects was rounded both at the top and bottom to express the idea that they had only superiors to obey. Such tablets were carried suspended from the girdle, and were used as memoranda or for jotting down notes. An official, when he had an audience at court, inscribed his report on the tablet and recorded the emperor's reply or command. At a later time they were reserved for the organs of government and became emblems of dignity.

Chopsticks were originally made of bamboo or wood, but in the time of the Chou dynasty (1122-247 B.C.) were also carved from elephant ivory. According to an ancient tradition, the man who first conceived this innovation, was Chou, the last emperor of the precedig Yin or Shang dynasty, notorious for his debauchery. He was remonstrated for this extravagance by one of his relatives who said, "He makes chopsticks of ivory! Next he will doubtless make a cup of jade, finally he will think of the precious and extraordinary objects of distant countries, and will have them carted to his place. From that moment he will crave in ever increasing numbers chariots and horses, mansions and palaces, and there will be no way of keeping him off."

Fortunately we now have at our disposal also a few ivory carvings of the archaic epoch. The veteran Chinese archæologist, Lo Chen-yü, made a few of these known in one of his illustrated works in Chinese. They are of the same character as the four objects illustrated here in Plate I. That in Fig. 1 is in the collection of the Metropolitan Museum of Art, New York, the photograph being due to the courtesy of Mr. Bosch-Reitz, curator of the Oriental Department. It evidently is an implement used for untying knots; corresponding implements were made from jade (cf. "Jade," pp. 238-242). It is firmly and handsomely carved with a running animal, a conventionalized animal's head and an eye, all set off from a background formed by a composition of square and triangular spirals. This implement was worn suspended at the girdle, being regarded as a token of maturity; it belonged to the equipment of one growing into manhood and indicated his competency for the management of business, however intricate; it accordingly symbolized a man's ability to solve knotty problems. The objects shown in Figs. 2-4 were obtained by me at Peking in 1923. The plaque in Fig. 2, though bone-like in appearance, is decomposed and calcined ivory; it is deeply incised at both ends with a band of geometric designs. The ivory character of the object in Fig. 3 is unmistakable; it is carved on both sides with designs which are identical with those found in the contemporaneous bronze vessels. This observation also holds good of Fig. 1. Fig. 4 shows the hilt or top of a knife or dagger engraved on both sides with a double row of angular spiral designs. These examples of ancient ivory carving differ radically in style and technique from any later works in ivory. They certainly do not go to prove that the elephant existed in ancient times in northern China; for it might be argued that the ivory of which they are made was imported as well. The ancient Greeks wrought ivory long before they became acquainted with the elephant. Considering, however, all available evidence, the conclusion may be hazarded that the objects in question were made of indigenous ivory; at any rate, they are good witnesses in confirmation of the ancient records.

While the ancient Chinese were acquainted with the elephant and used its ivory for various purposes, it must be stated, on the other hand, that they do not seem to have taken a deeper interest in the animal. It played no role whatever in their mythology and gave no rise to religious conceptions. It may even be questioned whether the elephant was hunted by the Chinese themselves. We have several ancient descriptions of hunting-expeditions, but none of these alludes to the chase of the elephant. The passage in Mong-tse, that Chou Kung, who died in 1105 B.C., "drove far away the tiger, leopard, rhinoceros, and elephant to the great joy of all people" is the only one from which occasional elephant-hunting on the part of the Chinese might be inferred. It is more probable that the elephant was usually hunted by the aboriginal "barbarous" tribes, who sold the ivory to the Chinese or with it paid their taxes to the imperial government, and that much of the ivory obtained by the Chinese was "dead" ivory (of animals which died a natural death in the jungle). Han-Fei-tse, a philosopher, who lived in the fourth century B.C., observes that men but rarely see a live elephant, but usually encounter the bones of a dead one. Above all,—and this observation bears out the point in question,—the ancient

Chinese never made any effort to tame or train the elephant, as was done by the nations of Indo-China, Java, Ceylon, and India. It was only in 121 B.C. that the first tame elephant was sent to the court of the emperor Wu of the Han dynasty from Nan Yüe; that is, the country in the southeast, at that time inhabited by tribes of Annamese origin. The commentator of the official Annals of the Han Dynasty, Ying Shao, feels obliged, with reference to this passage, to define what a tame elephant is, "It is docile, can make obeisance and rise again, and quickly grasps man's intentions." This, accordingly, was something entirely novel to the Chinese. Subsequently such gifts of trained elephants from the south are mentioned frequently; they made salaams and would even dance, or draw a carriage. Plutarch (*Of Fortune*) writes, "What is bigger than an elephant? But it also has become man's plaything and a spectacle at public solemnities; and it learns to skip, dance, and kneel."

There is an old tradition that when the emperor Shun was buried at Ts'ang-wu, elephants trampled down the earth around his tumulus, so that the land looked like a ploughed field. Ts'ang-wu then was a territory abounding in elephants; it was situated in what is now the district of Ning-yüan in the prefecture of Yung-chou, Hu-nan Province. In ancient times it was part of the state of Ch'u, which was inhabited by a non-Chinese population, presumably a member of the widely diffused Tai stock. It was a warlike and aggressive nation settled in the area now occupied by the two provinces of Hu-pei and Hu-nan on both banks of the middle Yang-tse. In opposition to the Chinese, the inhabitants of Ch'u seem to have tamed the elephant to a certain extent, and elephants were kept at the court of their king. There is one instance on record to the effect that they were even used for purposes of war. In 506 B.C. the kingdom of Ch'u was invaded and temporarily overrun by the army of the king of Wu, a state on the lower Yang-tse. Defeated in the field, the prince of Ch'u, in order to detain his pursuers, launched against the enemy a flock of elephants with lighted torches tied to their tails. This isolated occurrence does not prove that in ancient central China elephants were really trained and customarily employed for war: the act of the defeated king was rather a counsel of despair resorted to as a last stratagem; had he actually possessed war-elephants, he would have turned them to more effectual use right at the opening of the engagement. The fact, however, remains that in early times the Yang-tse Valley swarmed with elephants, that they were hunted for the sake of their ivory and hides, and also that they were caught, partially tamed, and kept.

The elephant must have survived in the Yang-tse Valley at least until the end of the tenth century A.D. Isolated occurrences of elephants in the ancient territory of Ch'u are still on record during the middle ages: thus we are informed in the Annals of the Sung Dynasty that in A.D. 962 elephants were seen in the district of Hwang-pei (lat. 30°56′, in the prefecture of Han-yang, province of Hu-pei), and subsisted on the crops of the people; at the end of the following year they were captured in the district of Nan-yang (lat. 33°06′, in the province of Ho-nan), and their teeth and skins were sent as a gift to the throne. Again, in A.D. 964, elephants appeared in the same locality, Nan-yang, and were slain by foresters; teeth and skins were dealt with as in the preceding case. In the same year elephants were observed in the districts of Li-yang (lat. 29° 37′) and An-hiang (lat. 29° 22′), in the province of Hu-nan; others were noticed crossing the Yang-tse and entering the district of Hwa-jung (lat. 29° 30′, in the prefecture of Yo-chou, province of Hu-nan), and others even reached the northern part of the city of Li-yang. In A.D. 966 elephants arrived spontaneously at the capital.

In the western part of the empire, the present province of Se-ch'wan, formerly styled Shu, elephants are noted in early records, and survived there at least into the period of the two Han dynasties (206 B.C.-A.D. 220), during which they were sent as tribute by the native chieftains to the court of the emperor at Ch'ang-an, where they were kept in the imperial animal-park. The Han emperors were exceedingly fond of curious and exotic animals and plants, and organized a sort of natural history museum in their palaces.

The present province of Yün-nan was originally inhabited by a stock of peoples designated as Tai or Shan, the forbears of the Siamese. They formed a powerful kingdom which was destroyed by the Mongols in A.D. 1252. The Tai were a warlike and chivalrous nation, and had a highly organized army. Military service was compulsory, and every adult was a soldier. The capital, Ta-li, was the centre of the military industry, where harness and helmets were manufactured from elephant skins.

As early as the second century B.C. an unsuccessful Chinese mission, sent out for the exploration of the southwest, received a dim knowledge of an "elephant-riding nation" living farther to the south and west. This was the ancient Tai kingdom, where the elephant played an important part in the life of both rulers and people, in court pageantry, as a riding and draught animal, and as a beast of burden. The elephant was native to this region and plentiful. Fan Cho, who in A.D. 860 wrote an interesting account of the aboriginal tribes of Yün-nan (*Man shu*), says that elephants occurred there in large numbers, and were caught by men who kept many of the animals to draw their ploughs. The same is also reported by subsequent authors, for instance, by T'an Ts'ui in 1799. Liu Sün, who lived toward the end of the ninth century and who wrote an interesting work on the products of southern China (*Ling piao lu i*), observed in Yün-nan that every family kept elephants for carrying loads over long distances, exactly as oxen and horses were used in China.

The Chinese received their first knowledge of India when Chang K'ien, during his memorable mission to the western countries, sojourned a year in Bactria in 128 B.C., and was informed that the people of India rode on elephants to fight in battle. Subsequently the Chinese also learned the fact that war-elephants were employed in Persia and Camboja, the latter country being reported to have two hundred thousand of them. The introduction of fire-arms put an end to the use of elephants in war, and the Chinese themselves demonstrated the futility of this mode of military tactics. In A.D. 1388, while Mu Ying was governor of Yün-nan, he gained an overwhelming victory over the Burmese, his cannon and powerful crossbows proving too much for the mailed elephants; and in the following year Burma acknowledged the suzerainty of China. In Yün-nan the elephant survived longer than anywhere else in China, and it may still occur here and there in outlying jungles. The native tribes use bracelets and large ear-rings of ivory.

In the two southeastern provinces, Kwang-tung and Kwang-si, elephants have always been numerous and persisted through many centuries. The provinces of Kwang-si and Yün-nan are still given in the *Ko ku yao lun* (written by Ts'ao Chao in A.D. 1388) as producing ivory. The same work lists Tonking and the countries of the Southern and Western Barbarians (Siam, Burma, India) as sending ivory to China. The ivory of the Southern Barbarians is extolled as long and large; that of Kwang-si and Annam is described as small and short, and a kind yielding a red powder when cut by a saw was regarded as very excellent.

In the seventh century A.D. the animal was still plentiful in Tonking, as well as in the prefectures of Ch'ao-chou, Hui-chou and Lei-chou of Kwang-tung Province, and was captured by the natives who regarded its flesh, especially that of the trunk, as a great delicacy. The tusks of the Kwang-tung variety are described as small and red, very suitable for ivory tablets. Chinese writers, further, emphasize the fact that their elephants were all dark or black in color, while white elephants are ascribed to the distant lands of the Arabs, of Fu-lin (Syria), and India. A white elephant was sent to China from Gandhāra in A.D. 509. It was kept in a special building near the capital Lo-yang in Ho-nan Province. A white elephant was sent from Burma to Hui Tsung, emperor of the Sung dynasty, in A.D. 1105.

At Canton elephants were employed as late as the tenth century in putting criminals to death. P'eng Ch'eng, who lived in the first half of the eleventh century, writes in his *Mo k'o hwi si,* "In the district of Chang-p'u (lat. 24° 07′) in Chang-chou fu (Fu-kien), which is adjoined to Chao-yang fu (in northern Kwang-tung), there are still numerous elephants usually encountered in herds of ten, yet they are harmless. Solely the rogue elephant pursues men and tramples them down till their flesh and bones are reduced to a pulp. Of all elephants, the rogue elephant is the most ferocious."

Chou Ta-kwan visited Camboja in A.D. 1295-97, and in his Memoirs on the Customs of Camboja writes, "The ivory from the tusk of an elephant killed by means of a pike is considered best; next in quality is the ivory of an animal which was found shortly after it died a natural death, while least esteemed is that discovered in mountains many years after the animal's death." This observation, which the Chinese learned from the Cambojans, is quite correct; and the Chinese have adopted this rule until the present day. Even in their materia medica the tusks of slain elephants are preferred to those who have died of a disease or otherwise. It has been noted that the "dead ivory" (taken from a dead animal some time after its natural death) is always dull, and when used, will be covered with brown spots of irregular size and very opaque.

The Chinese have also preserved much curious folk-lore in regard to the elephant. It was believed that the designs in elephant tusks were formed when the animal was frightened by a peal of thunder, while the patterns in the horn of the rhinoceros were supposed to be produced when the animal was gazing at the moon. This notion has reference to the "engine-turned" pattern (similar to that on the back of a watch-case) which ivory displays in cross section. It is probably due to this peculiarity of internal structure that it possesses the high degree of elasticity which forms one of its most valuable properties.

Pliny writes that the elephants, when their tusks have fallen out either accidentally or from old age, will bury them in the ground. The ancient Chinese told a similar yarn, according to which the animal would shed its tusks regularly and hide them in a hole dug by itself for this purpose; in order to take them away, it was necessary to leave a pair of wooden teeth in their place, so that the animal would not notice the theft. In regard to the rhinoceros it was also believed that it annually sheds its horn and that a wooden horn must be deposited in lieu of the real one when picked up. These notions were naturally prompted by the observation that detached tusks and horns were occasionally found in the wilderness, which suggested to the people a natural process akin to the shedding of cervine antlers.

During the middle ages ivory was imported into China, chiefly by the Arabs, from several states in the Malay Peninsula, from Java, Borneo, the eastern coast of Sumatra, southern India, and from the Somali Coast of eastern Africa. The Chinese of the twelfth century knew perfectly well that the African ivory was best of all, and speak of African tusks as reaching a weight of over a hundred pounds. Ivory then was a sort of government monopoly in China inasmuch as the merchants who desired to import it required an official license for tusks weighing thirty pounds or over. The tusks imported by the Arabs are described by a contemporary observer as being straight and of a clear, white color, with patterns displaying delicate lines. In weight they varied from fifty to a hundred pounds, whereas the tusks coming from Tonking and Camboja were small, weighing only from ten to twenty or thirty pounds, and had a reddish tint. The African ivory was designated as "great ivory" (*Ling wai tai ta*, written by Chou K'ü-fei in 1178). In the African species both sexes are furnished with tusks of large size, while in the Asiatic species they are generally restricted to the males, and even then are but poorly developed. Masudi, an Arabic geographer (A.D. 983), informs us that Arabic ships brought the ivory of the Zenj, as the Negroes were called by the Arabs, into the country of Oman on the east coast of Arabia, whence the traders transhipped it to India and China, adding in a tone of regret that ivory would be plentiful in the Musulman countries were it not directed to foreign ports. He further states that the tusks entirely straight or but little curved are held in high esteem by the Chinese and that these serve for the manufacture of palanquins for persons of high rank;

no important official would dare present himself in the palace of the king in a chair made of another material than ivory. Masudi writes also that the Negroes themselves made no use of ivory and did not understand, like the Hindu, to tame the elephant. It is noteworthy that the large straight tusks command the highest price in China even at the present time.

Kubilai (A.D. 1214-94), the great Mongol sovereign of China, was famed for the large number of elephants in his possession. The nucleus of his stock was formed by two hundred animals captured in a fierce battle of the Mongols against the Burmese in A.D. 1277. The king of Burma, as Marco Polo informs us, opposed the invaders with two thousand elephants, "on each of which was set a tower of timber, well framed and strong, and carrying from twelve to sixteen well-armed fighting men." The elephants could not withstand the force of the Mongol arrows, but turned tail and fled. In another chapter Polo relates that the Great Khan's elephants amounted fully to five thousand and that they were exhibited on the New Year's festival, all covered with rich and gay housings of inlaid cloth representing beasts and birds, while each of them carried on its back two splendid coffers filled with the emperor's plate and other costly furniture required for the court on the occasion of New Year. On his hunting expeditions the Great Khan was carried upon four elephants in a fine chamber made of timber, lined with plates of beaten gold, and covered with tiger skins.

The Manchu emperors still maintained an elephant stud, and the emperor K'ien-lung (1736-95) had sixty of them. When the emperor, on the evening before the winter solstice, proceeded to the Altar of Heaven to offer sacrifice at dead of night, he mounted a carriage drawn by an elephant.

E. Ysbrants Ides, envoy of the Russian czar to the emperor of China in the years 1692-95, reports, "The emperor's life-guards were clothed in red calico, printed with red figures, and wore small hats with yellow feathers. They were armed with scimitars and lances. There were eight white saddle-horses for show. In the third court of state were four extraordinarily large elephants, one of which was white. They were all covered with richly embroidered cloth, and their trappings were ornamented with silver gilt. On their backs was a finely carved wooden castle spacious enough for eight persons. Being taken out of the court, I mounted one of the emperor's two-wheeled carts, and was drawn to my apartment by an elephant. There were ten persons on each side with a rope in their hands fastened to the elephant's mouth to lead him; and on his neck sat a man with an iron hook to guide him. He walked at his ordinary rate of speed, but this obliged the men to run, in order to keep up with him. In the emperor's stables there were fourteen elephants: they made them roar, sing like a canary, neigh, blow a trumpet, go down on their knees, etc. All these elephants were extraordinarily large, and the teeth of some a full fathom long. The mandarins told me that the king of Siam annually sends several by way of tribute."

John Bell of Antermony, when he was at Peking in 1721, observed, "After dinner we saw the huge elephants richly caparisoned in gold and silver stuffs. Each had a driver. We stood about an hour admiring these sagacious animals, who, passing before us at equal distances, returned again behind the stables, and so on, round and round, till there seemed to be no end of the procession. The plot, however, was discovered by the features and dress of the riders: the chief keeper told us there were only sixty of them. The emperor keeps them only for show, and makes no use of them, at least in these northern parts. Some of them knelt and made obeisance to us; others sucked up water from vessels, and spouted it through their trunks among the mob, or wherever the rider directed."

The Earl of Macartney, when sent as ambassador of Great Britain to the emperor of China in 1792, still saw the elephants in the imperial palace, and remarks that they were brought to China from the neighborhood of the equator, and a few of them were bred to the northward of the Tropic. The Chinese elephants, he says, are smaller than those of Cochin-China, and of a lighter hue; they are literally granivorous, being generally fed with rice and millet, though the food of that animal in its wild state consists more frequently of the tender leaves of trees and shrubs than of the seeds or blades of corn or grass.

In a description of Peking inserted in the *Chinese Repository* for 1834 it is said that at that time not more than eight or ten elephants were kept in the Siang Fang ("Elephants' Palace") and were used to increase the pomp of some processions and ceremonies of the emperor. When I visited the building in question in 1901, there were no more elephants there.

While it is not the object of this article to survey the whole development of the elephant motive in Chinese art, which would require a profound study of Indian-Buddhistic subjects, a few remarks may be added here in order to assist the reader in a better understanding of some representations of the elephant in the Museum collections. The *Po ku t'u lu* of Wang Fu (chap. 7, p. 8) and the *Si ts'ing ku kien* (chap. 9, pp. 25, 26), the catalogue of the bronzes of the emperor K'ien-lung published in 1749, which follows Wang Fu's authority, illustrate and describe bronze figures of elephants carrying a vessel on their backs and assign these to the Chou period. This date, however, is merely prompted by the fact that "elephant-vases" (*siang tsun*), as already mentioned, are spoken of in the ancient Rituals. The art of the Chou, in fact, represented the elephant only in a strongly conventionalized, hieratic form, but never in that realistic manner manifested by the elephant-vases of the two Chinese catalogues. These obviously exhibit the style of the Indian-Buddhistic elephant with smiling eyes and harnessed with neat trappings. These objects, therefore, cannot be older than the age of the T'ang (A.D. 618-906), and this type has ever since been favorite with the bronze founders and potters. The Museum has a good elephant figure of this type of cast bronze coming down from the Sung period (Fig. 14). For comparative purposes are added two elephant designs of the T'ang period (A.D. 618-906) in Figs. 11 and 12,

Fig. 11 Fig. 12
Brass Chessmen with Designs of Elephants.
T'ang Period (A.D. 618-906).

Fig. 13
Ivory Chessman bearing Written Symbol of Elephant (Modern Form).

which appear on brass chessmen; these are as large and as flat as coins and, on the obverse, bear the character *siang* ("elephant"), in the same manner as in the modern ivory chessmen (Fig. 13), which are solely provided with the characters for the men, not with their figures as was customary under the T'ang.

In the K'ien-lung period (1736-95) the elephant was a favorite art-motive. Many good examples of its application to bronze vessels may be seen in Case 24 of the Blackstone Chinese Collection. In a censer shown there both the three feet and the two handles are formed by realistic figures of elephants; the cover is surmounted by the figure of a recumbent elephant on which astride is a turbaned Mohammedan. In another censer the elephant on the cover carries on its back a basket filled with coral branches, jewels, and rhinoceros-horns intended as gifts for the emperor. There is also a set of five altar-vessels in which the

Fig. 14
Bronze Figure of Elephant Followed by Mahout. Sung Period (A.D. 960-1278).
Specimen in Blackstone Chinese Collection.

elephant is the leading artistic motive; the ornaments in the animal's caparison and trappings are indicated by inlaid coral and turquois beads. In Fig. 15 the bronze figure of an elephant of the period is shown.

In India the elephant was modelled in art at an early time. In the Museum's collection of Gandhāra sculptures (Case 37, Hall 32) may be seen a small stone figure representing an elephant of naturalistic style (first or second century A.D.).

Live elephants were transported from India to Samarkand and Khotan and thence overland to China.

Fig. 15
Bronze Figure of Elephant (K'ien-lung Period).
Specimen in Blackstone Chinese Collection.

Images of elephants were brought along the same trade route, and were distributed over Central Asia, Siberia, and Russia. In this manner peoples who had never before seen an elephant became familiar with its likeness through models distributed from India. Small elephants of metal have been found in Russian soil, one of bronze in the Government of Yekaterinoslav and another of silver beyond the Ural, worshipped by the Ostyaks as an idol. The former is decorated with a purely Indian ornament, the so-called çrivatsa, an emblem of Çiva, which has become widely known also in China. It must hence be inferred that this bronze elephant found in Russian soil is of Indian workmanship, and was imported into Russia.

The elephant is quite capable of standing cold climates. The trip of an elephant to the northernmost part of Sweden (lat. 64°) is on record in the *Smithsonian Miscellaneous Collections* (Vol. 47, 1905, p. 517).

FOLK-LORE OF THE MAMMOTH AND TRADE IN MAMMOTH IVORY

The ancient Chinese had a certain knowledge of the mammoth (*Elephas primigenius*), though blended with marvelous details and embedded in the ground of folk-lore rather than based on correct observations. The interesting point is that Chinese traditions regarding the animal show a striking resemblance to those of Siberian tribes. The *Shen i king*, a book of wondrous tales, traditionally ascribed to Tung-fang So, minister to the emperor Wu of the Han dynasty (140-87 B.C.), contains the following passage: "In the regions of the north, where ice is piled up over a stretch of country ten thousand miles long and reaches a thickness of a thousand feet, there is a rodent, called *k'i shu*, living beneath the ice in the interior of the earth. In shape it is like a rodent, and subsists on herbs and trees. Its flesh weighs a thousand pounds and may be used as dried meat for food; it is eaten to cool the body. Its hair is about eight feet in length, and is made into rugs, which are used as bedding and keep out the cold. The hide of the animal yields a covering for drums, the sound of which is audible over a distance of a thousand miles. Its hair is bound to attract rats. Wherever its hair may be found, rats will flock together.".

Another term for the mammoth was *fen,* a name which refers properly to a species of mole (*Scaptochirus moschatus*); it was also called "the hidden rodent" (*yin shu*). Hence T'ao Hung-king (A.D. 452-536), a distinguished physician and celebrated adept in the mysteries of Taoism, and Ch'en Ts'ang-k'i, who wrote a materia medica in the beginning of the eighth century A.D., speak of two animals of the name *fen* and discriminate between *fen* as a small mole and the *fen* of the size of a water-buffalo, which may be identical with the mammoth. The earliest definition of the animal *fen,* as given in the ancient dictionary *Erh ya,* is that of "an animal which moves in the ground." As the same was supposed in reference to the mammoth, it is easy to understand how the name was transferred from a burrowing mole to a creature apparently resembling it in subterranean habits. It is striking, however, that in none of the Chinese traditions any allusion is made to the ivory-furnishing tusks.

To some extent the Chinese were also acquainted with fossil ivory. Their materia medica registers two famous articles known as "dragon's bones" and "dragon's teeth," offered for sale by drug-stores. The former have been examined microscopically by D. Hanbury (Science Papers, p. 273) and proved to be, at least in some cases, fossil ivory. Fossil bones of *Stegodon orientalis* of Swinhoe are brought from Sech'wan Province in large broken masses, showing the cancellous structure of the large fossil bones of proboscidians. Portions of limestone matrix bearing the impressions of these bones are sold together with these genuine fossils. They are powdered and used in ague, fevers, hemorrhages, and fluxes. The "dragon's teeth," usually found in marshy ground of Se-ch'wan, also in Shen-si and Shan-si, consist of fossil teeth of *Rhinoceros tichorhinus, Stegodon sinensis* and *Stegodon orientalis*, horns of *Chalicotherium sinense*, teeth of *Hyla sinensis*, and molars of horses, mastodons, elephants, and hippotherium. They are supposed to act on the liver and to be of great service as cordial or sedative remedies. In a lot of dragon's teeth obtained by me for the American Museum of Natural History, New York, in 1902, were found one tooth of a mastodon, five teeth of a rhinoceros, two molars of an hipparion, and one tooth of an undescribed hipparion. Dragon's bones from T'ai-yüan in Shan-si and Tsin-chou in Chi-li are mentioned as early as the T'ang period (A.D. 618-906).

The Chinese, moreover, possess a certain number of accounts which allude to the discovery of fossil animal-bones, particularly in Yün-nan and Se-ch'wan, and which are explained by the people as the remains of saints or fairies, in the same manner as we have the giants' bones in European folk-lore. These notices are so vague, of course, that it is impossible to determine the character of these bones. There are other Chinese descriptions of fabulous animals found in Chinese soil which led some European writers to the conclusion that the Chinese of historical times were personally acquainted with the mammoth. This, however, is extremely doubtful and, at any rate, is of no interest to archæology, as it is not known that the Chinese ever made any industrial use of mammoth or any other fossil bones. The "dragon's bones" and "dragon's teeth" were employed medicinally, but for no other purpose.

In the years 1712-15, a Chinese embassy traversed Siberia on its way to the Volga for the purpose of inducing the Torgut, a Kalmuk tribe who had settled there under Russian protection, to return to their old homes on the Chinese frontier. The Manchu Tulishen, the envoy, writes in his Memoirs in reference to Yenisseisk, "In the coldest parts of this northern country is found a species of animal which burrows under the ground, and which dies when exposed to the sun and air. It is of enormous size and weighs ten thousand pounds. Its bones are very white and bright like ivory. It is not by nature a very powerful animal, and is therefore not very ferocious. It generally occurs on the banks of rivers. The Russians collect the bones of this animal, in order to make cups, saucers, combs, and other small articles. The flesh of the animal is of a very cooling quality, and is eaten as a remedy in fevers. The foreign name of this animal is *mo-men-to-wa* [i.e. mammoth]; we call it *k'i shu.*" The transcription is based on Russian *mamontowa* (scil. *kost,* "bone"). It is interesting to note that this Manchu official, apparently well read in the literature of China, had wit enough to identify the earth-wanderer of ancient lore with the mammoth of whom he heard in Siberia. The *Ts'e yüan,* a modern Chinese cyclopædia published by the Commercial Press of Shanghai, quotes a portion of his text in proof of the fact that the word *fen* denotes the mammoth.

In 1716 the emperor K'ang-hi, who was fond of natural history, wrote, "The books say that in the very cold regions of the north ice forms to a thickness of a hundred feet and melts not even in the spring or summer. This region is now known actually to exist. Again, the *Yüan kien lei han* contains the following statement: 'The *k'i shu,* which is described as reaching the weight of ten thousand pounds, is found even at the present day. In shape it resembles the elephant, and its tusks are like those of the same beast, but the ivory is yellowish in color.' In both these points, the ancient books are confirmed." Again, in 1721, in the sixtieth year of his reign, the venerable sovereign recurred to the same topic in an address to his ministers, "While all the assertions found in books are not to be implicitly believed, there are, on the other hand, statements which, however false and absurd they may seem, are nevertheless perfectly well founded. Thus, for instance, Tung-fang So relates that in the regions of the north ice is formed to a thickness of a thousand feet, and does not melt either in the winter or summer. When the Russians presented themselves at our court this year, they stated that in their country, at a distance of something over twenty degrees from the Pole, there is what is called the Polar Sea. The ice lies frozen there in solid masses and prevents the access of human beings. Thus, for the first time, the truth of Tung-fang So's assertion has been confirmed. Again he states that in the northern regions, under layers of ice, is found a large animal of the kind of a rodent, the flesh of which weighs a thousand pounds. Its name is *fen shu.* It burrows under the ground and dies when it sees the light of the sun or moon. Now, in Russia, near the shores of the northern ocean, there is a rodent similar to an elephant, which makes its way under ground and which expires the very moment it is exposed to light or air. Its bones resemble ivory, and are used by the natives in manufacturing cups, platters, combs, and pins. Objects like these we ourselves have seen, and we have been led thereby to believe in the truth of the story."

In 1666, the learned Hollander Nicolaus Witsen, who subsequently became mayor of Amsterdam, paid a visit to Moscow, where he collected the materials for his work "Noord en Oost Tartarye," which appeared in 1694. This work introduced for the first time the name *mammoth* to western Europe. Witsen describes how elephants' teeth are found in large numbers on the banks of Siberian rivers, and adds, "By the inlanders (the Russian settlers in Siberia) these teeth are called *mammouttekoos* (for Russian *kost*, "bone"), while the animal itself is called mamout." Ludolf (Grammatica russica, p. 92, 1696) writes that the Russians believed the teeth of the mammoth to belong to an animal living underground, larger than any above ground. They used it in physic in lieu of and for the same purpose as unicorn's horn (narwhal tusk). The more sensible among the Russians affirmed these teeth to be of an elephant, brought thither at the time of the deluge.

The etymology of the word *mammoth* is obscure. The Russian form is *mamont* or *mamut*. It has been suggested by Strahlenberg that the word is derived from *behemoth* (see below, p. 63) through the medium of an Arabic *mehemoth*. Howorth has accepted this theory, but it is by no means convincing. Byron (*The Deformed Transformed*, III, 1) has confronted the two animals:

When the lion was young,
  In the pride of his might,
Then 'twas sport for the strong
  To embrace him in fight;
To go forth with a pine
  For a spear, 'gainst the mammoth,
Or strike through the ravine
  At the foaming behemoth;
While man was in stature
  As towers in our time,
The first born in Nature,
  And, like her, sublime!

According to the conception of the Samoyeds, the mammoth is a gigantic beast which lives in the depth of the earth, where it digs for itself dark pathways and feeds on earth. They call it "stallion of the earth" or "the master of the earth." They account for its corpse being found so fresh and well preserved by assuming that the animal is still alive. Death, however, they contend, is the fate of any one who has the misfortune to meet on his way the bones of the master of the earth; and if he is to ward off this penalty, he must sacrifice a reindeer to the demons. This entitles him to the possession of the bones and to using or selling them as he pleases.

In the country of the Ostyaks on the Irtysh mammoth-bones are sometimes found in the slopes of steep banks after a landslip. Some of the Ostyaks look upon them as water-sprites, others regard them as sacred animals living under ground and call them "earth-oxen." They cannot bear the daylight and must die when reaching the surface of the earth (cf. the similar Chinese notion above). Pieces of mineral coal which occur on the banks of some rivers are regarded as the livers of mammoths. They subsist on tree-roots and hence dig up the earth, so that they undermine the river-banks and finally cause their collapse. They are also fond of residing in the depth of streams and lakes, and their presence is announced by the agitation of the water and whirlpools. Such places in rivers which are looked upon as abodes of mammoths are considered sacred, and nets must not be cast into the water. They do not even like to draw there water for drinking. In the winter the animals sometimes rise to the surface of the water, break the ice, and cause a tremendous noise. They are harmless to man and cannot grant him luck in his enterprises or health; yet the offerings made to them during a journey insure its successful completion by guarding against landslips and ice-breaks. Other animals when grown old may undergo a metamorphosis into mammoths; elks and reindeer, even bears, may change their status for a life in the depth of the waters when mammoth-horns will grow on them. Old pikes are said sometimes to choose the deepest spots of lakes where moss will grow on their heads and horns on their front, whence it is concluded that old pikes also are gradually transformed into mammoths. In this form they are called pike-mammoth.

The Samoyeds designate the mammoth "earth-bull" or "earth-stag." The former epithet is applied to it also by the Wogul, a Finno-Ugrian tribe. The Mongols and Manchus speak of the "ice-rodent."

The Buryats, a branch of the Mongols living around Lake Baikal, call the mammoth *arslan* ("lion") or *arsalyn zan* ("lion-elephant"). They believe that its bones represent a smashed dragon (*lu*). When the dragons have grown old, they take refuge in the earth. The animal is further connected with the Biblical story of the flood: it boasted that it could not perish on account of its size, and refused to enter the ark. It swam around for several days, but was finally drowned. Hence its bones are now found in the ground. The Russians of Transbaikalia have a similar story, adding that while the mammoth was floating, birds perched on its "horns," as they could not find a dry place; for a long time it struggled against the flood, but the birds increased to such a number that its strength finally dwindled, and it perished after a few days.

In 1611 an English navigator, Jonas Logan, visited the land of the Samoyeds and returned to London with an elephant's tooth which he had bought from them. This presumably was the first mammoth tooth that came to England. The Scotch traveller, Bell of Antermony, observed in most of the towns which he passed between Tobolsk and Yenisseisk many mammons' horns, so called by the natives. Some of them were very entire and fresh like the best ivory in every circumstance excepting only the color, which was of a yellowish hue; others of them mouldered away at the ends and, when sawn asunder, were prettily clouded. The people made snuff-boxes, combs, and diverse sorts of turnery-ware of them. "I brought a large tooth or mammons' horn with me to England," he adds, "and presented it to my worthy friend, Hans Sloane, who gave it a place in his famous museum, and was of opinion also that it was the tooth of an elephant. This tooth was found in the River Obi at a place called Surgut." The Russians developed a lively trade in mammoth ivory from the latter part of the eighteenth century when Liakhoff, a merchant, discovered vast stores of mammoth bones between the rivers Khotanga and Anadyr and obtained the exclusive right to dig for them. The quantity of fossil ivory which was sent from Siberia to the European markets was enormous. In 1821, an ivory-hunter from Yakutsk brought back twenty thousand pounds of ivory, each tusk weighing on an average about a hundred and twenty pounds. In the London market as many as 1,635 mammoth tusks were sold in a single year, averaging 150 pounds in weight; of these 14 per cent were of the best quality, 17 per cent inferior, while more than half were useless commercially. The total number of mammoths represented by the output of fossil ivory since the Russian colonization of Siberia is estimated as not being far from forty thousand.

Vague allusions to the mammoth loom up also in the writings of the Arabs of the middle ages. Thus al-Beruni (A.D. 973-1048), in his discussion of the substance *khutu,* which, as will be seen in the following chapter, applies in the main to walrus ivory, remarks that it is the frontal bone of a bull who lives in the country of the Kirgiz. In an Arabic chronicle, written in A.D. 1076, are mentioned teeth resembling the tusks of elephants which were obtained in the country of the Bulgars who at that time lived on the Volga. These teeth were thence exported to Kharizm (now Khiva), where they were wrought into combs, boxes, and other objects. Abu Hamid, who visited the country of the Bulgars in A.D. 1136, observed there "a tooth four spans long and two spans wide and the cranium of the animal resembling a dome; teeth were also found in the ground like elephant's tusks, white like snow, one weighing two hundred *menn;* it was not known from what animal it was derived; it was wrought like ivory, but was stronger than the latter and unbreakable."

It is possible, as is assumed by several scholars, that the question is here of mammoth tusks; but it is equally possible that it was simply elephants' tusks. P. S. Pallas (Reise durch verschiedene Provinzen des russischen Reichs, Vol. I, p. 140, 1801) found in the Volga territory several bones and even the cranium of an elephant on the banks of a rivulet; at Simbirsk he saw some objects wrought from the ivory found there and which could not be distinguished from "healthy" ivory; only the tip of the tooth had undergone a certain degree of calcination. Another elephant's tusk found on the bank of that rivulet he describes as having assumed an intensely yellow color. Similar discoveries may have formed the source of supply for the Bulgar ivory.

The Chinese were acquainted with Kharizm as early as the T'ang dynasty and emphasize the point that it was the only country of western Asia, where carts drawn by oxen were to be found and that the merchants travelled around in these vehicles. In A.D. 751 a prince of that country sent an embassy to China with gifts. The ancient capital of the country, Urgenj, was captured and destroyed by the Mongols in A.D. 1221. Khiva, the present capital of Kharizm or the Khanate of Khiva, is situated about a hundred miles southeast of ancient Urgenj. If Kharizm was a centre of the mammoth-ivory industry during the middle ages, we receive in this manner a clew as to how this material may have found its way to China.

In the work of the learned Philipp Johann von Strahlenberg, which appeared at Stockholm in 1730, we are informed that great quantities of white mammoth tusks were carried from Siberia for sale to China. Strahlenberg was a Swedish officer in the service of Charles XII, was taken prisoner by the Russians in the battle of Pultava, and resided in Siberia for thirteen years. It is only surprising that Strahlenberg characterizes these teeth exported to China as white, while in the same breath he describes mammoth tusks as yellow, some as brown as coconuts and even black-blue. It has therefore been suggested by Ranking that the teeth in question were walrus teeth.

"In the northern part of Siberia, so great is the abundance of mammoth tusks, that for a very long period there has been a regular export of mammoth ivory, both eastward to China and westward to Europe" (N. N. Hutchinson, Extinct Monsters, p. 183). This view, however, must be adopted only with certain reservations. As far as the last three centuries are concerned, it is without any doubt correct, but going beyond the seventeenth century, the matter becomes one of uncertainty, and we have no definite evidence either archæological or historical. The archæology of Siberia is fairly well known, and no ivory has as yet been discovered there in any grave or otherwise. The notices of the Arabs given above refer to tusks found in the territory of the Volga, but the Arabs never had any knowledge of the Siberian mammoth. As will be seen in the following chapter, what the Arabs traded was chiefly walrus ivory. In regard to the Chinese we are confronted with a puzzling fact: they have ancient traditions and certain notions of the mammoth as an animal, but they never allude to its tusks or ivory. Only as late as 1716 the emperor K'ang-hi learned from the Russians that the animal furnishes ivory, and about this time the same fact dawned upon the envoy Tulishen when he was in the heart of Siberia. Previously, however, the Chinese, if we rely on their literature, possessed no knowledge of mammoth ivory. Howorth argues that "from early times mammoth ivory was sent from Siberia to China, that the Chinese had a knowledge of Siberia and its products from a very early time, a fact not otherwise improbable," and he adds that the throne which Carpini describes as having been made for the great Mongol Khan Kuyuk by the jeweller Cosmas out of ivory was doubtless made from fossil ivory, showing it was well known in Mongolia in the thirteenth century. Carpini, in his Latin text, uses merely the word "ivory" (*ebur*), and considering the vast number of elephants kept in the empire of the Mongols (above, p. 18), there is no reason why Kuyuk's throne could not have been made of plain elephant ivory as well. True it is, as Howorth says, that from an early date the Chinese were well acquainted with the peoples and products of Siberia; in fact, nearly all we know about this subject in times prior to the Russian occupation we owe to the official annals and other records of the Chinese. They tell us a great deal about the fine Siberian peltry like the sable, marten, fox, lynx, beaver, but never mention any ivory used by a tribe of Siberia or imported from there into China. This silence surely is not fortuitous, it is ominous. Considering the movements and migrations of the former peoples inhabiting Siberia, it is most unlikely that the northernmost parts of the country were inhabited in very early times. These inhospitable regions were only the refuge of weaker tribes who were gradually pushed northward by more powerful neighbors. It was in the tundras and along the littoral of northern and northeastern Siberia that most of the remains of the mammoth were discovered, and this was possible only in times after the northward movement of the tribes.

There is another reason why a history of the trade in mammoth ivory cannot be written with an absolute degree of certainty or cannot be given, so to speak, a clean bill of health. The accounts we have are confused, and in many of them the tusks of mammoth, walrus and narwhal, and even fossil rhinoceros-horn, are hopelessly mixed up. These various products are all comprised in Siberia under the commercial term "horn." The Yakut, for instance, indiscriminately designate mammoth and walrus ivory as "horn" (*muos*). To us who have a clear notion of the animals this lack of discriminating faculty may seem strange at first sight, but looking into the conditions under which the said animal products are found in Siberia we find it easy to gauge the situation. Immense deposits of mammoth and rhinoceros bones are accumulated together with masses of stranded walruses and fossil narwhal tusks along the northern littoral, and are collected promiscuously by the treasure-hunters. Walrus and narwhal as live animals are familiar solely to the maritime tribes, and totally unknown to the inland peoples. Again, the mammoth and the rhinoceros, which occur there only in fossil remains, are unknown as animals to any of them, and their bones accordingly are not distinguished. It must further be taken into consideration that in many cases it is not a complete tusk or horn which is traded by the ivory hunters, but merely fragmentary pieces; rotten and hollow portions are cut off as useless, as soon as the best-preserved pieces have been picked out, and the remainders if still of a considerable size may again be sawed into smaller portions to be rendered fit for transportation on pack-horses. The distant trader who will buy up this cargo and the consumer still more remote from the place of provenience hardly have any means of obtaining a clear idea of the true origin of the product, still less of the character of the animal from which it may have come. The door was thus thrown widely open for fabulous speculations of all sorts in regard to the "horn." This term is encountered everywhere in Europe, among the Arabs, in China, Korea, and Japan, with reference to walrus and narwhal tusks, as the following chapter will demonstrate.

TRADE IN WALRUS AND NARWHAL IVORY

The trade in walrus and narwhal ivory is a veritable romance in the history of commerce, and it is not generally known that in times long prior to the discovery of the Arctic shores of North America and long before the two animals were described in our natural history, a lively traffic in this kind of ivory was carried on all over Asia and Europe. As this subject has never been clearly set forth in any book and is based on researches almost entirely my own, I hope that a somewhat detailed digest of the matter will be welcome. Those desirous of knowing the original sources and the exact texts of the documents may fall back on my previous studies of the subject cited in the Bibliographical References at the end; on the other hand, much new information is given here.

In the zoological system the walrus belongs to the order Pinnipedia which consists of the three families, Otariidae (eared seals), Trichecidae (walrus), and Phocidae (seals). The genus *Trichecus* (walrus) consists of two species,—*T. rosmarus* and *T. obesus*. The former occurs on the coast of Labrador northward to the Arctic Sea, alòng the shores of Greenland, and in the polar areas of the eastern hemisphere to western Asia. The latter inhabits the northwest coast of America, the Arctic Sea and Bering Strait, as well as the northeastern coast of Asia. The most striking characteristic of the animal is formed by the pair of tusks corresponding to the canine teeth of other mammals; they descend almost directly downward from the upper jaw, sometimes attaining a length of twenty inches or more and a weight of from four to six pounds. The tusks do not form a solid mass throughout, but are hollow about two-thirds of their length, so that large objects and billiard-balls cannot be carved from them. The outer layer of the tooth has a dark coloration, and is not smooth as in elephants' tusks, but is fluted and hard as glass. The tips of the tusks yield a hard and solid mass which is evenly yellowish white, and in a cross-section, displays speckled designs. The lateral portions, likewise yellowish white, are crossed by fine yellow lines, or are interspersed with large, yellow, flamed spots. When exposed to the atmosphere or to moisture for a long time, the tusk will lose its whiteness and assume a yellow tobacco color.

The walrus was formerly styled "sea-horse" (Latin *Equus marinus*), its tusks "sea-horses' teeth" (e.g. John Ray, Synopsis methodica animalium quadrupedum, p. 193, London, 1693). Likewise such descriptive terms as sea-ox, sea-cow, sea-elephant were in use. In earlier literature also *morse, mors* (derived from Russian *morzh,* a word of unknown origin, through the medium of French *morse*) appears occasionally: thus William Baffin ("Relation of his Fourth Voyage for the Discoverie of the North-West Passage, performed in 1615") speaks of "peeces of the bone or horne of the sea unicorne, and divers peeces of sea mors teeth." Jonas Poole (in Purchas), in 1610, writes *mohorses,* with adaptation to *horse.*

The narwhal (*Monodon monoceros*) belongs in the zoological system to the order Cetaceae. Our word is derived from Old Norse *nahvalr,* Swedish-Danish *narhval; hvalr* or *hval* means "whale," the origin of the first element of the word is obscure. The animal frequents the icy circumpolar seas, and is rarely seen south of 65° N. latitude. It resembles the white whale in shape and in the lack of a dorsal fin. Its peculiar characteristic lies in the absence of all teeth, save two in the upper jaw, which are arranged horizontally side by side. In the male, usually the left tooth and occasionally both teeth are strongly developed into spirally twisted, straight tusks which pass through the upper lip and project in front like horns. They frequently reach a length of about seven feet; that is, half, and even more, that of the entire animal, which in the state of maturity may attain to fifteen feet. Its life-history is but little explored, and the biological function of the tusk is conjectured rather than accurately ascertained; it is supposed to serve as a weapon of defence, for breaking ice in order to breathe, and for killing fish. The ivory yielded by the tusk, which is hollow in the interior, possesses extreme density and hardness and in this respect surpasses elephant ivory; it is of a dazzling whiteness, which does not pass into yellow, is easily wrought, and easily receives a high polish. Along the northern shores of Siberia are also accumulated fossil tusks of the narwhal together with enormous masses of mammoth and rhinoceros bones.

In the eighteenth century a narwhal was observed cast adrift at the mouth of the Elbe, and another at the estuary of the Weser. Caxton, in his Chronology of England, has an entry under the year 1482, "This yere were take four grete fisshes between Erethe and London, that one was callyd mors marine." This is the earliest instance of the occurrence of the term *morse* for the walrus in English literature.

The first acquaintance of England with the walrus, however, was much earlier and dates from the latter part of the ninth century, and is connected with the daring exploits of the Norseman Ohthere from Helgeland in Norway, who in A.D. 890 undertook several voyages, rounded the North Cape, and reached the Kola Peninsula. He reported on this enterprise to King Alfred the Great of England (848-901), who embodied Ohthere's narrative in his Anglo-Saxon translation of Paulus Orosius' History of the World. The passage with reference to the walrus runs thus: "The principall purpose of his travelle this way, was to encrease the knowledge and discoverie of these coasts and countreyes, for the more commoditie of fishing of horse-whales, which have in their teeth bones of great price and excellencie: whereof he brought some at his return unto the king. Their skinnes are also very good to make cables for shippes, and so used. This kind of whale is much lesse in quantitie then other kindes, having not in length above seven elles. And as for the common kind of whales, the place of most and best hunting of them is in his owne countrey. Whereof some be 48 elles of length, and some 50, of which sort he affirmed that he himselfe was one of the sixe which in the space of 3 dayes killed threescore. Their principall wealth consisteth in the tribute which the Fynnes pay them, which is all in skinnes of wild beasts, feathers of birds, whale bones, and cables, and tacklings for shippes made of whales or Seales skinnes. The richest pay ordinarily 15 cases of Marternes, 5 Rane Deere skinnes, and one Beare, ten bushels of feathers, a wat of a Beares skinne, two cables threescore elles long apiece, the one made of Whales skin, the other of Seales."

The Anglo-Saxon word used in this text is *horshwael*, from Old Norse *hrosshvalr* ("a kind of whale") and *rosmhvalr* ("walrus").

In the sixteenth century when walrus ivory reached England from North America, it was paid for at double the rate of elephant ivory. Thomas James of Bristol, who visited the island of Ramea near Newfoundland in 1591 and who gives a description of the walrus he encountered there, writes that its teeth were sold in England to the comb and knife-makers at eight groats and three shillings the pound weight, whereas the best ivory was sold for half the money; the grain of the bone, he remarks, is somewhat more yellow than the ivory. He also tells a curious story about his friend Alexander Woodson of Bristol, an excellent mathematician and skilful physician, who showed him one of these beasts' teeth brought from Ramea and half a yard long, and who assured him that he had made trial of it in ministering medicine to his patients, and had found it as sovereign against poison as any unicorn's horn. Gerat de Veer ("The First Navigation of William Barents, alias Bernards into the North Seas," 1594) speaks of "sea-horses being a kind of fish that keepeth in the sea, having very great teeth, which at this day are used instead of ivory or elephants teeth." Martin Frobisher (Voyage in 1577) relates, "They found a great dead fish, round like a porpoise, twelve feet long, having a horn five feet ten inches long, growing out of the snout, wreathed, and straight like a wax taper; and might be thought to be a sea-unicorn: the top of it was broken. It was reserved as a jewel by Queen Elizabeth's commandment in her wardrobe of robes, and is still at Windsor to be seen."

From the ninth century onward walrus tusks formed an important article of trade in the north-eastern part of Europe, and this was the case long before the discovery of Greenland. In Russian history they are known as "fish-teeth," as in bygone days the walrus was classified among fish everywhere in Europe and Asia. In old Russian tales are mentioned precious chairs of fish-teeth, and these fish-teeth appear as highly priced objects. At Novgorod they were traded like marten and squirrel skins and accepted as monetary values. In 1159 the grand-duke Rostislav presented to the prince Svätoslav Olgovich sables, ermines, black foxes, polar foxes, white bears, and fish-teeth. During the period of Mongol and Tartar sway over Russia frequent demands for this product were made from Asia, and in 1476 Ivan Vasilyevich received a fish-tooth as a gift from a citizen of Novgorod.

S. von Herberstein, who, in 1549, published his work "Rerum Moscoviticarum Commentarii," a primary source for the history of Russia, and who was ambassador to the Grand Prince Vasily Ivanovich in the years 1517 and 1526, gives the following account: "The articles of merchandise which are exported from Russia into Lithuania and Turkey, are leather, skins, and the long white teeth of animals which they call *mors*, and which inhabit the northern ocean, out of which the Turks are accustomed very skilfully to make the handles of daggers; our people think they are the teeth of fish, and call them so. The ocean which lies about the mouths of the river Petchora, to the right of the mouths of the Dwina, is said to contain animals of great size. Amongst others, there is one animal of the size of an ox, which the people of the country call *mors*. It has short feet, like those of a beaver; a chest rather broad and deep compared to the rest of its body; and two tusks in the upper jaw protruding to a considerable length. The hunters pursue these animals only for the tusks, of which the Russians, the Tartars, and especially the Turks skilfully make handles for their swords and daggers, rather for ornament than for inflicting a heavier blow, as has been incorrectly stated. These tusks are sold by weight, and are described as fishes' teeth." Von Herberstein, accordingly, identifies the commercial label "fish-teeth" with the zoological term "morse"; that is, the walrus.

Richard Chanceleur, in "The Book of the Great and Mighty Emperor of Russia" (1553), writes, "To the north parte of that countrey are the places where they have their furres, as sables, marterns, greese bevers, foxes white, blacke, and redde, minkes, ermines, miniver, and harts. There are also a fishes teeth, which fish is called a *morsse*. The takers thereof dwell in a place called Postessora, which bring them upon hartes [reindeer] to Lampas to sell, and from Lampas carie them to a place called Colmogro, where the hie market is holden on Saint Nicholas day."

Farther on, he gives somewhat more detailed information on the same subject, as follows:—

"The north parts of Russia yeelde very rare and precious skinnes: and amongst the rest, those principally, which we call sables, worne about the neckes of our noble women and ladies: it hath also martins skinnes, white, blacke, and red foxe skinnes, skinnes of hares, and ermyns, and others, which they call and terme barbarously, as bevers, minxes, and minivers. The sea adjoyning, breedes a certaine beast, which they call the Mors, which seeketh his foode upon the rockes, climing up with the helpe of his teeth. The Russes use to take them, for the great vertue that is in their teeth, whereof they make as great accompt, as we doe of the elephants tooth. These commodities they carry upon deeres backes [reindeer] to the towne of Lampas: and from thence to Colmagro, and there in the winter time, are kept great faires for the sale of them. This citie of Colmagro, serves all the countrey about it with salt, and salt fish. The Russians also of the north parts, send thither oyle, which they call traine, which they make in a river called Una, although it be also made elsewhere: and here they use to boile the water of the sea, whereof they make very great store of salt."

Anthony Jenkinson, who travelled in Russia and Turkestan from 1557 to 1571, was well familiar with the life of the Russians and their use of walrus ivory.

"When he rideth on horse backe to the warres, or any iourney," he writes, "he hath a sword of the Turkish fashion, and his bowe and arrowes of the same manner. In the towne he weareth no weapon, but onely two or three paire of knives having the hafts of the tooth of a fishe, called the Morse" (E. D. Morgan, Early Voyages and Travels to Russia and Persia by A. Jenkinson, p. 40).

On his return from Persia in the autumn of 1564, Jenkinson's efforts were bent toward organizing a voyage to Cathay by the northeast passage; and in pursuance of this plan he addressed on the 25th of September, 1565, to the Queen of England a "petition relating to the north-east passage." In this memorable document he presents the following argument in which walrus teeth play a prominent part in favor of his contention that Cathay could be reached in that manner: "At my beinge in Scythia and Bactria, I divers tymes talked and conferred wit̄h dyvers Cathayens [Chinese] who wer there at that present in trade of merchanndyse towchinge the comodyties of their countrey, and how the seas aborded unto them, I learned of them that the said seas had theire course to certen northerly regions with whom they had traphyque by seas. Also havinge conferrence with th'inhabitantes of Hugarye [Ugria] and other people of Sameydes [Samoyeds] and Colmackes whose countreys lye very farr northerly (and nere whereunto I gesse the said passage to be) whiche people sayle alonge the saide coastes *fysshinge after the greate fyshe callyed the Morse* for the benefyte of his teathe. Of whome I have learned that beyonde them the sayde lande and coastes trenche and tende to the east and to the southwarde, and that the corrauntes and tydes runne east southeaste and west northweste very vehemently, whiche manifestly arguethe a passage. Further this laste yere at my beinge in th'emperoure of Muscovia his

Coorte, yt chaunced that there cam thyther certen of th'inhabitantes of the foresaid countryes to present unto the said prince a certen straunge hed with a horne therein, whiche they had fownde in the Ilonde of Vagatts [Vaigats, separated from the Siberian mainland by Yugor Shar, called Pet Straits], whiche is not farre from the river of Obbe and the mayne land of Hugarye. And for that th'emperoure neyther any of his people knewe what yt was for the straungenes thereof he commaunded that soche straungers as wer thoughte to have any judgement therin shold see the same, and be asked there opynion what they thoughte it to be. Amounge whome yt was my chaunce to be. And so was it fownde, by the reporte of them, that before had seane the lyke, to be the hedd and horne of an Unycorne, which is in no smalle pryce and estymacion with the saide prynce. Then I imagynyd with my self from whence the said hedd sholde come, and knowinge that unycornes are bredde in the landes of Cathaye, Chynaye and other the Orientall Regions, fel into consideration that the same hedd was broughte thyther by the course of the sea, and that theire muste of necessytie be a passage owt of the sayde Orientall Occean into our Septentrionall Seas, for how elles cowlde that hedd have come to that Ilonde of Vagatts."

This argument is alluded to by Martin Frobisher in his First Voyage of 1576, "That voyage was then taken in hand, of the valiant knight, with pretence to have gone eastward to the rich countrey of Cataya, and was grounded briefely upon these reasons. First, bicause there was a unicornes horne found upon the coast of Tartaria by the river Obij, which (said he) was like by no other ways to come thither, but from India or Cataya, where the saide unicornes are only found, and that by some sea bringing it thither" (R. Collinson, The Three Voyages of Martin Frobisher in

Search of a Passage to Cathaia and India by the North-West, 1576-78, p. 39).

Anthony Marsh, a factor for the Moscovie Company of England, wrote in his Notes concerning the Discovery of the River of Ob in 1584, "Not farre distant from the maine, at the mouth of Ob, there is an island, whereon resort many wilde beasts, as white beares, and the morses, and such like. And the Samoeds tell us, that in the winter season, they oftentimes finde there Morses teeth."

Giles Fletcher, who, in 1588, was sent as ambassador of Queen Elizabeth to Theodor, emperor of Russia, has the following report in his "The Russe Common Wealth," also entitled "The Native Commodities of the Contrey:"—

"Besides these (which are all good and substantiall commodities) they have divers other of smaller accompt, that are naturall and proper to that countrey: as the fish tooth (which they call Ribazuba) which is used among themselves, and the Persians and Bougharians that fetch it from thence for beads, knives, and sword hafts of noblemen and gentlemen, and for divers other uses. Some use the powder of it against poison, as the unicornes horne. The fish that weareth it is called a Morse, and is caught about Pechora. These fish teeth some of them are almost two foote of length, and weigh eleven or twelve pound apiece."

R. Stevens of Harwich ("Voyage to Cherry Island in 1608." This island, named in honor of Sir Francis Cherry, lies south of Spitsbergen) writes, "The ninth day we got one tierce of morses' teeth, besides four hundred other teeth. We brought a young living morse to court, where King James and many honourable personages beheld it with admiration. It soon died. It was of a strange docility, and very apt to be taught." In 1610 the Russia Company took possession of Cherry Island, and that year they killed a thousand morses and made fifty tons of oil (John Harris, Voyages and Travels, 1764, Vol. II, p. 389).

In 1652, Deshneff sailed down the Anadyr as far as its mouth, and observed on the north side a sand bank, which stretched a considerable way into the sea. A sand bank of this kind is called in Siberia *korga*. Great numbers of sea-horses were found to resort to the mouth of the Anadyr. Deshneff collected several of their teeth, and thought himself amply compensated by this acquisition for the trouble of his expedition. Another expedition was made in 1654 to the Korga, for the purpose of collecting sea-horse teeth. A Cossack, named Yusko Soliverstoff, was one of the party. This person was sent from Yakutsk to collect sea-horse teeth for the benefit of the crown (W. Coxe, Account of the Russian Discoveries between Asia and America, pp. 318, 319, London, 1780).

An important contribution to the subject is furnished by the Jesuit father Avril, who in the latter part of the seventeenth century gathered the following information from the Russians: "Besides furs of all sorts, which they fetch from all quarters, they have discovered a sort of ivory, which is whiter and smoother than that which comes from the Indies. Not that they have any elephants that furnish them with this commodity (for the northern countries are too cold for those sort of creatures that naturally love heat), but other amphibious animals, which they call by the name of *Behemòt*, which are usually found in the River Lena, or upon the shores of the Tartarian Sea. Several teeth of this monster were shewn us at Moskow, which were ten inches long, and two at the diameter at the root: nor are the elephant's teeth comparable to them, either for beauty or whiteness, besides that they have a peculiar property to stanch blood, being carried about a person subject to bleeding.

The Persians and Turks who buy them up put a high value upon them, and prefer a scimitar or a dagger haft of this precious ivory before a handle of massy gold or silver. But certainly nobody better understands the price of this ivory than they who first brought it into request; considering how they venture their lives in attacking the creature that produces it, which is as big and as dangerous as a crocodile." Farther on, Avril quotes a story told him by the Voyevoda of Smolensk about an island at the mouth of the great River Kawoina, beyond the Obi, that discharges itself into the Frozen Sea. "This island is spacious and very well peopled, and is no less considerable for hunting the *Behemot*, an amphibious animal, whose teeth are in great esteem. The inhabitants go frequently upon the side of the Frozen Sea to hunt this monster; and because it requires great labor and assiduity, they carry their families usually along with them." Avril, accordingly, confirms the fact that the Russians hunted the walrus along the shores of the Arctic Sea, and that the animal's tusks were conveyed to Moscow and traded to the Persians and Turks.

The Arabs, as we learn from al-Beruni (A.D. 973-1048) in a treatise on precious stones written by him, prized walrus ivory highly and called it *khutu*. They received it from the Bulgars, who then resided on the Volga and who brought from the northern sea "teeth of a fish over a cubit long," which were wrought into knife-hilts. The Arabs traded them even to Mekka. The Egyptians craved them and purchased them at a price equal to two hundred times their value. Maqdisi or Muqaddasi (about A.D. 985) mentions fish-teeth among the exports from Bulgar into Kharizm (Khiva). The Jesuit Avril, as quoted above, observes that the Persians and Turks bought up walrus teeth at a high value and preferred a scimitar or a dagger haft of this precious ivory to a handle of massive gold or silver. The Persians adopted both the foreign term *khutu* and the designation "fish-tooth" (*dandān māhī*, also *shīr māhī*, "lion-fish"), and turned combs and knife-hilts out of it, which were transmitted to India. In the second volume of his Memoirs, the emperor Jahangir tells how delighted he was when he received from Persia a dagger whose hilt was made of a fish-tooth. He was so much impressed by this hilt that he despatched skilful men to search for other specimens in Persia and Transoxania. Their instructions were to bring fish-teeth from anywhere, and from any person, and at any cost. A little later a fine specimen was picked up in the bazar of his own capital of Agra, and was brought to him by his son, Shah Jahan. Jahangir had the tooth made into dagger-hilts, and gave one of the craftsmen an elephant as a reward, and bestowed on the other increase of pay and a jewelled bracelet. The idea that this ivory was believed to be an antidote to poison, and also to reduce swellings, added greatly to its value. From a statement in the history of Akbar the Great, known as the Akbarnāma, it appears that about 1569 a Raja in Malabar, who probably was the Raja of Cochin, sent Akbar a knife which had the property of reducing or removing swellings, and that Akbar told his secretary that it had been successfully applied in more than two hundred cases. Probably this knife was made, wholly or in part, of walrus ivory, which could easily have been brought to Cochin by sea.

At present India still has a kind of ivory known as "fish-tooth" (*mahlīka-dant*). This is always of a dirty (oily) yellow color with the texture looking as if crystallized into patches, which is characteristic of the interior of the walrus tusk. The significance of being called in every language and dialect of India "fish-tooth" at once suggests a common and, most probably, foreign origin for the material. An inquiry made by

George Watt disclosed the fact that it was more highly valued for sword and dagger hafts and more extensively used for these purposes than is ivory. It is put through an elaborate and protracted process of curing before being worked up. The crude fish-tooth is wrapped up in a certain mixture and retained in that condition for various periods, the finer samples for as long as fifty years. The advantages are its greater strength, finer and smoother surface, and greater resistance (less liability to slip in the hand) than is the case with ivory. According to Watt, the fish-tooth of Indian trade is mainly, if not entirely, the so-called fossil ivory of Siberia—the ivory of the mammoth; but he thinks it equally possible that a fair amount of walrus ivory finds its way into India by passing, like the Siberian ivory, over land routes to India. And from the antiquity of some of the swords, found in the armories of the princes of India with "fish-tooth" hafts, it would seem possible that there has existed for centuries a traffic in carrying this material to India.

A Turkish work on mineralogy, written in A.D. 1511-12 by Ibn Muhammad al Gaffari, contains the following account: "On the Khutu Tooth. The *khutu* is an animal like an ox which occurs among the Berber and is found also in Turkestan. A gem is obtained from it; some say it is its tooth, others, it is its horn. The color is yellow, and the yellow inclines toward red, and designs are displayed in it as in damaskeening. When the khutu is young, its tooth is good, fresh, and firm; when it has grown older, its tooth also is dark-colored and soft. The padishahs purchase it at a high rate. Likewise in China, in the Magrib, and in other countries it is known and famous. It is told that a merchant from Egypt brought to Mekka a piece and a half of this tooth and sold it on the market of Mina for a thousand gold pieces. Poison has no effect upon one who carries this tooth with him, and poison placed near it will cause it to exude. For this reason it is highly esteemed."

Pierre Belon (1518-64), a prominent French traveller and naturalist, wrote in 1553, "The Turks have this custom in common with the Greeks that they carry their knives suspended from their belts. These knives are commonly made in Hungary with very long handles; but when the merchants of Turkey buy them, they turn them over to artisans to add to them a butt which is commonly made of Rohart tooth [walrus tooth]. There are two sorts: one is straight white and compact, resembling the tooth of the unicorn [narwhal], and is so hard that steel will hardly affect it unless it be well tempered. The other tooth of Rohart is curved like that of a boar: we might have believed that it was the tooth of a hippopotamus, had we not observed this animal alive which had no such teeth." The French word *rohart* (also *rohar, rohal*) refers to the walrus, and is connected with Old Norse *horshvalr* (Norwegian *rohal, roshal*). In the Latin translation of Belon's work prepared by the botanist C. Clusius (1589) the name "morse" for the animal has been added.

In the beginning of the seventeenth century, the Company of the Greenland Merchants of England shipped to Constantinople a "horn," as it was then called, found by an English sea-captain in 1611 in the ground on the coast of Greenland, and the sum of two thousand pounds sterling was offered for it. The Company, however, in the hope of a better price, declined to sell and sent it on into Muscovy, where approximately the same price was bidden. Hence the tooth was transported back into Turkey, where a much smaller sum was then proffered than before. The Company therefore decided that the tooth would sell more easily in pieces than entire, and had it broken up.

The single pieces were finally disposed of in different places, but the proceeds amounted to only twelve hundred pounds sterling (account of Pietro della Valle in 1623; the complete text is given in "Sino-Iranica," p. 567).

Quite independently of Europe, the Chinese received walrus ivory from the northeast of Asia through the medium of numerous tribes settled in this region. Beyond the boundaries of Korea, in the east conterminous with the ocean, the northern limit being unknown, there was from remote ages the habitat of the Su-shen, who have greatly stirred the imagination of Chinese and Japanese chroniclers, and who are frequently mentioned in the Chinese Annals. They were the Vikings of the East, raiding on several occasions the coasts of northern Japan and engaging in many a sea-battle with the Japanese in the seventh century. For a thousand years earlier, the Chinese were acquainted with this nation and its peculiar culture. They used flint arrowheads, usually poisoned, which were preserved as curiosities in the royal treasury of China. They lived through a stone age for at least fifteen hundred years down to the middle ages when they were merged in the flood of roaming Tungusian tribes. They availed themselves of stone axes which played a role in their religious worship, and of hide and bone armor for defence. In A.D. 262 they sent to China a tribute consisting of thirty bows, wooden arrows, three hundred stone crossbows, twenty suits of armor made of leather, bone, and iron, respectively, as well as a hundred sable-skins. This enumeration of objects brings us into close contact with the state of culture that partially still prevails in the northern area of the Pacific, and the main representatives of which at the present time are the Koryak, the Chukchi, and the Eskimo. In this area still occurs that peculiar type of bone plate armor composed of rows of overlapping ivory plates, and the plates in this type of armor are commonly carved from walrus ivory, possessing as it does a higher degree of elasticity than any ordinary kind of bone. The Su-shen, accordingly, appear to have been in possession of walrus ivory, at least prior to A.D. 262, and probably wrought it themselves into plates for armor.

A product of the nature of walrus ivory first became known in China during the reign of the T'ang dynasty (A.D. 618-906) under the name *ku-tu kio* ("horn of the *ku-tu*," the latter being a non-Chinese word derived from some native tribe of northeastern Asia). *Ku-tu* is given in the T'ang Annals among the taxes sent from Ying-chou in Liao-tung, and this was the domicile of the Kitan and other Tungusian tribes. It is also mentioned as a product of the Mo-ho, likewise a Tungusian tribe, whose country abounded in sables, white hares, and white falcons. The Mo-ho were settled to the north of Korea and extended east of the Sungari River as far as the ocean. They lived in close proximity and intercourse with the Liu-kwei, a people briefly described in the Annals of the T'ang Dynasty. The geographical position of the country of the Liu-kwei is clearly enough defined to lead us to Kamchatka. The culture of this people, as characterized by the Chinese, plainly reveals a type that is still found in the North-Pacific area. These cultural traits are, absence of agriculture, economy essentially based on the maintenance of numerous dogs, subterranean habitations, utilization of furs as winter costume, employment of fish-skins as clothing in the summer, and transportation on snow-shoes. The Mo-ho entertained a lucrative commerce with the Liu-kwei by way of the sea, the voyage lasting fifteen days; and when the latter in A.D. 640 sent a mission to China, their envoys travelled across the Mo-ho country. One of the three interpreters with whom they arrived at the Chinese

Court appears to have been a Mo-ho, and the extract in the Annals is doubtless based upon a report made by the Mo-ho. The latter, accordingly, were in intimate contact with a people that had the walrus and its product within easy reach; and from the descriptions of Steller and Krasheninnikov, which represent the principal sources for our knowledge of the ancient Kamchadal or Itelmen who are now almost extinct, we know surely enough that these tribes hunted the walrus and utilized its ivory for industrial work.

Hung Hao (A.D. 1090-1155) was sent as an ambassador of the Sung to the court of the Kin dynasty which belonged to the Jurchi or Niüchi, a tribe of Tungusian origin. He remained there for fifteen years (1129-43), and in his memoirs (*Sung mo ki wen*) has this note: "The *ku-tu* horn is not very large. It is veined like ivory, and is yellow in color. It is made into sword-hilts, and is a priceless jewel." In the History of the Liao or Kitan Dynasty (*Liao shi*, chap. 116), which ruled from A.D. 907 to 1125, the word *ku-tu-si* is defined as "the horn of a thousand years' old snake," the word *tu-na-si* being added as a synonym. These evidently are words belonging to the native language of the Kitan, although the Kitan on their part may have derived them from peoples living farther to the north. In the Annals of the Kin Dynasty (*Kin shi*, chap. 64) are mentioned daggers with hilt of *ku-tu-si* of the ancient Liao. Hung Hao also wrote in 1143 or shortly afterwards, "The Kitan hold the *ku-tu-si* in esteem. The horn is not large, but it is so rare that among numerous pieces of rhinoceros-horn there is not a single *ku-tu-si*. Unlike rhinoceros-horn, the latter has never been wrought into girdles. It has designs like those in elephant ivory, and is yellow in color. Only knife-hilts are made from it, and these are considered as priceless. The emperor T'ien Tsu (reigned A.D. 1101-19, died in 1125) had a girdle-pendant (*t'u-hu*) made from this substance." This was an exceptional case, for the girdles of the Kin dynasty were made of jade, gold, rhinoceros-horn, ivory, bone, and horn, and were graded in the order of these materials. This point bears out the fact that *ku-tu-si* represents a category of its own, and can have been neither elephant ivory nor rhinoceros-horn, which were the common articles for the girdles of the Kin or Jurchi. Moreover we know the Jurchi terms for elephant ivory and rhinoceros-horn, and these are distinct from *ku-tu-si*, which refers to walrus ivory.

Chou Mi (A.D. 1230-1320), a celebrated and prolific writer of the Sung period, alludes to *ku-tu* in two of his works. In one of these he cites the opinion of Sien-yü Ch'u, a poet and caligrapher, who possessed two knife-hilts made of this substance, to the effect that "*ku-tu* is a horn of the earth," which may possibly mean a horn found in the ground or underground. This might be construed to allude to mammoth tusks, although the evidence is not conclusive in view of the fact that walrus and narwhal tusks are likewise found in and under the ground along seashores. In another book Chou Mi writes that "*ku-tu-si* is the horn of a large snake and that, being poisonous by nature, it is capable of counteracting all poisons, as poison is treated with poison." This poison-curing property is a notion transferred to *ku-tu* from the ancient beliefs in the efficacy of rhinoceros-horn. The Chinese fondly entertained the idea that the rhinoceros feeding on brambles swallows all sorts of vegetable poisons which penetrate into its horn, so that in accordance with the principle that poison can neutralize poison, the horn or a cup carved from it becomes an efficient antidote. In 1320, Pi-ming, a son of Sien-yü Ch'u, was still the owner of the objects of which his father had spoken to Chou Mi thirty years earlier, and Ye Sen who saw them in his possession wrote an additional note on the subject which is embodied in Chou Mi's work. He observes that the natural designs displayed in the two knife-hilts of *ku-tu-si* resembled the sugar-cakes then sold in the markets; there also were white spots somewhat like those on candied cakes and pastry. "When touching this substance with your hands," he concludes, "it emits an odor like that of cinnamon; when after rubbing it no odor is perceptible, it is a counterfeit." Walrus ivory, on being rubbed, indeed emits a certain odor. Ye Sen's remark shows that the fakers were no less busy in China six hundred years ago than at present and that then also certain sleights were performed to test the genuineness of an article.

During the Mongol period the Chinese learned the fact that walrus ivory was found among the products of the western countries. In A.D. 1259 Chang Te was dispatched by the Mongol emperor Mangu as an envoy to his brother Hulagu, king of Persia. On his return to China he published a diary of his journey in which he mentions, among the products of the west, *ku-tu-si* as the horn of a large snake which has the property of neutralizing every poison. It is an interesting coincidence that the Kitan-Chinese term *ku-tu* has migrated westward and that for the first time it makes its appearance in a mineralogical treatise of the great Arabic traveller and scholar, al-Beruni (A.D. 973-1048); subsequently it recurs frequently in Arabic, Turkish, and Persian authors. Al-Beruni writes that *khutu* is much in demand, and is preserved in the treasuries of the Chinese who assert that it is a desirable article because the approach of poison causes it to exude, and that it was wrought into knife-hilts.

In the age of the Mongols we receive an interesting bit of folk-lore which has been recorded by Haithon, king of Armenia (1224-69), in the narrative of his journey to the Mongols written by Kirakos of

Gandsak. Haithon relates the universal legend of the country of the dog-heads, where the men have the shape of dogs, but the women have the human form and are endowed with reason. These fabulous creatures were located by the Chinese in an island of the northern Pacific, and Haithon adds, "There is also a sandy island there where is found a precious bone in the form of a tree, called fish-tooth; when it is cut, another bone will shoot forth at the same spot, in the manner of deer's antlers." The question is here of walrus tusks: the tusk was regarded as a "horn" (cf. p. 36); and in the same manner as the stag sheds its antlers, so also the "horns" of the marine mammals were believed to become detached and to grow again.

Toward the latter part of the seventeenth century, when the Russians established commercial relations with China, they traded chiefly two articles—seal-skins and walrus tusks, the latter being styled in the Russian documents of the time "bones of walrus tooth." A contemporaneous Chinese book (the *Pa hung yi shi* written by Lu Ts'e-yün in 1683) contains a brief description of Russia under the name A-lu-su (based on the Mongol name of Russia, Oros) and mentions the fact that in the reign of K'ang-hi (1662-1722) the Russians presented fish-teeth, black sables, gyrfalcons, a striking clock, glass mirrors, and other objects. It is perfectly obvious that the "fish-teeth" of this text, as corroborated by the coeval Russian documents, represent walrus tusks.

Gerbillon, one of the old Jesuit missionaries working in Peking (in Du Halde, Description of the Empire of China, Vol. II, p. 263, 1741), speaking of the trade of the Russians, mentions "the teeth of a sort of fish, which are much finer, whiter, and more precious than ivory. With these they drive a great trade to Peking, though scarcely any people but the Russians, who are poor, and inured to cold and fatigue, would take so much pains for so little profit." In a footnote it is added, "They are those called Mamuts teeth, found lately to be the teeth of elephants." But as already remarked by J. Rankin (p. 454), this is an addition of the translator, and the term "fish-teeth" used by Gerbillon, as well as the emphasis laid on the white color, demonstrates plainly that the question is of walrus tusks.

Since the beginning of the eighteenth century, in the era of K'ang-hi, the Chinese have gradually become acquainted with the walrus. In the dictionary *Cheng tse t'ung* it is designated "sea-horse" (*hai ma*), defined as "a fish or seal with teeth as strong and bright as bone and adorned with designs as fine as silk,—workable into implements." A curious description of the animal is also given in the *Hai lu* ("Records of the Ocean"), a small book published in 1800 by Yang Ping-nan and containing accounts of foreign nations from information received through a friend who had spent fifteen years voyaging to different parts of the world.

Besides the Russians, the Gilyak also, who are settled at the mouth of the Amur and on the island of Saghalin, traded walrus ivory to the Chinese on the Sungari. The Gilyak, on their part, received the Arctic product through the medium of the northern adjoining tribes in times prior to the Russian colonization of the Amur territory; the animal itself is known to the Gilyak solely by name. From 1853 they purchased its tusks from the Russian-American Company of Nikolayevsk and bartered them with the Chinese of the Sungari in a profitable trade for other articles. Vladivostok, prior to the war at least, received a share of walrus ivory from Gishiginsk and Baron Korff's Bay, a region inhabited by the Koryak. The Chukchi in the farthest corner of northeastern Asia, are great walrus-hunters, and formerly carried on an enormous trade in the tusks.

In the K'ien-lung period (1736-95) walrus ivory was carved into snuff-bottles, dishes, stems for tobacco-pipes, and covers for cricket-gourds. As a rule, the material was stained a bright green by means of verdigris to lend it the appearance of jade; but it must not be inferred from this that any ivory thus treated is that of the walrus and that ivory kept in its natural colors is necessarily that of the elephant. The hand of the back-scratcher shown in the case, for instance, is white, but of walrus ivory.

Finally, America came to the fore in the exportation of this article. It was during the nineteenth century that walrus ivory under the name *hai-ma ya* ("sea-horse teeth"), by which it is still commonly known in Canton, was imported into that city in large quantity from California, Sitka, and other parts of western America. The first American ship that reached China was the "Empress of China" which arrived at Canton in 1784, mainly with a cargo of ginseng. A company in Boston sent in 1788 two ships to the Northwest Coast, the "Columbia" and the "Lady Washington," which spent the spring and the summer of the following year in trading along the coast. At the close of the season all the furs collected were put on board the "Columbia," which then proceeded to Canton to dispose of the peltry, and with a cargo of Chinese goods returned to Boston by way of the Cape of Good Hope, arriving in August 1790 as the first American vessel which had circumnavigated the globe. The following years show a considerable growth in the American Northwest Coast trade, until in 1801 there were at least fourteen American ships on the coast. The normal voyage was to sail from the United States in the summer or early autumn, and to arrive on the Northwest Coast in the spring. The captains would then trade with the Indians from inlet to inlet, getting skins, preferably those of the rare sea-otter, in exchange for trinkets, knives, firearms, blankets, cotton and woollen cloths. In the autumn they would cross the Pacific to Canton, or if they had not yet obtained a cargo, they would winter at the Hawaiian Islands and trade on the Coast a second and even a third season before going to China. Once there they would exchange their cargo for tea and other goods and return to the United States by way of the Cape of Good Hope. The voyages, as a rule, were very lucrative. The original outlay was small, the furs sold in Canton at a large gain, and the teas and other goods purchased with the proceeds brought another profit in America or Europe. The voyages, however, were full of risk and required experience, and the trade was in the hands of a few large firms (cf. K. S. Latourette, History of Early Relations between the United States and China 1784-1844, Yale University Press, 1917). The beginnings of the trade in walrus ivory may be traceable to these early cruises.

In 1913 the Department of Commerce and Labor in Washington published the following communication from F. D. Cheshire, American Consul-General at Canton: "Before the revolution, about eighteen months ago, there was considerable trade in the manufacture from walrus ivory tusks of tobacco-pipe mouth-pieces, handles of fans, thumb-rings, and peacock-feather tubes for mandarin hats. These articles were sent to Peking, where they were dyed a green color, resembling the color of jade, but since the revolution there has been very little activity in the manufacture of such goods from walrus tusks. The demand has fallen off considerably, and the trade is confined to making cigarette holders, tooth-brushes, and chopsticks. The value of walrus tusks is $280 to $400 Hongkong currency per picul (133⅓ pounds). Elephant tusks are worth $700 to $1,200 Hongkong currency per picul. The elephant tusks are more serviceable and at the same time more valuable."

At the same time, Consul-General G. E. Anderson of Hongkong reported to the Department that inquiry among local importing and exporting firms and dealers in ivory of Hongkong failed to locate any importations of walrus ivory, but that elephant ivory was imported in large quantities, and was shipped mostly to Canton.

According to a communication of the United States Collector of Customs at the port of Juneau, Alaska, there was during the year 1913 exported direct from Alaska to China 4,000 lbs. of walrus ivory, to the value of $1,200, and from Alaska to the United States 7,763 lbs. of foreign walrus ivory, to the value of $2,717. The destination of the latter quantity was unknown to the office at Juneau, but it was believed there that the bulk of this ivory found its way to Japan and China. The shipment of ivory to China was made in that year by the Norwegian tramp steamer "Kit" from Nome en route to Japan; there is no regular transportation line direct from the Alaskan coast to the Orient, but occasionally tramp steamers call at different ports, bound for the Orient.

During the year 1924 there were shipped from Alaska to the United States 4,854 pounds of ivory valued at $6,602. This includes ivory of all kinds, but it is mostly walrus ivory (communication of the U. S. Collector of Customs at Juneau, January 19, 1925).

During my stay in China in 1923 I made inquiries among the ivory carvers of both Peking and Shanghai. They were perfectly well acquainted with walrus ivory and knew that the tusks, specimens of which were kept in their shops and readily shown me, came from America. In Shanghai the old term "fish-tooth" (*yü ya*) is still in use; both in Shanghai and Peking a new term has also sprung into existence —*ts'iu kio* ("horn of the *ts'iu*," originally the designation of some giant fish and the loach).

The Japanese likewise utilized (and still utilize) both walrus and narwhal ivory (besides elephant ivory) for their *netsuke* and other carvings. At the end of the eighteenth century shipwrecked Japanese sailors cast adrift on the Aleutian Islands acquainted their countrymen with the walrus by means of a somewhat grotesque, but unmistakable sketch. It happens that walrus sometimes get astray into the waters of Japan, and about 1890 one was caught near Hakodate in Tsugaru Strait, which must have passed along the Kurils from the north. The walrus is called in Japanese *kaiba* ("sea-horse"); its teeth, *kaiba no kiba*. It is curious that formerly also the term *unikōro* (our "unicorn") was used and written with two Chinese characters which mean "single horn." This hints at a trade in the product with Portuguese and Hollanders. Under the foreign name mentioned a pair of walrus tusks is figured in the Japanese cyclopædia *Wa-kan san-zai-zu-e*, first published in 1714, with the explanation that such tusks were imported on Dutch ships coming from Batavia; they were 6-7 feet in length and measured 3-4 inches in circumference, resembling elephant ivory. Among the temple treasures of Nikko, a narwhal tooth is still preserved in the temple of Iyemitsu. The rectangular box in which it is kept is inscribed with the words "a horn of the Barbarians" (*Ban-kaku isshi*). This tusk is said to have been presented by the Hollanders in 1671. According to Thunberg whose travels extended over the years 1770 to 1779, narwhal ivory was contraband in Japan before his time, and the Hollanders reaped immense profits by it, as the Japanese, who attributed to it all medicinal virtues, were willing to pay exorbitant prices for it.

IVORY SUBSTITUTES

Substitutes for ivory have been plentiful, both in ancient and modern times. The ancients used the teeth of the hippopotamus ("river horse") of the Nile like ivory: thus Pausanias mentions a golden statue of Demeter whose face was formed of hippopotamus teeth. The animal was known to the Hebrews under the name *behemoth* (Job, XL, 10), which is derived from the Egyptian *p-ehe-mau* ("water-ox"). We have seen that Avril availed himself of this Hebrew word for the designation of the walrus, and it has even been suggested that the word *mammoth* has been derived from *behemoth* through the medium of an Arabic form *mehemoth*. It is by no means certain, however, that this etymology is correct. The word *behemoth* was used rather flexibly, and was referred not only to the mammoth, but also to any large and strange beast, for instance, to the rhinoceros. Hekataeus, one of Herodotus' authorities, is the first who gave a description of the hippopotamus. The animal was hunted by means of harpoons, and large numbers were captured alive to be sent to Rome for the purpose of fighting in the circus with crocodiles. Both skin and teeth were used for industrial work. Arabic authors like Masudi and Damiri refer to the animal; and, in his notes on Egypt, Chao Ju-kwa, a Chinese author, who wrote in A.D. 1225, mentions water-horses in the Nile, which come out of the river to feed on the herbs growing on the banks, but which dive into the water at the sight of a man. Kubilai, the Great Khan, as Marco Polo relates, received from envoys he had sent to Africa "two boars' tusks, which weighed more than fourteen pounds apiece; and you may gather how big the boar must have been that had teeth like that! They related indeed that there were some of those boars as big as a great buffalo." These boar's teeth, as Yule comments in his edition of Marco Polo, were indubitably hippopotamus teeth, which form a considerable article of export from Zanzibar. Burton speaks of their reaching twelve pounds in weight.

Francesco Carletti, who travelled in America and in the Far East from 1594 to 1606 and whose very interesting book in Italian, entitled "Ragionamenti," was published at Florence in 1701, discusses the Mohammedan trade at Goa and mentions the importation of sea-horse teeth (*il dente del cavallo marino*), which he identifies with the hippopotamus. He further mentions another tooth, no less marvelous and of no less virtue coming from a fish called Fish Woman (*Pesce Donna*), so called from the resemblance it bears to a human creature. It is said that this fish has solely one tooth of marvelous virtue for stopping the flow of blood; yet of all these teeth they make without distinction crowns and rings, as likewise of the tooth of the hippopotamus to which they attribute the same virtue, but it is not so highly esteemed. This "fish" is the dugong (*Halicore dugong*), a cetaceous animal found in all parts of the Indian Ocean. Pliny and Aelian have written about it; according to the latter, these creatures partly resemble satyrs, partly human women. The Chinese also have their share of fables about this creature, believing that in the course of many years its teeth change into dragon's teeth, and in this state they call the animal "pig-woman-dragon." Its teeth were formerly imported into China from the southern seas to be made into knife-hilts and handles for fans. They were even imitated with elephant ivory which was subjected to an artificial treatment by means of chemicals.

At the first quarterly ivory sale held at Antwerp in 1912, 71 hippopotamus tusks were sold, at the second quarterly sale 262 hippopotamus tusks, and at the third quarterly sale 97 tusks. At one of these sales held in

1911, twelve kilos of rhinoceros tusks are mentioned, so that also these must serve as a substitute of ivory.

Bones of whale, crocodile, and large sea-fishes also are said to be used in lieu of ivory, particularly in Annam, where the material is exposed to the smoke of a charcoal fire; it is then gently rubbed in the sunlight, and is finally rolled for twenty-four hours in fresh tobacco-leaves of *Nicotiana rustica.* An ultimate energetic massage will produce a certain ivory-like appearance and a yellow tint which is not unpleasing. Similar bones are utilized in China for cheap ornaments, but they are invariably dyed a pink color and sold under the name "fish-bone" (*yü ku*), not as ivory. Teeth of the sperm-whale, lamantin, and other phocine animals are imported into China in limited quantities, and are also used like ivory.

In Japan, the large canine teeth of the sea-lion, some of which are nearly four inches in length and of the consistency of ivory, are sometimes carved into *netsuke.* In the same manner the Koryak of north-eastern Siberia employ the teeth of the white whale and the bear in carvings.

Finally we have also blessed the Chinese with celluloid which they euphemistically call "European ivory."

OBJECTS MADE OF IVORY

From the preceding discussion it becomes clear that the Chinese, in the course of their history, have utilized the ivory of the elephant, the mammoth, the walrus and the narwhal. In specimens of the archaic period, as illustrated in Plate I, we may not err in tracing the ivory to the native elephants of ancient China. From the period which marks the end of antiquity down to the middle ages (that is, from the Han to the close of the T'ang dynasty, 206 B.C.—A.D. 906) we are bound to assume that the bulk of ivory used by the Chinese came from Kwang-tung and Yün-nan, Annam, Camboja, Siam, Burma, and India. From the tenth century onward the chief importers of ivory into China were the Arabs, who obtained the material from Java, Sumatra, India, and the east coast of Africa. At present it is chiefly imported from Siam, India, and Africa, and the African ivory will presumably preponderate. The export of ivory from Siam for the fiscal year ending March 31, 1910, amounted to 4,301 pounds, valued at $8,489, and this is regarded as a fair average of the export for the preceding five years; this ivory is obtained from domestic elephants that have died a natural death, as the animal is not hunted in Siam for its ivory; the number of tame elephants kept in Siam is roughly estimated at about three thousand. From Bombay also much ivory is at present exported to China.

To what extent mammoth ivory was utilized in China is a question difficult to answer. It is said that the furniture-makers of Ning-po used it for inlaying tables. The desk-ornament illustrated in Plate IX, Fig. 3, is possibly a piece of mammoth ivory; it is deep brown and yellow in color, and is left in its natural state, being only sawed off and polished along the base. The Chinese collector from whom I obtained this object in 1923 was unable to give any information about it.

The articles most commonly made of walrus ivory are combs and back-scratchers both of which are kept intact in their natural colors, handles for fans (see exhibition of fans in Blackstone Chinese Collection), dice for gambling, ear-rings and other small ornaments usually dyed green with verdigris, and, above all, chopsticks. It is a curious fact that, although the wholesale price of raw walrus ivory is lower than that of elephant ivory, chopsticks of the former material are higher priced in Peking than those of the latter. A pair of good elephant-ivory chopsticks may be bought anywhere at a price of Chinese $1.80-2.00, while a pair of green walrus-ivory chopsticks, like that placed on exhibition, which is a perfect specimen, retails in Peking at Chinese $12.00, and a pair but partially green at Chinese $5.50. Chopsticks are a remarkable Chinese invention, and, as stated, were used as early as the Chou period; in the sculptured bas-reliefs of the Han period many banquets are illustrated with chopsticks in evidence. While all other nations took their food with their fingers, the Chinese were the first who introduced and practised good table-manners.

The small dish illustrated in Plate IX, Fig. 1, is likewise of walrus ivory stained green and carved all over into a swirl of waves rising into crests along the edge. The two covers for cricket-gourds in Figs. 4 and 5 of the same Plate are made of the same substance. One shows a boat in which a lady with a basket stands, speaking to a man seated in a mat-covered cabin. The other is carved in two layers, the upper one in green representing a bird on the wing flying toward a blossoming plum-branch (see, further, below, p. 74).

Archaic objects of ivory are figured in Plate I and have been referred to on p. 9. As to the T'ang period (A.D. 618-906), the Japanese Treasure-house of Nara furnishes us some good examples, as, for instance, a backgammon board of sandalwood decorated with ivory inlays (*Tōyei Shukō*, pl. 72), and two standard foot-measures of ivory colored red and green, respectively (pl. 82-83). As Omura Seigai informs us, ivory at that time (eighth century) was colored crimson, indigo, green, or some other shade, and on this colored surface floral designs were so engraved that the unstained part of the ivory stood out. He states that this process has altogether been lost, and was not used in later times. This may hold good for Japan, but in China it survived at least down to the K'ien-lung period from which we have ivory snuff-bottles with paintings of the same technique. The fact that ivory was painted in China under the T'ang may be gleaned from an ivory fragment found by Sir Aurel Stein (Serindia, p. 779) in the Limes of Tun-hwang; it bears traces of a painted leaf-scroll in green. Ivory dyed by means of purple in Asia Minor was known in the Homeric age (*Iliad,* IV, 141). The Hawaiians colored whale ivory yellow by smoking it with green banana leaves.

Under the Mongols who ruled China as the Yüan dynasty (A.D. 1260-1367), a bureau for carvings in ivory and rhinoceros-horn was established. In this court-atelier couches, tables, implements, and girdle-ornaments inlaid with ivory and horn were turned out for the imperial household. An official was placed in charge of it in 1263, and the force consisted of a hundred and fifty workmen. Again, toward the end of the seventeenth century, under the reign of K'ang-hi, an atelier for ivory works was founded within the palace at Peking in connection with twenty-six other establishments for the practise of all industrial arts. Experienced craftsmen for the various branches of work were summoned to Peking from all parts of the empire. These factories lasted somewhat more than a century, and were closed after the reign of K'ien-lung (1795). Authentic productions which could be safely identified with the output of the Mongol imperial works have not yet come to light; but there are authentic specimens of the K'ien-lung period, which have come from the imperial palace and without any doubt were fabricated in the imperial atelier, as, for instance, the fan illustrated in Plate VIII. This is a marvel of technical skill and harmonious beauty, being plaited from finely cut ivory threads held by a tortoise-shell rim and overlaid with colored ivory carvings of lilies, peonies, asters, and a butterfly. The handle, likewise of ivory, bears etched designs in colors of flowers and butterflies. The carved medallion on the dividing rod in the centre is of amber, and the ornament above the handle is of brass inlaid with blue kingfisher feathers. At the time of the Han, Wei and Tsin periods (first to fifth century A.D.), as we read in Chinese accounts, mats of ivory were made. The ivory plaiting in the above fan may give us a clew to the technique of such mats. John Barrow, who visited China in 1792, speaks of neat baskets and hats made at Canton from ivory shavings interwoven with pieces of quills, and as light and pliant as baskets and hats of straw.

Paléologue (L'Art chinois) said in 1887 that the fine Chinese ivories are excessively rare, and that the Buddhistic statuettes offer us the most interesting specimens of sculptured ivory. Unfortunately such statuettes never bear the signature or seal of the artist, nor are names of ivory-carvers preserved to us in any records, so that we are entirely ignorant of art-schools and artists working in this field.

The goddess of compassion, Kwan Yin, has been the most favored subject of the sculptors, who were particularly tempted by the task of presenting the drapery of her long, flowing garb in graceful sweeps and elegant lines. The statuette (Plate II, Fig. 1) representing her is a masterpiece of modeling animated by life and motion, and is a triumph of the spirit over matter. Her left hand supports a bowl believed to be filled with the nectar of immortality, and her right hand touches a ladle inserted in the bowl, ready to distribute her gifts among her devotees. Bracelets adorn both her wrists. Her face is refined and spiritual and astir with religious fervor. The lines of the figure and the drapery of the robe are exquisite, worthy of a Madonna. No less impressive is the statuette of Tung-fang So (Plate II, Fig. 2) whom we met as author of a book of marvels (p. 24). He lived in the second century B.C. as a poet, statesman, and adept versed in the mysteries of Taoism. He was reputed as being possessed of divine wisdom and supernatural powers, and is said to have thrice abstracted from paradise the famed peaches of immortality which ripen but once in three years. He was on intimate terms with the emperor Wu, amusing him with humorous sallies and earning for himself the sobriquet of the Wit. Our figure shows him as a genial old man with long whiskers and deep furrows over his eyes; he handles a palm-leaf fan in his right hand. The carving of the figure is cleverly adapted to the natural curve of the elephant tusk. Both these statuettes may be confidently ascribed to the Ming period (1368-1643). Both have developed fine patinas of dark brown and deep yellow.

A somewhat different style is represented in the statuette of a Buddhist monk of pure-white ivory (Plate III), apparently a portrait modelled from life in the era of K'ang-hi (1662-1722). He is obviously shown in the act of preaching a sermon of Buddha's gospel based on the text that is written on the roll of paper which he grasps in his left hand. Baldheaded, as the Buddhist monks are, with bright, intelligent eyes (outlined in black), high forehead, and his lips in motion, he stands there a worthy disciple of Çakyamuni, humble and modest, sincere and fully conscious of the truth of his convictions.

The Arhats (in Chinese Lo-han), the celebrated disciples of the Buddha, form the subjects of the ivory figures in Plates IV and V. The two grouped in Plate IV were evidently turned out by the same artist in the K'ien-lung period (1736-95); in style and attributes they closely approach the Arhat paintings of that time. Both figures are characterized by the same massive head, the same high helmet-shaped craniums, bulging eyes, large noses, heavy mustaches and beards. The Arhat in Fig. 1 sets his foot on the back of a lion (symbolizing the saint's power over the wild animals); in his right hand he holds a branch with fungus of immortality (*ling-chi*) which, strictly speaking, is a Taoist emblem, and in his left hand a fly-brush or chowry, an ancient emblem of royalty or rank, usually made of yak-tails or coir. The tip of the chowry tickles the lips of the lion who devotedly looks up to his master. A rosary is slung around his neck. His companion (Fig. 2) sets his left foot on the head of the three-legged mythical frog. He is represented in the act of conjuring a dragon from his alms-bowl, pointing at him a bead of the rosary which he holds between the thumb and index-finger of his right hand.

In each of the two figures illustrated in Plate V two Arhats are grouped together, each pair, including the base, being carved from a single piece of ivory. In the first group a monk hurls a dragon's head into the face of his frightened companion, who is suddenly interrupted in his prayers during which he was running off the beads of his rosary and burning incense. The counterpart of this figure presents a monk clasping his arm around his brother's shoulder and showing him a snake. Terror-stricken he screams aloud and presses hard his left foot on the tiger below who feels the force of his master's emotion. The conception of both groups is highly dramatic and emotional.

In A.D. 484 Jayavarman, king of Fu-nan (Camboja), sent Nagasena, an Indian monk, with a long letter to the emperor of China, offering as presents an elephant carved from white sandal-wood and two topes (stūpas) of ivory. The Museum is in the possession of an ivory seal from Siam presented by Miss C. Wicker; it is carved in the form of a tope, and such seals are still used by Buddhist monks. The design in the seal is a cat amid plants; the cat was sacred to the monks as the animal exterminating the rats which threatened their sacred books with destruction; the domesticated cat was hence introduced by the Buddhists into China and all other countries of the Far East.

The esthetic needs of the scholar are cared for by the ivory-carver in the production of handsome writing-brushes provided with ivory handle and encased in a sheath of ivory. He is also fond of ivory foot-measures etched on the back with delicate floral designs and birds in colors, but, above all, delights in brush-holders (*pi tung*) as a suitable decoration of his desk. These are carved out of the central portion of the tusk in the round, and are decorated with designs in high relief of a plum-tree growing out of a rock and surrounded by bamboo-leaves (Plate VI), or are adorned with a genre-scene in flat relief, as that in Plate VII, which shows a horseman at night in a mountain-pass followed by a flag-bearer, a boy with a lantern lighting the path; the scenery is enlivened by pine-trees and clouds. Another ambition of the literary man is to possess an arm-rest of ivory (usually carved from bamboo) with elaborate designs; objects like these are used on the desk for resting the forearm while wielding the writing-brush.

The ivory objects of which the ancient Chinese were proudest are flat tablets used for ceremonial purposes and call *hu*. Six fine specimens of these coming down from the Ming dynasty (1368-1643) are placed on exhibition. These tablets carved from elephant tusks formerly played a prominent part in official life. In very early times they appear to have been made of bamboo, being suspended from the girdle which belonged to the dress of every young gentleman. They were used as memoranda for jotting down any notes. At a somewhat later epoch they were made of ivory and reserved for the organs of government, developing into insignia of rank. When a high official had audience at court, he respectfully held such an ivory tablet, clasping both his hands around the broader base, the upper narrow part being at the height of his mouth, so that his breath might not touch the imperial face. He had inscribed on the tablet whatever business he wished to report and submit to the emperor, and recorded on it the imperial replies or commands. The ancient Book of Rites (*Li ki*) contains this passage, "When the great prefect had washed his head and bathed, his secretary brought him the ivory tablet to write down his thoughts, his replies, and the orders of the prince."

Friar William of Rubruck, who sojourned among the Mongols from 1253 to 1255, relates as follows: "Whenever the principal envoy came to court, he carried a highly-polished tablet of ivory about a cubit long and half a palm wide. Every time he spoke to the Khan or some great personage, he always looked at that tablet as if he found there what he had to say, nor did he look to the right or to the left, nor in the face of him with whom he was talking. Likewise, when coming into the presence of the Lord, and when leaving it, he never looked at anything but his tablet." Under the T'ang dynasty (A.D. 618-906) the ivory tablets were round above and angular below, and were used by officials down to the sixth rank; those below the sixth rank had bamboo or wooden tablets. Under the Ming the ivory tablets were angular at both ends, and were granted to officials above the fourth grade; those of the fifth grade and below had wooden ones with painted designs. They were abolished under the Manchu. In Korea such tablets were used down to quite recent times. Yüan Shi-kai, toward the end of his presidency, is said to have attempted to introduce them again in connection with his scheme to restore the monarchy, but in this he failed.

Girdle-pendants of which the Chinese were formerly very fond were made of ivory also in the K'ien-lung period, but are rather scarce. The one illustrated in Plate IX, Fig. 6, is carved in the shape of the so-called "wooden fish,"—a sort of wooden drum used in Buddhistic temples to mark time in the recitation of prayers, the handle being formed by two dragon-heads. Another pendant (Plate X, Fig. 3) represents two bean-pods with tendrils and leaves, and that in Fig. 5 two boys, so arranged that the complete figure of a boy (altogether four) may be seen from every angle. A scent-box in the shape of a flower-basket is shown in Fig. 2 of Plate IX; it is carved in open work with peaches and pomegranates and consists of two halves joined together. It is filled with perfume and worn in front of the dress during the summer.

In ancient Rome parrots were kept in cages of gold, silver, and ivory (Statius, *Silvae*, II, 4, 12), but none of these has come down to us. From China we receive bird-cages entirely made of ivory rods and adorned with numerous small carvings of the same material. No other nation has been more considerate of the welfare of its pets and lavished on them the most precious substances and the most exquisite work that art could offer.

Crickets are kept by the Chinese for two purposes—to enjoy their melodious chirps and to train them as fighters. A cricket-fight is a great event, and large sums are staked on the champions. In Peking a special kind of gourd is raised to keep the insects during the winter. Many of these gourds are elaborately decorated and provided with finely carved lids of jade and ivory. Five covers of such cricket-gourds are reproduced in Plate X. That in Fig. 1 shows an open-work composition of plum-blossoms with two birds; that in Fig. 2, leaves and tendrils of a gourd, with a butterfly. The cover in Fig. 5 is surmounted by three full figures of lions carved in the round and playing with a ball in the centre; the lower band is decorated with a row of peonies and leaves. A floral composition is spread over the cover in Fig. 6, and a dragon striving for the flamed pearl is carved in Fig. 7.

Ever since Europeans came into contact with the Chinese, their ivory fans have elicited unbounded admiration. John Barrow, private secretary to the Earl of Macartney on his mission to China in 1792, has the following interesting notice on this subject: "Of all the mechanical arts that in which they seem to have attained the highest degree of perfection is the cutting of ivory. In this branch they stand unrivalled, even at Birmingham, that great nursery of the arts and manufactures where, I understand, it has been attempted by means of a machine to cut ivory fans and other articles, in imitation of those of the Chinese; but the experiment, although ingenious, has not hitherto succeeded to that degree, so as to produce articles fit to vie with those of the latter. Nothing can be more exquisitely beautiful than the fine open work displayed in a Chinese fan, the sticks of which would seem to be singly cut by the hand, for whatever pattern may be required, or a shield with coat of arms, or a cypher, the article will be finished according to the drawing at the shortest notice. The two outside sticks are full of bold sharp work, undercut in such a manner as could not be performed any other way than by the hand. Yet the most finished and beautiful of these fans may be purchased at Canton for five to ten Spanish dollars."

Ivory beds were a prominent feature in many oriental countries. Mong Ch'ang-kün, a Chinese minister of state, who lived in the third century B.C., was famed for his extravagance and had as many as three thousand retainers, all of whom wore shoes embroidered with pearls; he is reputed to have been the owner of an ivory bed which he presented to the prince of Ch'u. A certain Yü Yang, who lived at the time of the Liang dynasty (A.D. 502-556), was no less noted for his love of luxury; he possessed a bed inlaid with ivory, gold, and silver. On Java, the Chinese Annals of the T'ang dynasty report, princes and people had ivory beds; and the same is on record in regard to India, where couches and seats were inlaid with ivory. With reference to the city of Cambaya (now Cambay), the Portuguese traveller, Duarte Barbosa (A.D. 1518), writes, "A great quantity of ivory is used here in very cunning work, inlaid and turned articles such as bangles, sword-hilts, dice, chessmen and chess-boards; for there are many skilful turners who make all these, also many ivory bedsteads very cunningly turned, beads of sundry kinds, black, yellow, blue and red and many other colors, which are carried hence to many other places." Wooden beds with ivory inlays are still made at Ning-po in Che-kiang Province.

The opium-smoker has a particular veneration for ivory: he may use a pipe with ivory mouth-piece and ivory boxes to contain the drug (cf. Leaflet 18, pp. 24, 35); he may also avail himself of an ivory spatula for taking a pill of opium out of the box, and he may worship the "god of opium" in the form of an ivory image (figured by A. de Pouvourville, L'Art indo-chinois, p. 189). Tobacco-pipes of ivory are described and figured in Leaflet 18 (p. 22). Snuff-bottles were also made of this material, carved with designs or painted.

The concentric ivory balls which have attracted much attention and which are still turned out at Canton were manufactured as early as the fourteenth century under the name "devil's work balls." There is a tradition also that they were made in the palace of the Sung emperors. They are the result of patient toil, the balls being carved one within the other. Good, old specimens are difficult to get; the modern ones are usually intended for the foreign market.

In India chessmen and backgammon were made of ivory at an early date (account of Masudi, A.D. 983). The Chinese make of ivory chessmen (Fig. 13), dice, dominoes, and many other games derived from the latter. The foreign craze for ma-jong has now caused nearly all available ivory to be absorbed for the manufacture of ma-jong sets, which has disorganized the whole ivory industry and unfortunately stopped the production of artistic carvings.

The main seats of the ivory industry are Canton, Amoy, Shanghai, Suchow, and Peking. As a rule, the objects are carved and sold in the same shop. Ivory is now preferred in its pure white state, and Canton workmen are successful in removing yellow tinges from ivory and restoring it to its pristine whiteness and brilliancy. On the other hand, there is no lack of methods of lending ivory a yellow-brown patina and making it appear old: for this purpose it is placed in a decoction of tobacco or tea leaves, or exposed to the fumes of burning incense.

Canton, for at least a century or more, has catered to foreign taste and produced immense quantities of ivory ware for export. Although many of these articles are marvels of patient workmanship and technical skill and ingenuity, they lack artistic feeling and finish; the carved concentric balls, models of boats, houses, temples and pagodas belong to this class. Others like brooches, chains, glove-boxes, etc., are entirely foreign to the Chinese, and are solely intended for the European or American market. Articles like these were strictly excluded from the ivory collection of the Museum, and only those made for and used by the Chinese were selected.

B. LAUFER.

BIBLIOGRAPHICAL REFERENCES

BEVERIDGE, H.—The Emperor Jahangir's Treasures of Walrus and Narwhal Ivory. Indian Magazine, February, 1914, pp. 37-39.

BISHOP, C. W.—The Elephant and Its Ivory in Ancient China. Journal of the American Oriental Society, Vol. XLI, 1921, pp. 290-306.

DAWKINS, W. BOYD.—On the Range of the Mammoth in Space and Time. Quarterly Journal of the Geological Society, 1879, pp. 138-147.

HOWORTH, H. H.—The Mammoth and the Flood. London, 1887.

KUNZ, G. F.—Ivory and the Elephant in Art, in Archæology, and in Science. New York (Doubleday, Page and Company), 1916.

LAUFER, B.—Arabic and Chinese Trade in Walrus and Narwhal Ivory. T'oung Pao, 1913, pp. 315-364, with Addenda by P. Pelliot, pp. 365-370.

Supplementary Notes on Walrus and Narwhal Ivory. T'oung Pao, 1916, pp. 348-389.

Sino-Iranica, pp. 565-568 (with special reference to Persia).

LULL, R. S.—The Evolution of the Elephant. Yale University, Peabody Museum of Natural History, Guide No. 2, reprinted from the American Journal of Science, Vol. XXV, 1908.

LYDEKKER, R.—Mammoth Ivory. Annual Report of the Smithsonian Institution for 1899 (Washington, 1901), pp. 361-366.

MAYERS, F. W.—The Mammoth in Chinese Records. China Review, Vol. VI, 1878, pp. 273-276.

RANKING, J.—Historical Researches on the Wars and Sports of the Mongols and Romans: in which Elephants and Wild Beasts were Employed or Slain. And the Remarkable Local Agreement with the Remains of Such Animals Found in Europe and Siberia. London, 1826.

WATT, G.—Indian Art at Delhi, p. 173. Calcutta, 1903.

The Commercial Products of India, pp. 695-699. London, 1908.